# Ethics for the Built Environment

Built environment professionals are expected to meet high standards of client care as well as maintain their own professional codes. Much closer relationships are being formed in the development of the built environment and cultural changes in procurement methods are taking place so that the need has grown to re-examine the ethical frameworks required to sustain collaborative trust and transparency. Young professionals are moving between companies on a more frequent basis and taking their personal values with them, but need guidance. What can companies do to support their employees?

The book looks at how people develop their personal values and it tries to set up a model for making effective ethical decisions. It exposes areas of weakness that may inhibit better relationships in projects and partnerships and suggests decision-making frameworks.

Part I establishes some theory and critically evaluates a number of topical areas through the life-cycle of development to give a holistic view of built environment ethics. Part II illustrates good practice by using case studies and questions for resolving dilemmas. Chapters cover the following issues:

- Ethical foundations and dilemmas
- Human rights, moral integrity and employment law and equal opportunities
- Professional conduct and integrity and good faith
- The ethics of development control
- Trust, business morals and relationships
- Corporate social responsibility and governance, and their implementation in construction and development organisations
- Corruption, procurement and competition
- The role of education and development
- Responsibility for sustainability, health and safety and the environment.

This book will be invaluable to professional institutions, students, contractors, clients and young practitioners in all stages of the development cycle from planning, property management, design, project management and facilities management.

**Peter Fewings** is Programme Leader, MSc Construction Project Management, University of the West of England, Bristol.

# Ethics for the Built Environment

Peter Fewings

Taylor & Francis
Taylor & Francis Group

LONDON AND NEW YORK

First published 2009
by Taylor & Francis
2 Park Square, Milton Park, Abingdon, Oxon OX14 4RN

Simultaneously published in the USA and Canada
by Taylor & Francis
270 Madison Avenue, New York, NY 10016

*Taylor & Francis is an imprint of the Taylor & Francis Group,
an informa business*

© 2009 Peter Fewings

Typeset in Sabon by
RefineCatch Limited, Bungay, Suffolk
Printed and bound in Great Britain by
CPI Antony Rowe, Chippenham, Wiltshire

This publication presents material of a broad scope and applicability.
Despite stringent efforts by all concerned in the publishing process,
some typographical or editorial errors may occur, and readers are
encouraged to bring these to our attention where they represent
errors of substance. The publisher and author disclaim any liability,
in whole or in part, arising from information contained in this
publication. The reader is urged to consult with an appropriate
licensed professional prior to taking any action or making any
interpretation that is within the realm of a licensed professional
practice.

*British Library Cataloguing in Publication Data*
A catalogue record for this book is available from the British Library

*Library of Congress Cataloging in Publication Data*
Fewings, Peter.
  Ethics for the built environment / Peter Fewings.
    p. cm.
  1. Construction industry–Management.  2. Construction workers–
Professional ethics.  3. Construction industry–Corrupt practices.
  4. Social responsibility of business.  5. Business ethics.
  6. Construction industry–Employees–Health and hygiene.  I. Title.
  HD9715.A2F49 2008
  174′.969–dc22
                                                    2008002199

ISBN10: 0–415–42982–X (hbk)
ISBN10: 0–415–42983–8 (pbk)

ISBN13: 978–0–415–42982–5 (hbk)
ISBN13: 978–0–415–42983–2 (pbk)

This book is dedicated to my wife Linda who has always been there for me.

# Contents

*List of figures*                                                                xv
*List of tables*                                                                xvii
*Acknowledgements*                                                              xix

**Introduction**                                                                  1

*The built environment  2*
*Responsibilities and stakeholders  2*
*Methodology and structure of the book  4*
*Ethical decision-making  5*
*Notes  6*

**PART I**
**Theory and application**                                                        7

**1  Development of an ethical framework and the built environment**               9

*Development of building and its impact on ethics  9*
*Integrated professional behaviour  12*
*The basis of ethics – good and bad, right and wrong  12*
*Right and wrong  20*
*Application to the built environment  24*
*Conclusion  27*
*Notes  28*

**2  Ethical dilemmas and decision-making**                                       29

*Definition and context  29*
*Kohlberg's stages of moral reasoning  30*
*Background of ethical perception  33*

*Ethical perceptions – business and professional practice  35*
*Classic ethical dilemmas  38*
*Ethical dilemmas in the built environment  39*
*Planning dilemmas  39*
*Property development and management dilemmas  41*
*The designer's dilemma  43*
*Construction dilemmas  47*
*Assessing 99 per cent situations  49*
*Labour dilemmas  52*
*Ethical decision-making models  53*
*Conclusion  58*
*Notes  58*

3   **Business ethics and corporate social responsibility policy**          61

*Business ethics  61*
*Objections to an ethical code and moral responsibility  64*
*Code framework  65*
*A business ethic model  67*
*Corporate social responsibility (CSR)  68*
*Rationale for social and environmental reporting  74*
*Sustainable construction  77*
*CSR in construction  78*
*CSR reporting survey  80*
*Trading  83*
*Conclusion  84*
*Notes  84*

4   **The development of professional ethical codes**          87

*Definition  87*
*Professional exclusiveness  88*
*Professional competence  90*
*Professional rules of conduct in the built environment  93*
*Built environment and the rules  100*
*Transparency in construction trading relationships  101*
*Ethical leadership  108*
*The Considerate Contractors Scheme  111*
*Conclusion  112*
*Notes  114*

5   Discrimination and human resource ethics in the
    built environment                                                    116

    *Diversity and equality 116*
    *Human rights 117*
    *Employment and the psychological contract 125*
    *Recruitment, selection and retention 133*
    *Developing economies 136*
    *Learning organisations, training and development 138*
    *Absenteeism and presenteeism 140*
    *The labour force 140*
    *Conclusion 145*
    *Notes 146*

6   The ethics of construction quality, safety, health and welfare       149

    *Fit for purpose 149*
    *Learning from the statistics 150*
    *Occurrence of accidents, dangerous and health incidents 151*
    *Ethical approach to health and safety 152*
    *Responsibilities for health and safety in the construction
        life cycles 154*
    *Building managers 158*
    *Risk assessment and moral hot air (virtual morality) 160*
    *Quality and moral imagination 162*
    *Changing culture 165*
    *Conclusion 169*
    *Notes 170*

7   The planning ethics                                                  171

    *Questions 172*
    *Development planning in the UK 172*
    *Ethics of development control decisions 175*
    *Basis of planning decisions 178*
    *Different planner roles 181*
    *Facing moral problems 184*
    *Housing supply and sustainability imperatives in
        the UK 186*
    *Conclusion 187*
    *Notes 188*

8  Ethics of sustainability: a UK example                                189

    *Sustainable development 189*
    *Environmental accounting and motivation 193*
    *Urban planning and sustainability 201*
    *Sustainable procurement 203*
    *Sustainable construction 205*
    *The environmental management ethic 209*
    *Measuring sustainability 209*
    *Conclusion 212*
    *Notes 214*

9  Trust and relationships                                                216

    *Contracts and trust 216*
    *Definition 217*
    *The case for trust in construction 220*
    *Trustworthiness, values and ethics 228*
    *Trust in professionals 232*
    *Trust in practice 233*
    *Trust and risk 236*
    *Conclusion 240*
    *Notes 240*

10  Bribery and corruption                                                243

    *Business and professional environments 243*
    *Dealing with corruption 250*
    *Corruption in the construction business and competition 253*
    *Competitive bidding 260*
    *Competition and collusion 264*
    *Corruption in property deals 269*
    *Money laundering 272*
    *Achieving reform of corrupt value 273*
    *Conclusion 274*
    *Appendix Corruption and bribe payers indices 275*
    *Notes 278*

11  Delivering ethical improvement through contractual good
    faith                                                     281
    JIM MASON

*The construction context 281*
*The wider context 282*
*Partnering and good faith 283*
*The newer contract forms 284*
*The duty of good faith 285*
*Other stimuli towards the introduction of a duty of
    good faith 286*
*Judicial hostility? 288*
*How best to deliver what the parties want? 289*
*Conclusion 290*
*Notes 291*


**PART II**

**Case studies of good practice**         **293**

*The case studies 293*

12   Comparison of CSR between a developer and a contractor     **297**

*Introduction 297*
*The contractor 297*
*The developer 297*
*Conclusion 300*
*Note 300*

13   Partnering trust and risk management     **301**

*Introduction 301*
*Transparency and risk 301*
*Value for money 302*
*Trust and working together 303*
*Quality, integrity and achievement 303*
*Change management 304*
*Risk management 305*
*Sustainability and whole life costs 305*
*Site efficiencies 306*
*Ethics 306*
*Conclusion 310*
*Acknowledgements 310*
*Notes 310*

14   Roofing contractors collusion case study                           311

*Introduction 311*
*Corporate penalty 311*
*Individual penalty 312*
*The group cases 312*
*Discussion 314*
*Notes 315*

15   The Heathrow T5 major projects agreement vs
     false employment                                                   316

*Introduction 316*
*Self-employment in the UK construction industry 316*
*Heathrow Terminal 5 (T5) 319*
*Acknowledgement 321*
*Notes 321*

16   Health and safety systems in a large PFI hospital                  322

*Introduction 322*
*Health and safety policy and ethics 322*
*Management involvement 325*
*Achievements 327*
*Conclusion 328*
*Acknowledgement 328*

17   Stroud District Council planning case study                        329

*Introduction 329*
*Development control policy frustrations and ethics 330*
*Stroud District Council (SDC) 331*
*Conclusion 335*
*Notes 335*

18   The use of training to establish small-scale organisations
     in construction                                                    336

*Introduction 336*
*Enabling an ethical and strategic entrepreneurial outlook 336*
*The centrality of health and safety ethics 337*
*The training structure 338*
*The E scheme 338*

*Reflection on the ethics of the project 341*
*Conclusion 342*
*Acknowledgements 342*
*Notes 343*

19 **Manufacturing quality and trading relationships**                    **344**

*Introduction 344*
*Manufacturing 344*
*Building up trust with the contractor and client 346*
*Conclusion 349*
*Acknowledgement 350*
*Note 350*

20 **Educational partnership and sustainable contracting**                **351**

*Introduction 351*
*The agreement 351*
*The programme 352*
*Conclusion 354*
*Acknowledgement 355*
*Note 355*

21 **Trust and relationships in a mega property development**             **356**

*Introduction 356*
*Purpose of the development and stakeholder management 356*
*Procurement and management 357*
*Objectives and stakeholder consultation 358*
*The development of trust during the construction stage 358*
*Conclusion 364*
*Notes 365*

22 **Making it work**                                                     **366**

*Education and ethical dialogue 366*

*Index*                                                                   **368**

# Figures

| | | |
|---|---|---:|
| 1.1 | The development of ethical drivers of construction | 11 |
| 1.2 | Plato's individual souls (psyche) | 14 |
| 1.3 | Aristotle's account of the soul | 15 |
| 2.1 | The Business Ethics Synergy Star (BESS) | 42 |
| 3.1 | A business ethic model | 67 |
| 3.2 | The triple bottom line | 68 |
| 3.3 | The CSR stakeholder model | 71 |
| 3.4 | The five guiding principles of sustainable development | 77 |
| 4.1 | The continuous improvement cycle | 104 |
| 4.2 | UN sustainability and anti-corruption principles | 106 |
| 4.3 | When is a gift acceptable? | 110 |
| 5.1 | Declaration of Human Rights (clauses relevant to employment) | 118 |
| 5.2 | Matching the candidate to the position | 134 |
| 6.1 | Risk assessment process | 160 |
| 7.1 | The development control process | 174 |
| 8.1 | Three areas of sustainability | 192 |
| 8.2 | Projected temperature rise compared with $CO_2$ concentration | 197 |
| 9.1 | Individual delegation and trust | 223 |
| 9.2 | Supply chain trust | 225 |
| 13.1 | Target sum make-up | 302 |
| 13.2 | Sparkleometer comparison | 304 |
| 17.1 | Stroud College | 329 |
| 17.2 | Comparison of a traditional way and the method used in the case study | 333 |
| 17.3 | SDC reduced cycle time for college application, win–win way | 334 |
| 18.1 | The four strings of the 'Essential E' plan | 338 |
| 19.1 | Manufacturing process | 345 |
| 21.1 | Alternative balanced credit approach to build trust | 361 |

# Tables

| | | |
|---|---|---|
| 2.1 | Kohlberg's stages of moral reasoning | 31 |
| 3.1 | Social and environmental values of corporate and potential entrants | 82 |
| 4.1 | The most trusted professions | 89 |
| 4.2 | Comparison of professional conduct rules | 96 |
| 4.3 | Measures for Considerate Contractor Scheme | 112 |
| 8.1 | Retrofitted measures and cost comparison | 207 |
| 9.1 | Tests for partnering | 226 |
| 10.1 | CIOB survey on corruption perception of different activities | 257 |
| 10.2 | The determination of a bribe, gift, tip or price | 258 |
| 10.3 | TI's Bribery Payers Index, 2002 | 276 |
| 10.4 | Countries in range 8–10 on the TI Corruption Perceptions Index, 2005 | 277 |
| 10.5 | Countries in range 0–2 on the TI Corruption Perceptions Index, 2005 | 277 |
| 12.1 | Corporate differences | 299 |
| 13.1 | Comparison of traditional and BECC costs and time | 302 |
| 14.1 | Comparison of Bull Ring bids | 314 |
| 20.1 | The Student Village collaborative activities programme | 353 |
| 21.1 | Broadmead city centre expansion – does it achieve its goals? | 359 |

# Acknowledgements

Without the help and encouragement of many people, this book would not have been completed. I have been most fortunate to have an encouraging mentor, Ron Harte, who would discuss some of my thoughts and bring them into perspective whenever I needed it. He was also responsible for organising two important ethics events that brought together people from within the University of the West of England who share an interest in ethics. These events were a good place to share and develop ideas. I have also had encouragement from the Chartered Institute of Building. Jim Mason has generously contributed an excellent chapter on the specialist area of good faith in ethics (Chapter 11) and I thank him for that. I thank others who have generously given their time to provide material for the case studies. These have added a practical dimension to the best practice dimension of ethics. Included in this list are Bill Haley, Colin Rooney, Quentin Leiper, Paul Donoghue, Steve Iddon, Ray Brown, Barry Wyatt, Nicola Kingaby, Joanna Davies and Jerry Swaine. I am very grateful to them for their patience in explaining things to me and drawing my attention to an ethical approach. This has been my gain, but I take full responsibility for any mistakes. I thank Carol Graham for helping me to improve a needy script for a second time; she has been enormously helpful. And of course I thank my long-suffering family who have encouraged me and put up with my distraction away from them. I am also grateful to the University who have helped finance my time and research, and to Tony Moore my editor.

Peter Fewings
January 2008

# Introduction

Ethics is very hard to define and it should have a definition understandable or relevant to more than a few, as in philosophical ethics. In the context of the built environment, professional ethics are often quoted in the form of professional codes of practice, but there is some criticism as to the effectiveness of these codes and some public suspicion that they might be self-serving. The discipline of business ethics has many overlaps, though not exclusively so. *The Journal of Business Ethics*[1] describes ethics simply as 'all human action aimed at securing a good life', putting the responsibility on us to work out our own solutions. The context of business is understood 'to include all systems involved in the exchange of goods and services', which is relevant also to the supplier–customer relationship in many of the transactions that take place to create and maintain the built environment. This book will also consider the broader public policy issues which also influence the quality of our lives attributable to our physical surroundings.

Business ethics is divided between a stockholder view – propounded by Milton Friedman[2] who argues that ethics and economics primarily intersect at the market level – and the wider stakeholder view where it is argued that ethics and economics also intersect at the organisational and individual levels (Dienhart 2000).[3] Both views are argued from an ethical, economic and legal perspective. Friedman's view is that a manager's prime ethical duty is to increase profits and to use these profits to increase stockholder value. The manager should respect their rights, and that wealth creation will promote the general social good, according to Friedman. It is Dienhart's more holistic view that drives this book and broadens the argument. The business dimension therefore includes management ethics, human and employment rights, organisation, corporate governance and social responsibility reporting and feedback, stakeholder theory, trust and collaboration, professional decision-making, research and development, company values, competition, corruption, risk and value management (risk–reward relationships), occupational health and safety and quality, entrepreneurship and property development and contract law.

For the purpose of this text, the underlying approach is to define and apply ethics in the area between legal compliance and moral expectations in the built environment.

## The built environment

The built environment describes the man-made environment in which we live, and in recent years there have been particular ethical and moral issues that are connected with the development and maintenance of the built environment, which includes buildings, engineering structures and the spaces, mix and juxtaposition of such structures. The development life cycle of an asset can be described as the urban planning, inception, design, development control, construction, occupation/use and deconstruction of development, when the cycle might then restart. During this period many different professionals are involved in making decisions which have a long-term effect on the well-being of all of our lives, whether we are involved as clients, suppliers or as bystanders affected by the development of the built environment. Sustainability of built assets is just one example of where choices have to be made. Decisions in this area might affect the use and quantity of the asset and its components, the design, the planning constraints, the best construction method and the way in which the building is operated. In making these decisions it is often postulated that a stakeholder approach is appropriate where the needs of many different parties are considered. However, the existing way of doing things allows most of the decisions about the building to be made by the developer or client and the designer. They in turn have the best access to the planning authorities who are charged to act as the community's final arbiter in assessing the acceptable social and political impact.

In addition, the business application is supplemented by personal ethics, sustainability, politics, education and knowledge transfer, access, professional conduct, transactional analysis, legal and voluntary codes/contracts/best practice, developmental economics and psychology. All these have a bearing on our actions, decisions and behaviour. This book is essentially a book about people's interactions to support the technical understanding we have that informs our disciplines and adds additional dimensions to organisational decision-making. Some of the general principles discussed could be applied to situations outside the built environment and relevant business moral issues.

## Responsibilities and stakeholders

Responsible and integrated behaviour is important to present a holistic ethical view and to expose the ethical dilemmas that easily crop up for professional players in the built environment who have responsibilities to their employers, to the community in general and for the credibility and trust of their professional expertise. Ethical management in this context requires a cool head and an ability to balance a wide range of stakeholders while recognising a special duty to consider and understand the values of the party paying for your services. Choices are made by both sides about the compatibility of the client and supplier values in doing business together. It is also clear that ethical and moral outcomes emerge from the personal values and experience of the decision-makers and their integrity, transparency and fairness and their ability to work with others. As the ethical response is revealed, employers/employees may make some mutual assessment about who they want to work with.

The construction process has been criticised for poor ethical trading relationships and for corrupt practices (TI 2005).[4] Clients have felt that the product has not been up to the expected standard and that the service has fallen short of the standards they have been used to. Various reports on the industry have dealt with matters such as customer satisfaction, delayed supplier payments by dominant players, substantial unrewarded tendering costs, poor client advice, sub-standard health and safety measures leading to unacceptable accident levels and wasted resources, lack of innovation and research to reduce costs, and overruns on cost and time with unacceptable levels of disruptive defects. Although there are some mitigating factors for these criticisms, and there has been a sustained effort to improve the level of service, with some attention by government to legislate to force change, there still seems to be some deep-seated ethical inconsistencies in the process.

There is also a tendency for best practice to be ignored because of the fragmented nature of the industry. In the UK, but it is the same in other Western economies, 25 per cent of the total work is carried out by the top 50 major contractors, many of whom have worked hard to improve their customer satisfaction ratings. Some 40 per cent of the work is carried out by a large number of small contractors (c.190,000) who are able to tackle the smaller jobs and come face to face with the vast majority of clients who commission work often as inexperienced one-off individuals. These individuals are often subject to the traditional building process which can be quite a complex, fragmented and confrontational process, where they do not understand the rules and consequently they fall into common contractual traps. Where this happens, they often have no one to help them through the 'maze' and this leads to compromised designs, extended contract periods and over-extended budgets often ending in expensive lawsuits. Of course, one side will blame the inadequate budget or the poor design, but there clearly are ethical issues where what was promised or the budget agreed in the time agreed was not attained. This could be blamed on poor communications, unmanaged resources and a lackadaisical attitude to the needs of the client. The particular need of the client is to be informed.

### Exclusions

In order to contain the extent of this book I need to skate over much ethical theory and specifically exclude in-depth philosophical history (though some is appropriate), religious ethics (except for its influence in a cultural setting), accounting and financial process ethics. This might upset some, but I hope that the book will inform those wishing to understand the subject application to the built environment, and will be challenging for those who have wider experience in the development cycle, whether in planning, commissioning built assets, property development, design, construction and project management or facilities management.

## Methodology and structure of the book

The book will try to describe possible choices for making decisions rather than taking a moral stance. Inevitably I will be conditioned by my own values and environmental background. It is also useful to make comparisons between the general status quo perceived to exist in different contexts and cultures. In this context an international perspective will be introduced in Chapter 10. In general, the ethical case studies are ones applied to Western cultures and international trading.

The book is divided into two parts and deals with the theory and application context relevant to each development cycle stage in Part I, including a framework for decision-making, and seeks to apply them appropriately in Part II. Chapter 1 will look at the development of ethical and moral thinking. Chapter 2 presents the moral dilemmas created generally and applies them to built environment issues. Chapter 3 looks at the organisational framework for corporate social responsibility as an integrated whole and as a company's ethical response to its stakeholders. Chapter 4 looks at professional ethics and the distinct responsibilities of a professional. Chapter 5 deals with the organisation and will concern itself with employment and employee responsibilities and the whole area of diversity and equality in the property and construction industries. Chapter 6 discusses the ethical issues that connect the concept of health and safety with quality and considers the idea of managing risk in the process and outcomes of development projects. Chapter 7 looks at the beginning of the implementation stage in development and at the philosophy which is behind the development control system in different contexts, and considers possible implications. Chapter 8 looks at the future of development and the concept of sustainability ethics in the built environment. This is an integrated concept, balancing the economic, social and environmental aspects of the built environment and developing the project-related aspects of sustainability rather than an organisational response. Chapter 9 is a study of a key ethical aspect of project and trading relationships – trust – and considers the claims for collaborative working relationships in the industry. Chapter 10 looks at the dark side of the industry in terms of unethical action through bribery and corruption. It has both a local and an international dimension and considers key aspects of the Transparency International reports and some of the global action to combat corruption as it appears in built environment procurement. Chapter 11 is a focused consideration of how good faith can successfully be incorporated into construction contracts and be supported by the courts. Part II presents 10 brief case studies illustrating aspects of the application of good ethical practice in the construction industry.

## Ethical decision-making

Ethical decision-making is an important procedure, because most of us at some time have to make decisions on behalf of others which can be judged as good or bad or right or wrong decisions that must satisfy our standards and values. This might end up being a compromise, so how do we judge what is acceptable? Harrison (2005)[5] distinguishes five types of ethical investigation. *Philosophical* or *meta-ethics* (about ethics) may define the dilemma or even give us a moral answer, but may be quite theoretical. *Normative ethics* may give us one answer, depending on which view we follow. *Practical ethics* may provide us with more rules of application and practical models often combine more than one normative approach to suit particular organisational values. *Descriptive ethics* uses empirical studies and experiential learning to guide understanding and give principles or precedents for decisions. This makes it important for the observations to be relevant to the context and may take a long time to evolve. In practice, more than one approach may be used which reflects organisational objectives and values. International cultures are almost certain to differ and so complicate multinational trading and consultancy for the built environment. This is why it is so important to communicate organisational values and objectives. Surprisingly some construction organisations do not have positive ethical guidelines (Fewings 2006).[6]

Typical questions we can ask are:

- What is the role of trust, is it right to ever tell a lie? If yes, under what circumstances? Do we need to protect our commercial interests or are they protected better in the long term by building open relationships?
- What is a sustainable policy? Is the planning system fair? In the UK, 50 per cent of clients believe the planning system should be improved to help regeneration (Puckett and Stocks 2006).[7]
- Should we use low priced labour from abroad to reduce the price of our product? Many manufacturers are using cheap labour abroad which they say boosts the economy of a developing country and their global competitiveness and ability to meet demand. Unions at home say that their members are losing their jobs and this is short-termism.
- Do we have a fair tendering policy and contract? Is competition or collaboration fair? Many major contractors argue for fewer regular, long-term partnerships as an efficient method, but the EU Public Procurement Directive presses for open tendering to be the norm.
- How should corruption be defined? Can we trust others? TI finds a lot of corruption in the way construction and consultancy contracts are awarded (TI 2005).[8]

On top of these political and organisational policy questions, operational ethics affect our everyday relationships with employees and contractors and clients.

The development of trust is not really an option, but the degree to which we operate it may be important to the quality of the product.

## Notes

1 *The Journal of Business Ethics*, Home Page. Available: http://www.springer.com/west/home/philosophy?SGWID=4–40385–70–35739432–0 (accessed 18 December 2007).
2 Friedman, M. (1967) Presidential Address to the American Economic Association.
3 Dienhart, J. W. (2000) *Business, Institutions and Ethics: A Text with Reading and Cases*. New York: Oxford University Press.
4 Transparency International (2005) *Global Corruption Report: Corruption in Construction and Post Conflict Reconstruction*. 16 March.
5 Harrison, M. R. (2005) *An Introduction to Business and Management Ethics*. Basingstoke: Palgrave Macmillan.
6 Fewings, P. (2006) 'The Application of Professional and Ethical Codes in the Construction Industry: A Managerial View', *International Journal of Technology, Knowledge and Society*, 2(7): 141–50.
7 Puckett, K. and Stocks, C. (2006) 'Jolly Green Clients', *Building*, 29, Survey among clients by Carmague.
8 Transparency International (2005), op. cit.

# Part I

# Theory and application

# Development of an ethical framework and the built environment

The built environment has been planned since the time of ancient civilisation and groups of humans started working together. History provides a rich source of anecdote and case study with plenty of opportunity to consider other cultural ways of delivery. Some ethical impacts can be assumed by comparing the parallel development of philosophical ethics and the success of the built environment.

The aim of this chapter is to consider the ethical principles which have evolved to the present day and their impact on delivering our built environment. The main points are:

- ethical principles in the development of the built environment;
- good and bad, right and wrong – various ethical approaches, virtue theory, natural law, justice and morality, proportionalism, utilitarianism, and relativism/situation ethics;
- applications to the built environment.

## Development of building and its impact on ethics

Vetruvius (27 BC)[1] wrote one of the earliest surviving texts on building. *De Architectura* consists of 10 books covering the subjects of architecture, engineering, town planning, landscape architecture, mechanical engineering, water supply, and material science, coming up with the fundamental building principles of durability, convenience, and beauty (often interpreted as firmness, commodity, and delight), and the design principles of order eurhythmy, symmetry, propriety and economy.[2] He writes of the ethical responsibility of an architect:

> As for philosophy, it makes an architect high-minded and not self-assuming, but rather renders him courteous, just, and honest without avariciousness. This is very important, for no work can be rightly done without honesty and incorruptibility. Let him not be grasping nor have his mind preoccupied with the idea of receiving perquisites, but let him with dignity keep up his position by cherishing a good reputation.

The idea of engineering was born and developed from the ancient Greek passion for science 300–400 years before Vetruvius. Civil engineering developed separately from military engineering, and applied science to roads, buildings and other permanent town structures, particularly under the Romans who were well known for their road and wall building. The Roman style was distinctive, and they built structures which accentuated civic pride and orderly government. The ethic of both these styles was to indicate the common good of society and the authority of government.

Vetruvius shows an amazing modern relevance in his perception of building purpose rather than self-promoting building. He referred to building economics as well as the values that different types of clients place on their buildings. He understood the nature of 'place' and the context in which the buildings stood, with special reference to public buildings. His chapter on defence may not be outdated in the security crisis that the developed world is facing today. He also understood well the principles of environmental science and the physics of building climate and hygiene.

The gothic style, mainly used in majestic religious structures, emerged in Europe from the eleventh to the sixteenth centuries and heralded the next great cycle of the master builder-designer. The gothic style emerged from the desire to let light into the great cathedrals and used the pointed arch, partly inspired by Islamic architecture, to enlarge the window openings and increase their slenderness. The ecclesiastical style of Islamic and gothic architecture were inspired by the ethic to glorify God and huge projects were devised and built over many decades.

Architects re-emerged under the Italian Medici-inspired drive for artistic differentiation in the sixteenth century with Leonardo da Vinci and Michelangelo. The symbolism and beauty of their work combined religious fervour and a more down-to-earth egoism, thus promoting elitism among leading Italian families, who were sometimes dubbed the Italian godfathers.

The Industrial Revolution started with the invention of steam engines at the end of the eighteenth century and rapidly changed the face of Western Europe and North America with the advent of power for manufacturing machinery for mass production and transport. This was the new golden age of steam with engineers such as Watt, Brunel, Whitney and Singer, who with restless energy lobbied the industry barons to invest in their schemes which championed the might and power of the entrepreneur. This was an ethic of industrial egoism to the glory of profit and achievement. Ecclesiastical and civic buildings were grand and represented an ethic of compensation and duty, with entrepreneurs using public buildings to provide beneficent charity. The development of mass production further organised and divided labour with entrepreneurs like Ford and F.W. Taylor who were the fathers of scientific management and reinforced the functionality and productivity motive at the end of the nineteenth century.

The Cadbury brothers and the Quaker movement in general provided a softer face to the welfare of the workers at the beginning of the twentieth century. In contrast to Ford and Taylor, they had a much more holistic approach in the ethic of Garden Cities where better living conditions within the metropolis were used to improve the work ethic and achieve a better work–life balance. During the nineteenth century, the design process split off from the construction process, and engineering and architectural practices began to build up professional societies which offered guarantees to the public of their competence and public responsibilities.

The age of the formal town planning regime slowly evolved from the beginning of the twentieth century to deal with slums and control large-scale rebuilding programmes. It produced opportunities for the private property developer after the Second World War when large tracts of urban land needed to be redeveloped. However, there was a tension between the egoism ethic of the developer and the Kantian (duty) ethic of the control regime represented by the planning authority. These acts were modified after the Second World War with the development of new towns and the setting aside of green-belt in the 1950s to reinforce a modern virtue ethic in the face of rising private development. Sustainability and public health are perhaps two ongoing issues that we need to face in the twenty-first century, which bring the need for more disciplinary integration into the building process, and they have re-ignited what might be called the corporate social responsibility (CSR) era, where large corporate organisations now require to pay much greater attention to the impact of building upon the environment and the social fabric of more sustainable built environments. All planning applications now also require a greater attention to sustainable features in buildings and property development.

Buildings have evolved through these series of ethical justifications (Figure 1.1), but later came the driving socialist contribution of providing jobs and places to live. This has taken different forms internationally. There is, however, a sense that this system has compromised the needs of development and the compromise arrived at may not satisfy all of the parties.

Kant's deontological (duty-based) approach says the outcome is immaterial if the advice is entirely impartial and that we have fulfilled our moral duty to society. This satisfies certain moral behaviour towards each other, including

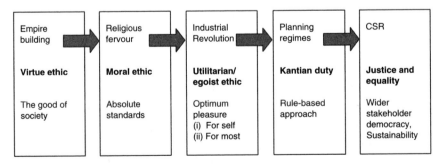

*Figure 1.1* The development of ethical drivers of construction.

honesty and fairness in an equal way – in short, our intentions need to be right. However, it is quite possible in the planning system for inequalities to appear and for one party to be favoured over another, depending upon their influence, their financial incentives or the lack of knowledge of others of the proposed impacts. Are planning systems tending towards a politically correct Kant-like duty-based approach emphasising the impact on others? If so, is this the ethical approach we want, or do we need to consider the quality of life factors of those who need to live and work in the buildings more than we currently do?

## Integrated professional behaviour

Groak (1992: 61–62)[3] reminds us how the building process brings together the treatment of design and building principles and further suggests that we distinguish falsely between white- and blue-collar workers (the thinkers and the doers) and that the building process is best delivered by recognising an integrated creative effort in design and production, which has not gone unnoticed in history. This is an ethical matter at the root of much division and has stilted development of the building process. Its resolution has the potential to lead us into more suitable forms of delivery for the twenty-first century. For example, boundaries now between the design, manufacturing supply and site production assembly processes are more blurred than previously, and new forms of procurement require us to break down the old divisions.

He also refers to the need for more building research to match the move from independent inventor to organisational research and development plans. Perhaps it is the slowness of the industry to move to a developing rather than a precedent-based governance of the building process that has earned it a reputation for being inefficient and giving less than optimal value to the client (Egan 1998;[4] Latham Report 1994[5]). It is the later Egan (2002)[6] and Fairclough (2002)[7] reports that have triggered a sustained drive in the UK to more efficiency and a more customer-orientated approach.

Ethically the two later reports put an obligation on the suppliers of design and production services in the built environment to integrate and get to know their customers' needs and to work with them to increase transparency and commitment. Fairclough advocates more innovation in the industry to match needs rather than imposing old methods. It is a call for a combined outcome and process approach to delivering the project.

## The basis of ethics – good and bad, right and wrong

Essentially ethics is actions that exceed a legal compliance. There is much debate in the philosophical arena about the definition of ethical behaviour. Many philosophers have sought to clarify the position of morality and ethics and have come up with many theories on the difference of emphasis between the concepts of good or right. Happiness or fulfilment have also been strong contenders in differentiating these theories and they can be easily split into consequentialist and non-consequentialist theories. Kantian, social contract and natural law theories are rule-based and concerned with good processes. They apply, whatever the consequences. Virtue ethics, egoism, utilitarianism and justice theories are concerned with good outcomes.

### Virtue ethics

Virtue ethics is a moral approach with a concern for the community and the identification of desirable universal qualities. Virtues were formally espoused by Plato and were later developed by Aristotle who consolidated an ethical frame-work in his work *Nicomachean Ethics*, in 10 volumes. This is still a classic text for virtue ethics. In modern times, Alasdair McIntyre has developed virtue theory in a form called neo-Aristotleism and has helped to identify an applica-tion for virtue ethics in modern communities.

Plato, who lived in the fourth century BC, suggested that good was an absolute moral value and that morally no human being was capable of living up to the standard which was only possible for the gods. His theory of *forms* identified such things as beauty or justice or good and these forms were chosen as being timeless, spaceless, changeless and immutable. However, we live in a world dominated by time and space dimensions and compromise. His philosophy of ethical behaviour therefore referred to approaching the good and the perfect. He described activity and behaviour which was less than good as *shadows* which shielded the full view of the *forms*, but claimed that it was possible in a hierarchy of greater clarity to move into a purer light or, in his analogy, nearer the cave entrance.

In defining a good house, we might agree that there is a perfect match for the needs of the building users, but practically we might agree that some comprom-ises might need to be made to mitigate harmful effects on others. In defining a good builder, we might think of quality or value for money or minimum risk or courteousness. Good in practice, then, is hard to define, but more of us might be able to apply the continuous improvement theme of getting closer to the 'cave entrance'.

Plato lived in a stratified society called the State and distinguished three classes of citizen: the rulers, the soldiers and the people. He passionately believed in a behavioural ethic which was for the good of society. For each class he introduced the idea of a particular virtue or social quality which he believed should exist for the moral good of the state. For rulers, it was the virtue of *wisdom* which is described as the 'capacity to comprehend reality and to make impartial judgements about it'. For soldiers, it was courage which is their 'willingness to carry out orders in the face of danger' and for the people, it was necessary to follow their leaders and he prescribed *moderation* (or *self-control*) which is defined as 'the subordination of personal desires to a higher purpose' (*Republic*, 433e, as in Kimerling 2001).[8] Although these describe a situation a long time ago, there is a great deal of overlap with the ethical working of a corporation, and virtue ethics is now often adopted as a base for an ethical business approach.

Plato also applied these three virtues to individuals as having three 'souls' – their rational soul for thinking, their spirited soul for willingness and their appetitive soul for feeling (*Republic*, 436b) (Figure 1.2). This division of human nature has influenced Western tradition and provides a basis for human actions on a different level and complements the idea of an absolute good.

Plato was teleological which means he held that acting morally and suffering were better because this would not hurt your soul. Plato argues against a relativism where good was relative to the context in which it was carried out, so ugly was not beautiful. When compared with something even uglier, it was still ugly. The problem was, how do you know you have arrived out of the shadows and have sufficient maturity to judge? There is no guidance on the practical day-to-day situation.

Aristotle's virtue theory grew out of a secularisation of Plato and it tackles the question of what a virtue is, but he recognises the *evil* or bad that can come out of an excess or a deficiency. *Nicomachean Ethics* has provided the basis for a natural law approach to science and ethics to this day.

Aristotle started with *eudaimonia* in Greek, often translated as good, but meaning fulfilment and justice done and this was considered to be the key aim which supported a society that had wisdom, courage and self-control and who could act as guardians for the good.

Aristotle described twelve moral virtues such as courage and patience and each of these was a mean between the extremes of excess and deficiency. So courage would have rashness and recklessness at one end of the spectrum and cowardliness at the other. Patience was seen as the mean between too much anger and a lack of spirit. Generosity is a virtue, but an excess of it may put the benefactor in debt and a scarcity of it is miserliness. There were also nine intellectual virtues – technical, scientific, prudence, intelligence, wisdom, good deliberation, judgement, understanding and cleverness, which complemented the 'irrational' choice of wants and desires with the rationality of the virtues which assessed logic, fact and truth. This model is indicated in Figure 1.3.

Aristotle believed there was a secular purpose for everything we do and that we uniquely possessed reason which needs to be exercised for the right choices. The superior aim is the supreme good which is defined as happiness, which, for

| Soul | Function | | Virtue |
|------|----------|---|--------|
| Rational soul | Thinking | ⟶ | wisdom |
| Spirited soul | Willing | ⟶ | feeling |
| Appetitive soul | Feeling | ⟶ | moderation |

*Figure 1.2* Plato's individual souls (psyche).

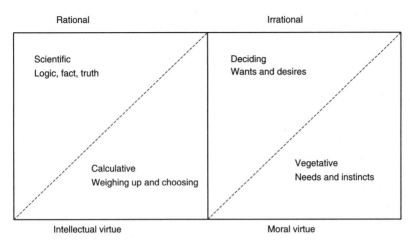

*Figure 1.3* Aristotle's account of the soul.

*Source:* Based upon Vardy and Grosch (1999)[9]

Aristotle, meant the well-being of all, and was not a temporary condition of one person, but applied to society as a whole.

> All men seek one goal, success or happiness. The only way to achieve true success is to express yourself completely in service to society. First have a goal . . . Second have the means to achieve that end . . . Third adjust your means to that end.
>
> (Aristotle 384–322BC, in Handy, p. 30)[10]

This quote clearly makes Aristotle's ethics theory a consequentialist approach, indicating the predominant emphasis on the objective, with some qualification on the means through the practice of the virtues.

There are problems in this approach which are based on the changing definition of virtue that might apply to modern-day thinking which may consider them sexist, ageist, speciesist and racist, given the pride Aristotle had in the virtues of his own society. However, the rationalistic nature of Greek society in expressing things mathematically provides some comparisons with a modern scientific approach. Currently we might consider it obscene for a beautiful building to have an excessive carbon footprint and we might consider it a virtue for superior energy efficiency which enhances its aesthetics in our mind. We might also consider it spartan to have a cold building.

An Aristotelian approach fits well with stakeholder theory where there is social concern for the *good* of the wider external stakeholders, including those affected by new developments and the good of users. Many architects have a philosophy regarding the beauty or aesthetic of the building in its context and

Aristotle interpreted beauty in the good that it will bring and how much good it can spread to other people. A building which is an eye-sore if it is functional is not necessarily bad, but a beautiful functional building would have a double effect if it also lifted the spirit.

### Neo-Aristotelian ethics

Later *consequentialist* theories look at maximising pleasure, and range from egotistic, 'What effect will my actions have on me or my group?', to utilitarianism, which widens the question to 'How will my action affect others?' Alasdair McIntyre (1985)[11] calls for a return to a more absolute theory of morals, moving away from what today is called liberalism and the acceptance of a relative standard of tolerance based on the context and society norm. Practically, this means a reflection on a wider consideration of human practice and traditions that have evolved over history in order to establish a single standard of behaviour. This is radical thinking and echoes some of the Aristotelian virtues in a modern context. Harrison (2005)[12] likens this to a 'narrative quest' in the development of not just an understanding of 'good', but the development of virtues such as wisdom and self-confidence through experience and practice. This is relevant to a mature application of ethics in a tough setting where there are plenty of conflicting demands and a need for mature guidance based on significant experience and reflection on lasting principles.

### Egoism

Egoism is when each person does what maximises his/her own interest. This is a hard concept to derive as an ethical approach as what is done for the good of self might be regarded as a dangerous precedent which might harm others or, at the very least, not consider their needs. This is called simple egoism. Some actually argue that everyone would be better off if they acted in their own interests, which is a sort of universal egoism. This is based on the idea of liberating behaviour so that the number of controls, rules and regulations is minimised, helping everyone to better happiness with the freedom it brings. This is discounted by the prisoner's dilemma (see Chapter 2). If both prisoners confess, doing the right thing, they both get less than if both do not confess and if they were found out, they get the worst sentence. If one confesses and the other is not found out, then the one who confesses is worse off (the most unfair!). It is clear that for this type of egoism not to dissolve into chaos, then some sort of rule favouring a few would be applied and this is generally called 'unethical'.

However, there is a sense in which egoism can be more altruistic in that it can be applied also to 'my group' which means that you will take action which is beneficial to more than oneself. This is ultimately still for our own good in that we look out for each other.

It can also refer to a deontological approach where it 'feels right to me' which might also be a personal notion, but it is an attempt to provide a good and fair approach for others. This is also altruistic and much behaviour is based on this, such as giving to charity and treating our customers well. However, it is still interpreted, maybe cynically, as a sort of psychological egoism to gain benefit such as self-esteem or as long-term egoism where benefit eventually accrues. The difference from non-egoism is that you have not agreed it with others as an ethical solution.

The golden rule which says 'do unto others only as you would have done to yourself', also has an egoist approach where you are comparing others with your own standards. This golden rule is very helpful in moderating behaviour and 'lifting' blind spots. It helps to think, for example, if I saw my actions printed on the front page of the newspaper, would I be ashamed? Again, it is based on our own values and may still upset others which may or may not be a problem to us.

David Hume talked about the subjectivity of making a judgement of what you ought to do in the absence of a religious or other standard. For example, if you say, 'you ought to value a property at the market value for sale is good', this is simply the same as saying 'I like to value the property at its market value'. Thus our moral contention is in fact a subjective feeling because others would say 'get what you can' or 'sell it at different prices depending on who it is'. Hume claims that in order to do a moral action, we need to do the right thing for the right motivation, which makes morality more objective.

### Utilitarianism

Utilitarianism can be formally defined as 'to maximise utility aggregated over all those affected [in a good, positive, bad, negative way] by an action' (Harrison 2005).[13] This is better known as the greatest happiness to the greatest number although there are variations on the inclusivity of the group. A more esoteric approach would look at the greatest good of society and a narrow view would look at the greatest good of the family or the client or the organisation. In business terms, this is a consequential approach distinguished by the share-holder or the broader range of the stakeholder. The problem with this approach is that different things give pleasures to different groups and to different levels of intensity. Stakeholder groups might conflict – the narrower the group, the easier this might be to define. This approach which was developed by Jeremy Bentham (1748–1832) and John Stuart Mill (1806–1873) motivated many of the great social reformers of the nineteenth century. Bentham articulated seven criteria to measure the greatest good: intensity, duration, certainty, extent, remoteness, richness and purity. These potentially provided a more transparent and objective measurement of pleasure and pain to different groups and a way of adjusting the mix to suit the known issues.

**Case study 1.1  Cribbs Causeway shopping mall**

For example, what level of pain will justify the benefit given by planning consent for a shopping mall and a leisure centre? This will cause pleasure to the unemployed who will get work, and will cause pain to neighbours who suffer reduced amenity, more pollution, and risk restricted views. When Cribbs Causeway was built, a small hamlet was displaced so that instead of being in the country, it was on the edge of an extremely busy retail and leisure development, with loss of amenity. However, Cribbs Causeway is extremely popular with shoppers and families and brings a lot of pleasure and richness of choice to thousands of people every day. There are often traffic jams because so many people use the facility and people are frustrated about being trapped in the jams. Others such as small shops in the hinterland of Cribbs have lost business to the larger retailers.

However, on balance, more shops gained and more people have greater or easier choice of shopping facility than they had before, and it is easy to shop out of peak hours and miss the jams. It is uncertain whether other shops in the hinterland have not simply refocused to regain lost business. So factors of duration, certainty, intensity, richness and extent have been considered on balance as positive.

The problem comes when this sum is more marginal, that is a *large* minority suffers, but on balance there is a gain. Is it worth doing?

*Question*

Consider what factors should have been considered in utilitarianism in the decision to go ahead with a third runway at Heathrow in order to increase the number of passengers by a third.

Another problem is that as a predictive measure, as in planning consent, it is quite difficult to know what results to expect. John Stuart Mill defined the higher (mind) and lower (body) pleasures and pain and these allowed the more extreme consequences on the minority which might cancel out any benefit. So if the Channel Tunnel could be built with a prediction of only 10 people being killed, then it might be refused planning permission, but these things are not really so certain, and promised measures to reduce accidents may not work.

There are two forms of utilitarianism:

1   *Act utilitarianism* which holds that certain acts are good, as discussed above.

2   *Rule utilitarianism* which holds that rules are good and, when broken, bring evil acts.

The latter form slightly muddies the category of a consquentialist approach as it seems to support the right way of doing things, but it still has its roots in the consequence of the act to bring about pleasure and to avoid pain.

The problem with utilitarianism is that happiness and pain mean different things to different people and it sometimes is quite hard to calculate the effect. Utilitarianism does not seek to deal with individual motivation and the final solution does not underline a good action for a single person and it does not refer to an end in the sense that it develops a person or produces a fruit, as in virtue ethics.

*Post-utilitarianism* has developed and Peter Sedgwick tried to rescue its tarnished image and proposed that: (1) a future greater good was better than a present lesser one; (2) that you treat others as you would wish to be treated; and (3) you regard the good of others as much as your own.

### Justice and morality

The other consequentialist approach is the ***justice and morality*** approach espoused by John Rawls (1971).[14] He talks most famously about acting behind a 'veil of ignorance' which refers to dispelling our preconceived ideas of position and influence. He has two principles of justice which would help a fair and equal distribution of social and economic advantages across society. These two principles are:

1   Equal rights to basic liberties compatible with others.
2   Equal opportunity for offices and positions to benefit the most disadvantaged.

This was based on the formation of a social contract which has an in-built concern for those who are least well-off, in an attempt to produce a level playing field in society. It tries to identify principles which will be the foundation for present and future generations in a local context and this is regardless of any adverse effect on those delivering the contract. This approach would be relevant to the establishment of a social policy which would provide a quality of life for all. In theory, it might mean that an agreement might be brokered for fair, on time payment between a large main contractor and a small sub-contractor that would protect the rights of both equally. In practice, a small supplier takes what they can get. Courts of justice are not an economically viable option for a small organisation and a more collaborative approach is practical on the basis of future work.

Nozick (1974)[15] distinguished a less benefactor-based approach with a personal contract committing to a just system for fair acquisition, transfer and rectification of wrongs. This did not include a proactive moral obligation on behalf of others, but a morality through personal example. This might be applied to the present division between organisations who act ethically to protect shareholder returns and those who broaden their social concern by reporting on their corporate social responsibility for a range of stakeholders and whose objectives are to put something back into the environment and social structure.

The justice system is not necessarily ethical in everyone's definition. Poor laws can produce injustices. In most cases the system provides a starting point for the establishment of an ethical standard. The legal process of having opposing advocate claims about the innocence or not of a person using language like 'lost your case' can itself raise questions about the morality of protecting a known guilty party, but also speaks volumes about the difficulty of finding a perfect moral solution.

## Right and wrong

Right and wrong have different connotations and refer to *non-consequentialist* ethical approaches which are based on acting in the right way so that you want others to follow your example. They are often given a moralistic tag. Thomas Aquinas's natural law and Immanuel Kant's moral law approach are duty-based and do not consider the individual circumstances or consequences. A duty-based approach is termed deontological. In this approach, the end does not justify the means and is therefore non-consequentialist.

### Natural Law

Thomas Aquinas (1225–1274) was a Christian theologian who believed that human beings, though fallen, had a leaning through their conscience and reason to understand an absolute law of right and wrong. He wrote from a religious perspective and believed there was an externally revealed moral standard and that the purpose of existence did not lie entirely in this life. His work has been taken as universally relevant because of his belief in the *natural law* tendency towards goodness which included those who were not religious. He believed that human nature, at its core, was a good guide to morality and through conscience understood natural law, which formed a basis for right and wrong actions. In Aquinas's language, lying would be wrong in any context and the goodness of the intention would not justify the means, e.g. stealing to help the poor. As no one is perfect, it depends on the premise that natural law can be deduced by experience from a study of human nature and its purposes. These judgements of right and wrong would therefore depend on the judgement of the current guardians of right and wrong. A lot of society's laws have developed under particular religious systems. Morality is considered to have weaknesses in an intercultural context if absolute or even relative standards are to be applied.

There is an argument in natural law for absolute standards based on an 'archetype' applicable across religious or humanistic doctrine. For example, love of our fellows and the acceptance that killing is wrong. This is a non-consequentialist approach.

An example of a natural law approach is the application of restricted Sunday open hours which would have emerged out of the Ten Commandments. The approach led to Kant's rather more agnostic approach of moral law.

### Moral law – Kantian ethics

Immanuel Kant's approach[16] was moral rather than religious. He set out a universal set of principles for ethical behaviour which he termed *categorical imperatives* (CI). A professional code of conduct is based on this approach, such as declaring conflicting interests. This allows for effective monitoring. It could include more open-ended commitments which describe the relevant process. This is a system that needs policing but provides early indications of unethical behaviour. He believed that truth was known independent of experience and he stressed moral duty rather than moral good. His theory is often termed deontological (duty-based) because it is a matter of doing the right thing.

The categorical imperative was a way of connecting rationality with morality and vice versa. In this way, a universal law could be created separate from religious morality on the basis of the in-built fairness of the universe. He claimed that moral truth could be known and that to act morally we must act in such a way so as to lead by our example. He also believed that truth was independent of experience and that we had a duty to *act* morally whether or not it benefits us to do so. Statements like 'God is good' depended upon experience and therefore were not a moral statement. The categorical imperative was a universal law. He believed that truth is known independent of experience. The universal law was that people should always be treated as ends and not as means. Eliminating self-interest was critical to the success of the theory and he believed moral duty was good and not the outcome. He proposed three formulas:

- Act as if your action was a universal law of nature.
- Act in such a way for the action to be an end as well as a means.
- Act as if you were a law-making member, i.e. for rational means that would integrate with others wanting to do the same.

This made Kant diametrically opposed to utilitarianism. All moral acts are chosen and not arbitrary and we act in a moral way when we act rationally in accordance with the formula. Kant recognised explicitly and implicitly that corruption existed.

An example of Kantianism is the application of a strict code of professional conduct to the actions of a professional regarding confidentiality. He believed that a rule such as 'keep the confidentiality of a client' was sacrosanct unless there was a case where the law was broken by the client. The rule should be maintained even if it meant some degree of discomfort or injustice to others who might be affected.

All of the judgement on morals Kant held as an *a priori* process, i.e. not depending on observations of results. The objection to moral law is a religious one, in that if all human beings act corruptly or irrationally, can they not be 'regenerated to moral acts'?

Moral and, to some extent, natural law takes no account of the consequences. However, it can easily be argued that different contexts need different approaches when dealing with different cultures. For example, a pre-payment may be seen as a bribe in one country and a commission in another.

## Proportionalism

Proportionalism takes the edge off Kantian ethics by recognising the responsibility of a manager to consider the outcomes. An act may be morally right, but you need to distinguish between good and right outcomes. A proportionalist will have the experience to recognise the situation in contravening the rule to mitigate the outcome. This may be very appropriate in medical ethics, as the duty of a doctor to save someone may put his own life or both lives at risk. In proportionalism, this would allow the doctor to consider whether he is in danger in doing so, or to weigh up the risks of other outcomes such as loss of quality of life. A 'last in first out' policy for redundancy may be adjusted where there is a critical effect on the efficiency of the remaining team. The legal system, which supports defending a known murderer and getting them off the hook on a technicality, is a Kantian approach and not a justice approach. It supports the un-proportionate interpretation of the law. Proportionate ethics would provide a place for the victim's family's protection and quality of life.

## Situational ethics

Situational ethics is a relative application of ethics and fits in well with post-modernist thought which puts a lot more emphasis on self-determination, rejecting some of the past certainties and positively allowing alternative thinking in the context of the individual. Postmodernism arose from the disillusionment after the Second World War with the ability of governments to establish a central ethic that matters to everyone, and therefore tends not to reject moral principles, but tends to subject them to a less legalistic interpretation where the context of the principle is even more important, like a sort of Robin Hood approach. Thus, situational ethics applied in this context could be dependent on any, or all, of the reputation of those making the decision, the context of the act and the culture and past experiences of the people who are affected. It might come up with different ethical solutions and by itself would use the sensitivity of the people around to find clues for an ethical solution. In a very real sense, it is subjective, but business can adopt it as a way of personalising the corporate approach to their customers and employees.

There is also a body of situational ethics that uses *agape* love (love for the good of the neighbour) as the underlying principle to eclipse all others (Fletcher 1966).[17] Joseph Fletcher was a clergyman and his application finds its roots in the principle of love in the Bible and, although it has a relative outlook, it has a strong moral base. Therefore, the context is the 'rule of love', so truth is still an important principle, but a lie told in love is good, but a lie that is told unlovingly is wrong, so love rules. Fletcher claimed there was a 'coalition with utilitarianism'. This principle has been criticised because it is difficult to make the boundaries clear for those disadvantaged by the position, who believe that the 'love is served better principle' decision is discretional.

Thus, situational ethics might approve of the assassination of Hitler before 1939 because millions of lives would have been saved. However, it is easy to justify an action from your own perspective as Hitler and Stalin might well have done in justifying their actions.

### Moral scepticism

Moral scepticism may be a view held by some who wish to liberalise action, and really asks the question 'Why should I be moral?' According to Wikipedia, moral scepticism is the view that there are no objective truths in morality. It identifies three forms:

- There are no objective truths because morality does not exist (ethical nihilism).
- Moral truths exist only when they apply to a particular culture (ethical relativism).
- Moral truths are correct so long as you hold those beliefs (ethical subjectivism).

> The moral sceptic's conclusion is that supposedly objective values are merely useful fictions that function for such purposes as social preservation. Furthermore, it is possible to invent moral values that are more likely to further our actual desires and interests as human beings living in particular historical circumstances.[18]

Nihilism is the belief that all values are baseless and that nothing can be known or communicated. Its context is in a revolutionary Russian movement that rejected the authority of the state, the church and the family. Although its roots have popularly been associated with destruction and revolution (annihilation) through the anarchist leader Mikhail Bakunin (1814–1876), its philosophical roots are in seeing 'rationalism and materialism as the sole source of knowledge and *individual* freedom as the highest goal'.[19] Nietzsche helped popularise this theory because he believed that there were no true altruistic actions and that instead our actions have a motivation to influence others for our own good. He proclaimed: 'Every belief, every considering something true is necessarily false because there is simply no true world.'[20]

## Application to the built environment

Ethical judgements which are made about complex social interaction need to be robust, but should not be dismissed if they do not have universal application. They might in some of the theories be permissible as a precedent in similar circumstances. An integrity approach would demand, however, that justifications, particularly by influential or dominant organisations or institutions, should not paper over the cracks of the ethical objection.

### Case study 1.2   Customer care

In 1995, the Yorkshire Water Company[21] was criticised heavily for making excessive profits and at the same time not meeting its targets to reduce leaks in its infrastructure at a time when a drought was forcing customers to severely ration their water consumption. There was a public outcry which over a period of five years forced a company turnaround in its service approach to such an extent that it has won consumer awards for its performance and community commitment. A particular feature of the new way of doing business was the emphasis on customer care. In business terms, ethical service is good business, as their profits have improved.

In Kantian terms, this may be considered as driven by selfish interest – 'the company cannot survive without doing better'. In justice and virtue ethics terms, public pressure has led to a fair service and a responsible company which can bask in the virtue of a good reputation.

### *Situational ethics applied to business*

Situational ethics may well be suited to a business or project in managing stakeholders. The four working principles of business situational ethics, according to Bowie and Werhane (2005)[22] are:

1   *Pragmatism*, where an emphasis is put on satisfying as many people as possible to get the job done.

2   *Relativism*, where alternatives are reviewed and the least harm solution is chosen.
3   *Positivism*, where belief in the solution is a way of getting others on board.
4   *Personalism*, where people are put in first place.

These types of solutions can be quite individualistic with unacceptable compromises and may not be robust in the long term, but with skilled leadership and some consultation to create ethical frameworks, they may work out to be very effective in the fast-moving world of business and project management. With an extremely broad stakeholder interest, for example, in major public projects, a lot more democratic process and reflection may be needed.

Bowie and Werhane identify the need in most corporate situations for management to be delegated. Managers have their own personal ambitions and although they are restricted in the number of shares and options they can hold, they are likely to benefit from actions that they take that will maximise the share price at the expense of reinvestment in the company or in its community responsibilities. Both attracting investors and having sustainable and fair policies with some stability and fair returns for the efforts of employees are important, but this brings the need to make sure that in a conflict of interest, managers will not benefit themselves at the expense of others.

For managers to ensure that they can fulfil their ethical responsibilities, they must have a clear purpose and a good understanding of their role – who they are managing for and to whom they are accountable. Having determined their role, it is iterative and will help them to determine what a good purpose is.

### Professional ethics

Professional ethics comes from a different angle, policed by a national or international society to ensure that there is a standard of practice threshold below which there is an assumed ethical code. This may take the following forms:

*   a code of conduct;
*   an ethical code;
*   a set of rules.

This code of conduct is not always extrinsically connected with ethical behaviour, but further efforts have been made by some societies to have an ethical code and to boost public confidence by giving more transparency in investigating unprofessional behaviour. Some also interpret their code of conducts as rules that deal with practical situations, such as professional indemnity insurance cover, marketing the brand and situations for reporting others. A key concern is conflict of interest in giving professional advice or service.

Conflict of interest is the conflict where a single person has obligations to more than one party to the contract, or there is a clash of private or commercial interests with a person's public position of trust. A delicate balance can be maintained, but it is important to remove strong temptations to be partial at the expense of another. An example of the latter is the withdrawal of a councillor from the planning committee who is affected personally by the planning decision.

## Planning and urban space

From a community point of view, it is surprising that the construction of new buildings and structures in the centre of populated communities has not been more accountable to the neighbourhood. Communities have had an indirect influence on the planning process through the appeal system. However, the imperatives of urgent housing and commercial needs can dilute the ethical democracy in providing full consultation, and central government controls can restrict the room for movement locally. There is a perceived obligation now to widen the stakeholder ownership of urban space. In the UK, the Commission for Architecture and the Built Environment (CABE) has received government support to encourage a study of urban space and environmental impact even on quite small schemes. This includes money for training local councillors and raising public awareness. CABE (2005a)[23] developed 20 'building for life' criteria in an audit for developers, planners and designers. These criteria for housing cover character, roads and parking, design and construction and environment and community. They are a guide for planners, designers and developers in good practice. A silver standard is awarded for 14 points and a gold standard for 16 points.

### Case study 1.3  House owners' views

CABE (2005b)[24] in their study of the views of house owners on housing estates in the North and Southeast of England carried out 11 case studies. The resident feedback covered the issues developed in their housing audit, covering the five areas of dealing with cars (car access and parking), a sense of place, meeting the neighbourhood, public transport and community facilities. The result indicated some differences from the professional audit results for good practice although this varied from case to case. The sense of place was a strong factor for residents, according with the audit recommendations, and was achieved by the views afforded, the uniqueness and variety of housing mix and the refurbishment and conservation of existing buildings, trees and other special features. They differed inversely with dealing with cars and public transport provision as most residents discounted the relevance of public transport provision because of the poor quality of the service, but they did perceive cul-de-sacs and pavements as an essential way of making the areas safer for children. In some commuting areas, the community facility was also discounted due to the high priority for some privacy and the lack of time for social interaction with neighbours.

CABE concluded that a mix of resident and good practice planning and design for sustainability was an important factor. Residents' choice of living place was a short-term subjective mix of trade-offs to suit their current circumstances, but was relevant locally for quality of life. It was also based on limited experience and a pragmatic comparison with a limited number of other developments.

The ethical treatment of such a case is interesting as stakeholders may be selfish, but an overall compromise of complex factors may not make anyone happy at all. It also indicates the strength of diversity and context in the application of design factors and the need to treat people uniquely.

## Conclusion

Philosophers have given us a meta-ethic which gives an explanation of the practical ethical thinking that exists today. It is important to recognise how influential players in the built environment have developed their ethics and to understand that perspectives have changed in the application of ethics to development and the political actions which have supported it through the centuries. Civic building has always represented an ethical challenge as there is a feeling of collective ownership, but urban planning and building have varied in their impact and many have been disappointed by great promises to bring equality, beauty and a better society through the urban landscape.

Ethics is essentially the application of values to society and a range of different perspectives have been discussed. Ethical behaviour is important to the man/women on the street as is witnessed by the outcry when there is a hint of corruption or sleaze in public leadership. Many of these ethical blunders have been associated with procurement competitiveness and with professionals who have lost public trust through not declaring conflicts of interest. They may have otherwise taken advantage of their position in business or have pretended there is a partnership of trust that has proved only beneficial to the dominant partner.

Employers also are discovering that there are new expectations in the psychological contract that are beyond formal employment contracts, which influence their ability to retain the best young staff. The younger generation are developing a greater sensitivity to sustainability because their future is more affected by the outcomes of decisions made today, but at the same time they are looking for a lead to develop their ethical approach. They are looking for justice, but they are also looking for career development. The next chapter picks up some of these themes with particular attention to the solving of ethical dilemmas.

## Notes

1 Vetruvius (~27 BC) *De Architectura*. Trans. Morris Hicky Morgan, in 1914, ed. Tom Turner in 2000.
2 Wikipedia (2006) 'Eurhythmy is beauty and fitness in the adjustments of the members. This is found when the members of a work are of a height suited to their breadth, of a breadth suited to their length, and, in a word, when they all correspond symmetrically.'
3 Groak, S. (1992) *The Idea of Building: Thought and Action in the Design and Production of Building*. London: E & FN Spon.
4 Egan, J. (1998) 'Rethinking Construction', in *The Report of the Construction Task Force*. London: DETR.
5 Latham, M. (1994) 'Constructing the Team', in *Final Government Report of the Government Industry Review of Procurement and Contractual Arrangements in the Construction Industry*. London: HMSO.
6 Egan, J. (2002) *Accelerating Construction*. Strategic Construction Forum/DTI. London: HMSO.
7 Fairclough, J. (2002) *Rethinking Construction Innovation and Research*. London: DTI. Available at: http://www.dti.gov.uk/files/file14364.pdf (accessed 7 August 2006).
8 Kimerling, R. (2001) *Plato: The State of the Soul*. Online http://www.philosophypages.com/hy/2g.htm
9 Vardy, P. and Grosch, P. (1999) *The Puzzle of Ethics*. 2nd edn. London: HarperCollins.
10 Aristotle (384–322 BC) in C. Handy, *Understanding Organisations*. 4th edn. Harmondsworth: Penguin, 1993.
11 McIntyre, A. (1985) *After Virtue: A Study in Moral Theory*. Notre Dame, IN: University of Notre Dame Press.
12 Harrison, M. R. (2005) *An Introduction to Business and Management Ethics*. Basingstoke: Palgrave Macmillan.
13 Ibid.
14 Rawls, J. (1971) *A Theory of Justice*. Cambridge, MA: Belknap Press.
15 Nozick, R. (1974) *Anarchy, State and Utopia*. New York: Basic Books.
16 Kant, I. (1785) *Groundwork of the Metaphysic of Morals*. Trans. H.J. Paton. New York: Harper & Row, 1964.
17 Fletcher, J. (1966) *Situation Ethics*. Philadelphia, PA: Westminster Press.
18 Wikipedia (2007) 'Moral scepticism'. Available at: http://en.wikipedia.org/wiki/Moral_skepticism (accessed 10 December 2007).
19 Internet Dictionary of Philosophy (2007) 'Nihilism'. Available at: http://www.iep.utm.edu/n/nihilism.htm (accessed 10 December 2007).
20 Nietzsche, F. (1968) *Will to Power*, in *Collected Writings*. ed. L. Kaufman. New York: Vintage.
21 BBC Panorama (2006) *Whose Water Is It Anyway?* Investigative journalist report on the water companies, broadcast 6 August.
22 Bowie, N. E. and Werhane, P. (2005) *Management Ethics*. Oxford: Blackwell.
23 CABE (2005a) 'Building for Life Criteria' in *Delivering Great Places to Live*. London: CABE and HBF, November. Available at: http://www.buildingforlife.org/buildingforlife.aspx?contentitemid=384&aspectid=15 (accessed 20 September 2007).
24 CABE (2005b) *What's It Like to Live Here?: The Views of Residents on the Design of Houses*. London: CABE. Available at: http://www.cabe.org.uk/AssetLibrary/1144.pdf (accessed 20 September 2007).

# Chapter 2

# Ethical dilemmas and decision-making

The aim of this chapter is to look at the nature of ethical problems and to determine how to structure decisions that are ethical and will provide outcomes that are acceptable on a personal and organisational level. In some approaches it is the presence of an ethical decision-making process rather than the outcome which is dominant. This Kantian approach cannot completely ignore the outcomes and an unfettered utilitarian approach might rationalise itself too narrowly and justify unacceptable behaviour. A decision-making model is a blended approach to combine different ethical approaches which successively resolve dilemmas. The objectives are:

- awareness of the ethical state of health;
- to explore the process of moral development;
- to understand the nature of an ethical dilemma and provide some context, categorisation and general case studies;
- to make applications to topical issues in the management of the built environment;
- guidance for the different decision-making models and the inputs required for success.

## Definition and context

Ethical dilemmas occur because different people see their responsibilities differently. These are defined here as managerial, business and professional issues which are complex due to the conflicting impacts that decisions have on stakeholders. For a professional, there is an additional responsibility to uphold the values of their profession which may conflict with their client or employer. Each person will hold a set of personal ethics which influences their judgement between different options and stakeholders. For example, they might see a client's instruction for air conditioning as an unsustainable solution and wish to design a more naturally ventilated building to save carbon emissions.

For a business, the shareholders' good might conflict with the good of employees or the community. A sustainable solution may have a serious impact on short-term profits, or competitiveness, or on the current behaviour of employees. In the end, the company will always have been operating with some stakeholder compromises, but ethical behaviour may be seen as developmental and may alter the balance of the previous guidelines in order to maintain a good reputation. Companies with a more deontological (Kantian) outlook will seek to work out the right way of doing things and will operate on the basis of 'taking the market with them' and being selective in the trading relationships they have. Companies with a consequential outlook will be concerned less about the way they work but more concerned about the ethical outcome of their product or service. For example, they will seek to ensure a good quality product and that the products do no harm to others. Intentional acts will be taken to ensure that the 'greatest good is achieved for the least pain'. Companies may be able to do this with a minimalist approach. They will also expect that to provide good profits for their shareholders without causing harm to others is ethically good. This will sometimes mean that minorities will not be winners. The company worries less about *how* they provide the service and satisfies customers through the value of the product.

Chang (2005)[1] distinguishes between macro and micro ethics. Micro ethics is to do with our relationships with each other and covers things such as transparency, integrity and honesty, which often affects the professionals that clients choose to work with. Macro ethics deals with the wider impact of our actions on society, the environment or upon the reputation of the industry, and depends upon our collective conscience for doing good.

Gilbert (2006)[2] warns against non-sensitivity to ethical dilemmas where we act unaware that we have made an ethical decision. In this mode, he warns that we will most likely cover up our actions and justify our behaviour as only being a 'little bit' ethically wrong and 'slide into a deeper pit' from which it is harder to justify an ethical choice. For example, we could justify a small gift to someone, when we have won a small contract, as a thank you for 'getting us off the ground'. Next time the contract seems more important and we offer a larger gift, until we are oblivious to the proportionality and reason, so we justify gifts as important ways of getting and keeping business. The ethical dilemma for gift giving is that we will subconsciously influence the receiver so that they will favour us over others.

## Kohlberg's stages of moral reasoning

Lawrence Kohlberg (1958)[3] carried out research into the development of moral thinking and came up with a theory to verify how people develop their judgement. The theory categorises why people make decisions and not what action they take morally. Kohlberg believed that there were three levels of moral reasoning and six stages of development, with an irreversible development of approach through the stages from one to six (Table 2.1). He developed his work from Piaget's (1932)[4] work on the development of children's morals. Piaget

*Table 2.1* Kohlberg's stages of moral reasoning

| Level | Focus | Stage | Orientation |
|---|---|---|---|
| Pre-conventional morality | Self-interest | 1 | Avoiding punishment or harm |
| | | 2 | Self-interest and individualism |
| Conventional morality | Community interest | 3 | Community norms + relationships |
| | | 4 | Maintaining social order |
| Post-conventional morality | Universal justice | 5 | Just rules and consensus |
| | | 6 | Universal principles – autonomous |

notes two stages of development from a rule-based judgement which was sacrosanct to a later consequential judgement where a more flexible system of rules was recognised. Both were based on self-interest.

The stage of the moral reasoning was tested by Kohlberg and his research team by presenting theoretical moral dilemmas which were used in interviews to ascertain the level of moral reasoning that the subject was comfortable with. Each of these levels is described.

The *pre-conventional morality* level is a level of self-interest: stage 1 is characterised by the minimisation of personal harm and stage 2 by the maximisation of personal gain. Stage 1 works on the basis of obedience to rules which are perceived as right, but are obeyed for the purpose of minimising punishment. At this level, if it was clear that you could escape the consequences, then why not do it? At stage 2, it is recognised that there is not just one outcome and that people are independent agents who act out of self-interest. Stage 2 opens up a broader range of actions which might bring about a more comfortable outcome where the interests of others could be considered in order to gain mutual advantage. Some companies seem to act at this level with the use of creative accounting practices. It is a risky policy as these companies may be subject to penalties if found out, but these may have been calculated, on probability of discovery, to be less than the benefit of not paying tax. At stage 2, mutual advantages may be gained by paying subcontractors early, but extracting discounts in the delivery of work costs.

The *conventional morality* level recognises a more mature development with the consideration of living within the acceptable norms of the community and the recognition that developing relationships is a good strategy. Individual needs are subordinated to that of the group. A value is put on being seen as good people by that group and acts as a deterrent to unethical behaviour and dishonesty. Stage 3 might include a comparatively small group such as a workplace. The assumption is that it would be a moral adopted by the majority of the community, and would attract fierce loyalty. The idea of building without adhering to building regulations would be frowned upon because of the inherent danger to others in the community. Stage 4 stretches into the broader inclusion of the maintenance of the social order for society as a whole and the moral course of action is focused on compliance to laws and codes and policies. This stage goes beyond the lower stage 1 order in that there is compliance in order not to be punished – and compliance so that there is an absence of chaos and that there is equality of treatment. At this level there might be support for a landfill tax, even though it would make building costs inflationary, because it would be seen as a way of reducing wastage of scarce resources.

The *post-conventional morality* level was considered by Kohlberg to be attained by comparatively few as it distils the essence of just laws and treatment to improve a moral situation. This level involves stepping back from society and looking at individual rights and asking the question 'What makes a good society?' At stage 5, an individual would support a moral principle which had gained consensus. For example, corporate social responsibility (CSR) is a moral development, but only qualifies at this level where there is significant market leadership and long-term rationale for the good of others in the policy. Stage 6 includes the ability to view morality autonomously on the basis of the application of universal principles over and beyond the general consensus and is therefore radical. This might have described the position of a campaigner 20 years ago proposing carbon-neutral living to uphold the principle of living within global means. At this level there is unlikely to be a commercial commitment that is sufficiently altruistic because it may have adverse consequences on profit levels and even reputation for some time. Voluntary codes are recommended, but these will be insufficient if a common baseline is not created by a legal obligation, which takes it down to stage 4.

Kohlberg (1969)[5] applied these stages to individual moral reasoning and was later unable to reliably test a six-stage moral reasoning. However, it is generally agreed that there is a good correlation of the theory with the thinking of non-consequential moral philosophers such as Kant and Rawls and also the thinking of recognised moral leaders who would be rated at the top of these stages in their autonomous approach to freedom and justice.

### Evaluation of Kohlberg's theory

Kohlberg spent 20 years conducting longitudinal studies to test his assertion that development of moral reasoning in individuals did not regress. His and other studies found that between 6–15 per cent of participants who were re-tested did regress. The instability level mainly concerned those at the higher levels.

Some also thought that the theory was flawed in its application to gender and other cultures as most of the original work was carried out on white, western, educated samples. Gilligan (1982)[6] contended that the greater preoccupation of women with relationships tended to devalue their normative level to three.

Simpson (1974)[7] questioned whether the upbringing in Eastern cultures would affect the value of the hierarchy that Kohlberg had created.

## Case study 2.1 Moral dilemma for Kohlberg

A woman was near death from a special kind of cancer, but there was one drug that the doctors thought might save her. It was a form of radium that a druggist in the same town had recently discovered. The drug was expensive to make, but the druggist was charging ten times what the drug cost him to produce. He paid $200 for the radium and charged $2,000 for a small dose of the drug. The sick woman's husband, Heinz, went to everyone he knew to borrow the money, but he could only get together about $1,000 which is half of what the drug cost. He told the druggist that his wife was dying and asked him to sell it cheaper or let him pay later. But the druggist said: 'No, I discovered the drug and I'm going to make money from it.' So Heinz became desperate and broke into the drug store to steal the drug for his wife.

Should Heinz have broken into the laboratory to steal the drug for his wife? Why or why not?

This story was used not to discover an answer, but to score the reasoning against criteria for each of the above six stages.

## Background of ethical perception

INTERCAPE is an academic organisation for professional and applied ethics based at the University of Roehampton, Surrey. INTERCAPE (2006)[8] identified five developing trends which affect ethics at work:

1  Understanding other people's point of view is becoming increasingly important in a globalised knowledge economy.
2  People will have more responsibility at an earlier age for managing their own careers ethically.
3  Organisations are changing in ways that create new ethical challenges.
4  The world is becoming increasingly diverse in matters of values and faiths, creating an increased demand for tolerance.
5  Finally, the complexity of our skills and knowledge raises new ethical questions in respect to technology and practice.

These trends indicate new challenges when facing ethical dilemmas in the workplace. And there are many studies on particular professions to test the ethical perceptions. Surveys on ethics are very popular.

Is sensitivity to ethical action growing? Some of the existing construction and property practices are being questioned, such as the influence of hospitality and gifts, the degree of transparency with which we treat our customers and clients, and the definition of what is fair competition and trading is being redrawn. The degree with which we trust each other has become more important in so-called collaborative relationships. Ethical dilemmas mount as companies have to consider a wider pool of stakeholders.

### Ethical perceptions – personal

In assessing public perception, we should be aware of the cultural and the personal ethical standards. Patterson and Kim (1992)[9] found that in a survey of 5,000 Americans, 90 per cent would make up their own mind about morality. Some 74 per cent of those surveyed said they would steal and 64 per cent said they would lie when it suited them. This suggests quite an independent American public ethical standard. Kelly Services (2005)[10] suggest a higher standard of personal ethics for those living in the UK compared with other European countries, only 3 per cent admitting they approve of using unlicensed software at home. Vee and Skitmore (2003)[11] found in the Australian construction industry that 93 per cent believed that their professional ethics were driven by their personal ethics and 84 per cent believe that a balance of client requirements and public impact should be maintained.

Role conflict or ambiguity is a source of stress and is often caused by different ethical expectations. For example, a female executive is often caught between the expectations of a work role where work is pre-eminent, often at the expense of family life, and a mother role where the expectation is to be there in the home for the children's needs. This is also an ethical dilemma in terms of personal ethics to be a 'good' mother and business ethics to make provision for flexible working.

Conflict of interests may also occur in carrying out two roles at work or between personal values of complete integrity and business values of, say, creative accounting, or making a commitment to deliver when you know you cannot in order to get the job.

### The generation gap(s)

In ethical decision-making it is important to distinguish between published differences between the generations. Distinction is made between *the baby boomers* (the offspring of those who experienced the war years) born between 1946 and 1960, who have developed a freedom of choice and liberalisation; *the X generation*, born between 1960 and 1980, and heavily influenced by the Thatcher era, they are sometimes stereotyped as cynical and security conscious but this belies the huge technical developments they have been responsible for driving; and *the Y generation*, born between 1978 and 1998, the offspring of the baby boomers.

The Y generation see themselves as connected from the technological point of view. In a survey of 2,500, of which 95 per cent were aged between 21 and 28, they listed career progression ahead of salary package and many of them are expecting to progress up the career ladder within the first year. According to Gribben, '57pc see two to three years as a "reasonable length of time" to stay with an employer', most do not have a 'nine to five' mentality and they prefer teamwork.[12] Their approach can be summarised as refreshingly invigorated, with high expectations and with a 'high tech' familiarity. This 'new' generation is poised to take the lead in the economic sector. They are likely to have a particular approach to ethics and with their mobility and enthusiasm they are likely to develop and carry their own model with them. They want to have more flexible work hours and are likely to put a priority on becoming rich or famous and so they will not be afraid of responsibility in order to achieve status. They are likely to choose companies with a commitment to social responsibility, but there is also an opinion that this generation is likely to be deeply sceptical about the ethics of corporate generations, but because of their mobility they are likely to 'vote with their feet', by giving loyalty only when their trust remains.

## Ethical perceptions – business and professional practice

Public opinion of business behaviour varies between countries. Pew (2005)[13] found public perception in America thought that 35 per cent of businesses obey the law and that 54 per cent try to circumvent it. In the UK, the IBE (2005a)[14] found that 53 per cent believe that business behaves ethically, which is not quite the same question, but it does suggest more trust in business in the UK. In another survey, IBE (2005b),[15] on people's views about ethics in their own workplaces, found that 80 per cent are positive about it, and 66 per cent think that their organisations live up to their corporate policies and guidelines; people under 35 are likely to be less strict about their ethical standards; women are likely to be more strict; and the public sector are more aware of ethical standards, and support them more than their private sector colleagues.

The UK public have most concern for business ethics in the following areas:

- enabling employees to speak out about wrongdoing;
- preventing discrimination. Younger people felt particularly strongly about this (42 per cent of 15–24 year-olds).
- environmental responsibility;
- executive pay.

It is likely that these concerns vary in different countries as different events impact on the public's conscience. Excessive executive pay is a feature that has waxed and waned in the public mind. Young people have different concerns and

Poon's (2004a)[16] research (among UK construction and project managers) and Fan *et al.*'s (2001)[17] research (among HK quantity surveyors) indicate that age matters in considering the importance of different ethical matters.

Poon's (2004b)[18] survey of the ethical attitudes of construction managers indicates that what they dislike most are falsifying reports, a low level of personal honesty and over-claiming of expenses. In addition, they consider themselves better than their peers in the collusion of organisations to produce a common price and whistle-blowing.

## Employment

The International Labour Organisation (ILO) produces safe work[19] codes of practice which have been written separately for different industries since 1950, and these are excellent basic guides to safety levels which have been applied internationally. They have also brokered a number of tripartite agreements to tackle particular ethical issues that involved international trading, such as 'sweat' labour. An example of this is in the Textile, Clothing and Footware industry (FCT) where multinational businesses find they are supporting poor labour practices, such as uneconomic wages and child labour, by outsourcing to Third World countries. Public opinion has led to the signing of agreements and the monitoring of conditions to ensure compliance. An example of this is Clinton's 'no sweat campaign' which was instrumental in signing up companies to a trendsetters' register for companies who were prepared to lead the way in ensuring better conditions of work and better health and safety in the sweat shops of some emerging economies.[20] This enhanced their reputation and helped others to take the step.

### Case study 2.2  Purchasing timber

The Forestry Stewardship Council (FSC)[21] is a typical organisation offering a chance to use timber from properly managed forests, of which they accredit 90 million hectares in over 70 countries. They offer an independent accreditation service for either branded timber with an FSC stamp or a 'chain of custody' certification for a project's supply or for a company policy. Certification bodies such as the Soil Association are used to accredit the supply chain, through forest management, accredited supplies and contractor purchase and installation. This allows FSC stamped and non-FSC timber, but with agreed source and supply routes, to reach building sites. Forestry management is about ensuring that it is environmentally appropriate in terms of replanting, being socially beneficial for local communities and economically viable so that forests will continue to thrive and local communities will not be deprived of their way of life, and timber will be available for the future. There are many instances too where conservation of ancient woodland is ethical. The scheme allows a large number of organisations to be committed to ethical and sustainable supply.

A compliance-based company provides a basic compliance with regulations which, under our initial definition is pre-ethical, i.e. only compliant. An ethical cultural approach needs an ethical code of conduct to provide guidelines and a basis for training employees. The ILO (1991)[22] report indicates that a large majority of North American firms have ethical codes, but fewer have a code in Europe (57 per cent). Fewings (2006),[23] in a small pilot, found that only 30 per cent of UK construction and property companies surveyed admitted to such a code.

Justifying your ethical decision-making is important by being consistent and using a decision-making model. If a decision made by an authority or business leader is not considered fair by community values, then the decision will invoke unethical behaviour in order to avoid it. For example, this could apply to raising a price or paying an unfair tax. However, if the decision is justified in effective terms, it goes a long way in allaying the feelings of those who perceive unfairness in the decision made and who may retaliate. Time is not wasted in trying to solve ethical dilemmas fairly for all parties and describing the reasoning and balance of the decision.

### Professional behaviour

A professional has a special relationship with their client based on certain minimum standards of behaviour which they are expected to maintain, or they should be removed from the register of approved practitioners. These codes do not make professionals ethical but they deal with integrity, confidentiality and conflict of interest and make some statements about working for the common good. The public expectation of the ethical behaviour of professionals and effective decision-making is high, which makes their rationale and thinking processes all the more critical. Professional institutions have a minimum level of compliance with codes which must be enforceable on their members. The impact of professional codes will be discussed in Chapter 4.

### Conflict of interest

Conflict of interest is the conflict for a single person who has obligations to more than one party to the contract or the clash of private or commercial interest with a person's public position of trust. A delicate balance can be maintained, but it is important to remove strong temptations to be partial at the expense of another party.

## Classic ethical dilemmas

An ethical dilemma generally applies to choosing between two or more imperfect alternatives. Because the outcomes are likely to affect people in different ways, there is a need to apply some discretion and judgement based on experience and on past precedent and criteria used.

### The prisoner's dilemma

One well-known ethical dilemma is the so-called prisoner's dilemma. In this, the police have arrested two people and have enough evidence to convict them both of a minor offence. However, the police believe that one or both of the two people arrested have committed a more serious offence. To obtain the evidence of the more serious crime the police need a confession. The prisoners have already agreed that they are not going to confess to the more serious crime in order to avoid a long sentence. However, the police are offering individually to each prisoner a reduced sentence if they give evidence to convict the other of the more serious crime. The prisoners have no further contact with each other. The prisoner's dilemma is whether he can offer evidence to reduce his sentence at the cost of 'stitching the other one up' for a very long sentence or keep to their original agreement. The prisoner also does not know whether the other prisoner will keep to the agreement or will implicate him in order to reduce his own sentence. If they both confess, they will both get the maximum sentence. If only one prisoner confesses, that prisoner will receive the maximum sentence. If neither prisoner confesses, then they will both receive the minimum sentence. How will the prisoners act, and why?

As this dilemma comes from bidding theory, there is a connection with competitive rivalry between different firms equally determined to put in the highest price that will win the bid. We can see that this is the sort of argument that might attract the attention of contractors in a competitive tender. If one is desperate for the job and puts in a lower price and another is equally desperate, then they will expect a low price. However, if the client picks the lowest quote, he will find that in order to make a profit the lowest priced contractor has to 'cut corners' or make claims. This raises the question about the effectiveness of simple blind competitions.

### The doctor's dilemma

You are a doctor on the battlefield and arrive at a field hospital where there are a number of badly wounded soldiers. The hospital can treat the wounded effectively but you cannot. At the hospital you are told that the enemy is very close and they are not keeping to the Geneva Convention and so are very likely to shoot all the inmates of the hospital – patients and doctors/nurses, that is unless you are useful to them. Some of the able-bodied party suggest that as the wounded are in a pretty bad way, they should be left at the hospital where they can be treated. There is always a chance the enemy will not turn up, or there will be a more merciful than usual commander who will save them. They also suggest that all staff and walking wounded should run as it would be a shame if they were killed when they could live to fight another day, and the wounded would die anyway as they are too sick to run. A boat for disembarcation has been arranged by the army, 11 miles away. The patients are likely to suffer if left without medical help, but many patients encouraged the able-bodied patients to flee and to leave them with a lethal dose of poison to commit suicide if the enemy came.

You are a doctor sworn under the Hippocratic Oath. Should you run anyway and leave the wounded to their chances, or should you stay and treat your patients and risk that you are all killed, with a small chance you might be saved?

The answer to this lies in the requirements of the Hippocratic Oath which used to apply to all doctors to have a duty to alleviate pain even if they put themselves in danger, which would make the suggestion to run unethical. However, that was removed in 1957. In fact, this is the true story of Eleanor Jones, a nurse in 1942 in a hospital in Penang, Malaysia,[24] when the staff and patients were ordered to evacuate the hospital and walk to a ship 11 miles away. She chose to run, but the ship for the evacuation was torpedoed and all the escapees either drowned or were captured by the Japanese. Eleanor survived capture to tell the story, but had always felt troubled by her action.

How much do we have a duty to care? If one doctor was left, could that doctor have alleviated the suffering and had the chance to save a lot more people? If the doctor decided to flee, are we creating a situation where what we do is always based on the worst possible outcome? If we flee in a lesser situation, are we seriously eroding the trust of a patient in their doctor or do we all have a similar expectation of the boundaries of care? When the SARS epidemic reached Canada, doctors refused to treat the patients because they were afraid of the voracity of the disease. Is this an excessive or fair reaction? Is it likely that fear of litigation by patients who have lost trust in the profession has produced a vicious spiral of defensive medicine to protect the doctors?

## Ethical dilemmas in the built environment

This section covers the application of ethical dilemmas in the built environment at different stages of the development cycle. It discusses planning dilemmas, property development and management dilemmas, designers' dilemmas, construction dilemmas, and labour dilemmas.

## Planning dilemmas

Planning authorities are set up locally or centrally as a means of providing fair and just allocation of land use and building permission. In the UK and many

Western democracies, the final decision rests in the hands of a democratically elected local council with members from different parties. They vote on the basis of technical compliance with the agreed local plan and fair social treatment of those affected by it. The councils have a conflict of interest between the needs of their constituents whom they have pledged to support, the technical level of compliance with the local plan, and common sense connected with the local context. They are not expected to take their political affiliation into account at this point, but this may have an unconscious influence on their decision. To help them make up their minds, a compliance report with a recommendation has been produced by technical officers employed by the council. They may disagree with the recommendation or see contradictions between the technical and social issues. If they turn an application down, whether following recommendations or not, the applicant may go to appeal to the relevant government department and an appeal procedure will follow, which will take the decision outside the control of local democracy. It is very common for applicants to appeal on failure to get permission. Alternatively, the council may give approval with conditions.

The dilemma ethically is to be fair, not only to any conflicting demands between applicants who might be in competition, but also to consider objections from third parties fairly and to consider the overall sustainability of the community. This is not necessarily represented corporately, except by a single councillor who may not be on the committee or may be excluded because of a conflict of interest.

## Case study 2.3  Shopping amenity versus local nuisance

If a developer wishes to apply for permission to build a new shopping centre in the middle of a new large housing estate, this will be attractive to the council because it will provide facilities for the new population locally and sustainably, because users do not have to make long journeys, some can walk, and because facilities elsewhere are not overburdened. To make it economical, some facility for industrial units has been made. However, if this creates large buildings adjacent to existing residential properties, affecting rights of light and attractiveness of the residential property and reducing the value, it may affect a small number of people. Parking might spill over onto the streets causing nuisance, noise and clatter throughout the day, and unwanted night noise through youths gathering could reduce the amenity significantly. Some controversial industrial tenants are worrying locals with reference to the health impact on their children and others are worried about lorries hurtling through mainly pedestrian streets and this has created a lively debate. A council member may have even protested on behalf of local residents at their request, for acute loss of amenity.

On a Kantian basis, there is a right to consultation and the expectation that plans can be altered or aborted on the basis of feedback and also to comply with the technical rules in the planning guidance and the land use requirements. On a utilitarian basis, the increased amenity of closer shops for the majority with no loss of property value would rule in favour of the majority at the cost of a few. On a justice basis, there would be a case for the unjust outcome of a few, justifying more expenditure, which might not be economically viable. Those affected might blame the council for insufficient long-term planning to avoid problems with a local shopping centre. On a natural law basis, there would be concern for the sustainability of the development as long as it did not affect basic human rights or contained immoral or illegal activities. On a social contract basis, a deal could be struck between the main stakeholders, allowing an agreed level of alternative amenity for amenity lost.

## Property development and management dilemmas

The dilemmas experienced in property development are that a basic motivator is to make money out of the buying, selling and renting of property so that shareholders are satisfied. This requires a cheap solution which will be more stand-alone than a community-led system where long-term issues are taken into account and the building is designed to fit in with the social aims of the community. To what extent do you spend money to satisfy a wider stakeholder community?

### Ethical dilemma of the property developer

The temptation for a developer is to make a building attractive to buy, but not to build in durability as this is more expensive and less relevant if he is going to sell the building off in a relatively short time. This produces an ethical dilemma which means the developer needs to balance his needs for profits against the desire to be seen as a careful and reputable developer who provides good value for money and a building which is reliable.

### Case study 2.4 Business ethics model applied to building quality

The Business Ethics Synergy Star (BESS)[25] is a way of trying to set out the business motive B and ethics motive E with the objective O while at the same time identifying and trying to reconcile and balance the two opposing motivators Z and −Z which generally emerge because of the nature of a dilemma. In Figure 2.1, the developer's dilemma is set out and B and E depend upon Z and −Z respectively.

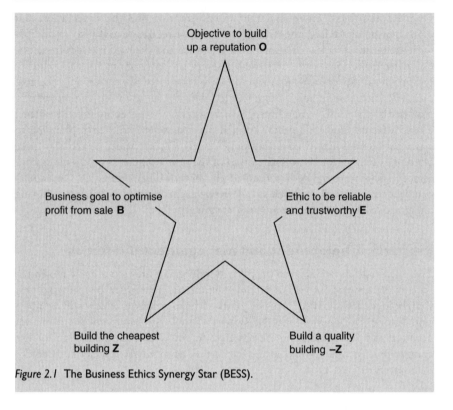

*Figure 2.1* The Business Ethics Synergy Star (BESS).

This dilemma can be resolved by looking at the axis B – –Z where the developer can build up a reputation because of the quality of the building and can look at ways of being able to charge more for the building because of their reputation, and thus make more of a profit in the medium-term as their reputation grows. The developer can afford to be trustworthy when making a profit and indeed would lose the ability to charge more if they let people down.

### Selling

There are a number of dilemmas for a property agent which are associated with a client's wish to sell a property at a maximum price and the professional duty to disclose information which the client may not divulge, but where one suspects a problem exists. There is also a responsibility to give a fair price and not just the price the client would like to sell at.

Another dilemma is associated with confidentiality and what you can tell a buyer in answer to direct questions and what is confidential to the client. You are anxious not to alienate the buyer. Advertising is often highly criticised by buyers as dishonest, but a property agent's desire is to get the buyers interested in viewing their client's property. On the other hand, to tell untruths to achieve that aim is dishonest and may put clients off, affect your reputation and contribute to the industry's disreputableness. What do you say to the vendor when they ask you about market value when you know it is overvalued in accordance with your client's wishes?

The next temptation is to accept two offers in order to have a back-up in the case of an offer falling through and then playing one offer against another to meet the client's desire to get the best price. In another scenario, if you have two offers, how should you treat the two fairly? Should there be a strict tender on the lowest price, or do you see one buyer as more worthy than the other, who is not out to make a profit, but to find a home? It is patently unethical to advise one on the offer of the other, but what do you do if your client asks you to favour one person over the other, and persuades them to raise the offer because they see the higher priced buyer as someone they do not wish to sell the property to. Is it fair to reject the offer of a client who cannot move immediately if they have a higher price than one that can, when you know your client needs to move quickly?

## Renting

Other dilemmas concern a fair rent for the tenant. In the rental of a building, an assessor is providing a service to the owner of the building and so will assess rents on a market range backed up with a typical assessment tool, making allowance for costs and market. Tenants are subject to a contract which specifies the level of rent for a given period and includes rent reviews.

Ethically, how much can you put up the rent of an existing tenant on review? Is this a supply and demand question or is there some loyalty owed to the client to raise rents in accordance with inflationary costs only? What level of service should you offer to tenants and what is a basic standard and what is worthy of a rent uplift? Should the tenant receive a basic level of services for a standard rent and what constitutes as a safe and healthy property? Should a tenant 'get away with' a low rent by threatening to move out of less desirable property in a buyer's market? From a Kantian point of view, rents are assessed on the basis of rules in place and the duty of a property professional to apply fair treatment, integrity and confidentiality to the client. In terms of utilitarianism, there is a desire to provide a reasonable compromise deal to the buyer and seller so that the degree of happiness between the two has a shared level of attainment. Justice-wise, the ethic would be to ensure a good deal for the less dominant party. Under natural law, there would need to be a moral solution where no one is being 'ripped off', i.e. paying above the standard price or rent.

## The designer's dilemma

The engineer or architect holds the strict trust of their client for the appropriateness and the safety of the design. A designer is also expected to advise the client on the best value solution and possibly provide separate options to give the client some choice. The ethical dilemmas are not the violation of codes, but the service which the client receives. It is quite possible too that innovative engineering and architectural designs are not covered by the codes and so a greater responsibility is put upon the designer in reassuring the client. A designer is going to need to protect their good name and also may be asked to design an innovative building. The building needs to be of good quality, safe as well as functional, and attractive. A designer could be over-protective of their name at the risk of value or cost.

## Cost vs functionality

One of the ethical dilemmas then is providing an economical solution without compromising the design. A designer needs to appreciate the client's values and also has a responsibility to society as a whole for the integration of the building into the community and, in the case of a public building, maximising the spending power of the tax payer's money. In another scenario, in order for the building to go ahead, a designer may be asked to put forward a more economical option to get the design through the approval stages. The risk for this may fall on the tax payer or upon the client's shareholders. In hindsight, it may seem unethical to be economical with the truth even if the building is seen to be a success afterwards, e.g. the Sydney Opera House which cost about six times the original budget, but would not have been built if this figure had been originally discussed.

### Case study 2.5  Citicorp Building

Citicorp New York is a 59-storey building designed by a renowned engineer called LeMessurier who had introduced a revolutionary new bracing system into the building which saved significantly on the weight of the steel in the building and therefore the cost. The design incorporated a large damper in the top of the building to counter the slightly greater tendency of the building to sway in winds. This damper operated on an electric hydraulic system. The building opened to great fanfare in New York in 1995 and performed well. LeMessurier was proud of his achievement until a student, backed by his professor, claimed there was a flaw in the design which produced a much greater stress on the joints if there was a diagonal wind which hit two sides of the building at the same time. The structural codes and LeMessurier's design only allowed for a perpendicular wind scenario.

LeMessurier did the calculations and realised that the greater stress was much greater than advised, putting intolerable stress on the joints, which would still be acceptable except that the structural contractor had calculated they could use bolts which were weaker than the total welding originally specified, but sufficient under the old calculations. Bolts would now be vulnerable; indeed, in a violent hurricane-strength storm, if one joint went and the electricity went off, stopping the counter-balancing effect of the hydraulic damper, it would cause violent swaying and a chain reaction which could cause the building to collapse.

With the hurricane season approaching New York, LeMessurier decided to approach Citicorp and recommend that strengthening work be done and that an evacuation plan for the building and the surrounds should be in place in case of a hurricane strike. Citicorp agreed to this and LeMessurier worked out a plan for doing the work at night in enclosed shelters to reduce disruption and not to allay panic among employees. Another engineer was also employed to check out the calculations and ensure all safety angles were now accounted for, and to make some suggestions for the strengthening. The work went ahead for an extra estimated cost of $4–8m. During work, it was reported a hurricane was heading towards the city, but as they were contemplating evacuation the hurricane swung away towards the sea. The work was eventually completed and Citicorp sent a bill for $4m to LeMessurier who was, however, able to negotiate a figure of $2m which was the amount he received from his PI insurance claim.

LeMessurier was relieved the building was now safe, but expected to pay a heavy additional insurance claim and to lose his good name in 'bungling' an innovative structural design. In fact, neither happened and he was able to convince the insurers that his early action saved them from the possibility of a collapsed building and the associated costs, and they lowered his premium for not sidestepping his responsibilities. He was also seen by his profession to have acted highly ethically and professionally in having discovered and treated an unforeseen problem promptly, competently and efficiently. So the story ends happily with his reputation enhanced.

(National Academy of Engineering 2006, part 8)[26]

The case in question shows the inherent increased design responsibility of an innovative design which may enhance building value, but induce unintentional design problems – a type of win–lose situation. In this case, the ethical stance transformed the win–lose to a win–win scenario as LeMessurier did not lose his reputation and also was able to set a good example of how to deal with a faulty design ethically.

### Architecture vs functionality

Ray (2002: 5, 6)[27] refers to why an award-winning history faculty building is also regarded as being a building that has not been appreciated by its users. There are several other reasons why this design might be unethical, such as poor functioning, as in inadequate temperature control, or because the layout and circulation do not match the habits of the users, or because aspects of the building have become out of date with the legislation.

Blockley (2005)[28] considers the problem of construction commissioned for unethical processes such as gas chambers. He calls this a duty of care. This extreme case did not deter the Nazi idealists, but it does point to other subtle differences of opinion in a democratic society, such as armament factories: used to produce weapons to defend or to attack; animal testing laboratories: prolonging cruelty to animals or enabling new cures for human disease; new bypasses across virgin countryside: upsetting fragile ecosystems or reducing urban pollution? It is not sufficient now as it once was to plan a major development without a sustainability plan. Design professionals are in the position to decide what is sustainable rather than ignoring their responsibilities. Contractors need to have a recycling plan and developers need an ecoHouse score not only because of legal requirements but also because of a greater public and client demand.

### Diversity of systems

In the global age there is pressure to standardise accommodation through lean production methods and forget about the users of buildings in the interests of capital building economics, including speedy completion. This may also favour certain business objectives. A diverse plan may conflict with an economic one, but in the deeper sense of client value it will add quality of life to those who are housed in standard buildings that are inappropriate for their unique purposes. It can reduce the sense of place when housing developments are built of repetitive boxes or buildings are globalised, losing the cultural context and history that have been developed geographically and historically. Innovation and imaginative design are another added value. Diversity is also produced by grouping and cladding and access differences and hidden structures may be standardised.

### Professional reputation

Blockley (2005)[29] is concerned about the effect of scandals, which have resulted from failed buildings and structures, on public confidence and the public's resulting loss of trust in professional advice. Typical examples are the Millennium Bridge over the Thames, which swayed dangerously when first opened and there were a large number of people on it, and the outbreak of legionnaires' disease because of the poor design of air conditioning coolers. There is a tendency to 'throw the baby out with the bath water' and to look to less informed sources which may only compound the problem of poor advice. The reputation of a profession needs to be protected by the professional institutions by policing the quality of professionals and making investigations and discipline transparent. Unfortunately, reputations that have taken years to build up take just a few days of negative press coverage to destroy and even longer to recover. The industry has also let itself down by not regulating and registering services which are the most commonly encountered by the general public, such as replacement windows, extensions and plastic fascias, so that painful horror stories abound.

## Construction dilemmas

Contractors' moral dilemmas revolve round their need to competitively bid for and win work which is sustainably profitable, and the responsibility to provide a good quality budget which is within the budget they have promised the client. Some 84 per cent of contractors in a US survey of contractors said that had encountered unethical acts or transactions in the construction industry, 63 per cent said they had also encountered illegal acts such as unauthorised use of equipment and mis-reporting of costs. Nevertheless, 24 per cent admitted that they might work with unethical contractors and the same percentage said they often ended up hiring them before they knew they were unethical.

### Design and scope

It is not unethical to claim for work which is extra to the original quote, but there is a dilemma for a manager who is expected to act to reach a fair profit level, and still to finish on time. In a traditional contract, when they can see a design is flawed and delays to get it right will cause disruption to progress, they have an opportunity to report it or to go ahead with a weaker design. This is a professional responsibility versus a commercial responsibility as a result of a combative contract condition. Where they have design responsibility, there is no dilemma as the design is their responsibility to be fit for purpose.

### Claims

In a tight competitive environment there is a temptation to win contracts as a loss leader. Claims are further made by using the contract to make claims on a marginal basis. These types of claims lower the reputation of a contractor, even if they can make a legal case.

Alternatively, contractors may be tempted to bid auctioning which involves going back to a favoured subcontractor and asking them to lower their price to match the lowest competitive bids. This is unfair competition, but is not illegal and allows the main contractor to claw back profit in a tight margin.

### Competition

Contractors are asked to bid free in most construction projects for what are quite time-consuming and extensive estimating costs, especially when a contractor's design is involved. When they have their resources tied up or feel less competitive or competent to do the work, they seek a cover price so as not to lose future custom with the client and to save estimating costs. This is strictly interpreted as a distortion of the competition and can lead clients to think that they have more price competition than they do.

### Cartels

Cartels are an area of temptation for an influential group of companies in an oligopoly to collude and artificially raise prices nationally or globally by agreement, effectively making excessive profits over and above what they could command in a competitive environment. Ethically, it is dishonest because clients believe they are getting value for money and they really are not. If they choose to negotiate a price with a company, then they will apply different principles to reach a contract of sale. National Competition Commissions are now working together internationally to limit the international scams that are easily hidden. These are apparent in the OECD Guidelines and the EU Procurement Regulations[30] to expose corporate perpetrators and make it possible for them to be prosecuted in their country of registration. One high profile case has been the Italian football clubs scam where they colluded to win and lose matches in order to gain financial advantage. In the construction industry, the Office of Fair Trading (OFT) in the UK has imposed fines of £1.38m (2006)[31] on four double glazing companies who were caught in a cartel. They have a campaign in the construction industry in the lead-up to the award of major contracts from the Olympic delivery authority.

Those who are victims of a cartel have recourse to compensation through the Competition Appeals Tribunal and only need to prove loss of earnings and not that there is a cartel. Those who are involved in a cartel and wish to 'blow the whistle' on it will be treated leniently if they are the first to report it. If there is a successful case, then cartel members are subject to heavy fines up to one-tenth of their worldwide turnover, and directors are subject to criminal investigation. The cartel members are also liable to pay damages to those affected and out-of-court settlements are common in order to 'stop the rush to the door' in the case of a widely distributed product. British Airways was a typical example of a company under investigation as part of a price-fixing cartel with other American airlines. In the UK construction industry, the roofing industry has been a cause for concern and there is much vigilance in the run-up to the 2012 Olympics because of the supply-led market for built facilities.

### Contractor image

The building industry has a poor image in that there is an unregulated 'cowboy' sector which is unashamedly unethical, and offers services they are not qualified to offer. Some have no idea about customer care and offer an unreliable service (time and quality), extortionate prices, or even fraudulent advice. These unethical services are good news for newspapers and destroy the image of the construction industry in general in the eyes of customers. Clients are also confused by the complex, fragmented and idiosyncratic contractual arrangements that are in place. Often they do not feel confident that they can check out advice and when they do take up references, they receive different accounts of the same organisations. Because the built environment requires complex unique solutions, things can easily go wrong. Often there is no easy and convenient recourse, such as guaranteed replacement or an immediate exchange, without a lot of disruption for a customer, such as a new house with a door that is jammed or a leak in the roof. These are symptoms of poor workmanship and design, and the ignorance of the client of the complexities of the product and its proper maintenance, and the reluctance/inability of contractors to give efficient, clean and cheerful after sales service. There is still a huge difference between a few excellent contractors and the large number of mediocre ones who think that the status quo is just the way the business is.

On larger projects, there is still a poor image due to the poor planning of complex resource management and wide and deep supply chains, lacking controls on quality, time or cost, and contractors who promise targets that are unintentionally broken, because of a lack of contingency for external factors. As the design, environment and logistics of construction sites are complex, then it behoves the contractor to use a better class of manager, clients to select only able contractors for their tender lists, and for the contractor and client to carry out risk and value management to enhance the pre-contract planning.

## Assessing 99 per cent situations

In a professional capacity there are never certainties. These uncertainties are often connected with the safety, quality or health factors that we think we have designed out, but because of the unique nature of the project, there are still some risks remaining. These risks are likely to have been assessed on a probability basis, but also on a severity or impact basis and it is the seriousness of the 1 per cent chance of something going wrong that is a concern. If someone gets killed or if the company is going to be exposed financially beyond its means so that many will lose their jobs, or if a product fault would affect a large number of people's safety or health, then the 'warning bells may ring' and we need to consider how ethical it is to take the risk. The ethical nature of the decision goes beyond efficiency and optimisation of resources, important as these decisions are, because it has a long-term effect which may ruin people's quality of life. Often, ethical sensitivity is needed to comprehend how serious a potential disaster might be. The evidence that supports a cautious approach is often provided very late, when delays or cancellations create extensive extra costs and/or loss of reputation and momentum so the problem becomes a management decision.

### Case study 2.6  The *Challenger* space shuttle

The *Challenger* space shuttle was a case in question when all the crew were killed in an unprecedented explosion, because a 'question mark' over an 'O' ring's unsuitability to work at certain temperatures was ignored as unproven and too disruptive to investigate at the late stage in the proceedings when questions were asked. Proper contingencies and risk mitigation and back-up systems need to be in place in the event of failure of critical components. This was patently not so and to go ahead on the following day had adverse implications for the whole shuttle programme. Management overruled the decision on the basis of a political and strategic, rather than an ethical, basis.

What is a suitable ethical model to adopt to avoid a future problem? St Augustine's model was not robust. The issue was legal. Checks had been made for all known problems. NASA would have supported the manager's decision to proceed with it and the newspapers would have made him more embarrassed if he had stopped the launch and, if it had been discussed with the astronauts they would probably have still been keen to continue. This issue revolved around the likelihood and impact of possible future outcomes.

### Case study 2.7  The Channel Tunnel

The Channel Tunnel suffered from a serious fire not long after it opened in 1998. This fire was in the centre of the tunnel and completely destroyed several carriages and their cargo of large trucks. However, the situation was judged to be a success because no one was hurt even though there was a full load on the train. The fire fighters were able to reach the passengers from a refuge off the third tunnel which had been designed to cope in such a contingency. The central tunnel and refuges were kept at a higher pressure so the fire that spread was contained in the main tunnels, and cargo in the lorries had been registered so that the correct fire fighting equipment was taken in to contain the fire without life-threatening spread.

These contingencies, together with training, provided sufficient back-ups, not to stop the risk of a fire and the damage to property, but to save serious injury and to ensure the reputation of the company so that people retained confidence in the service, so that many did not lose their jobs. In the event, freight trains were able to travel in the undamaged tunnel after three days and passengers after 15 days.

## Case study 2.8 The Mont Blanc Tunnel

A similar incident in the Mont Blanc Tunnel a year later (1999), in which 39 people were killed, closed the tunnel completely for three years. The tunnel was completely rebuilt and new safety features were built into the tunnel and special procedures for registration, monitoring and inspection were instituted.[32] The tunnel was built in 1961 and had an inferior ventilation system which could not remove the smoke quickly enough, even though there were pressurised emergency shelters. The case was taken to court after a long investigation and 16 people are facing manslaughter charges, including the tunnel operators and the lorry driver, the Mayor of Chamonix and the manufacturers of the lorry that burst into flames. The tunnel now employs more than 300 personnel[33] to ensure safety and as a measure to rebuild confidence, which was badly shattered.[34]

## Case study 2.9 Dell computers recall

Dell recalled 4.4 million Sony lithium batteries fitted in their laptop computers, because of the chance that they might catch fire and cause injury. A spokesman said: 'In rare cases, a short-circuit could cause the battery to overheat, causing a risk of smoke and or fire. It happens in rare cases but we opted to take this broad action immediately.'[35]

Dell advised people to only use the computer on the mains and set up a website so that a free replacement battery could be claimed. They said they were aware of six cases of catching fire in nine months, but the US Consumer Product Safety Commission said that they were aware of 339 such instances in the period 2003–2005 – not just in Dell computers. At first sight, Dell appears to be very safety conscious and provide satisfactory back-up, but in the light of the CPSC statement it seems only a short time before an accident could be more serious. Questions which come to mind are:

- Was a free replacement battery sufficient compensation for the risk?
- Was this an ethical slip-up which could have been avoided?
- Was the liability Dell's or Sony's for the compensation or any subsequent claims?

The BBC also reported a story a few days earlier that Dell had made false advertising claims for laptops sold in China which actually included cheaper components.[36] Dell said that advertising literature had been out of date.

It is difficult for a company to foresee every problem, but its fragile reputation is only as good as the people who work for it. However high its ethical standards, a company also needs to have contingencies which can spontaneously correct the errors. Enron had a 65-page ethical policy document Berenbeim (2002),[37] but unfortunately it appeared to employ senior people who ignored their own advice.

## Labour dilemmas

### Offshoring

This ethical dilemma covers the practice of using resources outside the country as a way of procuring services more efficiently or effectively, often where cheaper labour is available. The core business remains in the developed economy. It has become controversial as the practice has spread in the past 10 years, with the accessibility and efficiency of global communications, and many in production have lost their jobs because labour abroad is cheaper. The rationale for more efficiency is to offer better value of money to customers, allow access to leading-edge technology with flexible and adaptable solutions, decrease product cycle times and of course the economic arguments for being more competitive and surviving where, if they do not do it, their competitors will, and they will not have a job anyway. Companies could also argue that in the case of expansion, this encourages quicker returns and leaves room for the remaining employees to progress their careers in a post-industrial developed world.

There are some residual ethical issues: to whom, if anyone, do the company owe their allegiance most? If they outsource remotely, they are likely to have more difficult quality and production control problems in their supply chains which may affect their reputation and the health, safety and sustainability of their products. If they keep the product at home, it will become uncompetitive. These actions might be seen as helping the economy of the developing country with employment opportunities or the distortion and 'milking' of the economy by large multinationals leaking profits out of the economy.

The practice is also controversial because labour conditions are perceived to be unfair, for example, 'sweat shop' labour and child labour. The company may pay less regard to pollution of the local environment or create an imbalance in the developing economy. Their reputation may be affected by customers perceived drop in the product quality or the ethics of the service provided. In short, it is important to make sure that there is a product fit which is effective in the long run. In their defence, larger businesses justify the creation of employment opportunities which can alleviate poverty in developing countries and through their intervention improve working conditions and quality of life and knowledge transfer through training programmes.

In construction, there are particular issues to do with the design contracts being outsourced abroad in India. Construction itself is not easy to outsource as it is geographically based. However, it is obvious that a lot of the manufactured goods are moving their manufacture to other countries where production is cheap enough to justify bigger transport costs. This is not sustainable, particularly with heavy materials which are heavy on fuel costs and are not carbon neutral. Globalisation has led to consolidation of manufacturing and large multinationals are formed to remain competitive and centralise production facilities to make use of economy of scale.

## Using migrant labour

Migrant labour is an ethical dilemma because it is a solution that provides cheap good quality labour to help offset the continuing shortage of labour in the industry. Polish and other East European immigrants are taking over some of the jobs that may have been available for locals. A survey (2004)[38] indicated that 10 per cent of all construction labour have English as a second language. Some 100,000 migrant workers work in the construction industry alone and many of these are illegal workers. This has occurred because of the shortage of labour. These statistics indicate that there could be major health and safety problems due to a lack of knowledge of the language and training of a large influx of workers unaccustomed to the safety regime. Brumwell (2004)[39] pointed out that migrant labour is helping the construction industry to survive, but a boom in the accession countries would attract immigrant workers back to their homes and make the shortage even worse.

## Ethical decision-making models

Badaracco (1998)[40] sees character as being formed at the point where responsibilities clash with values. He calls these occasions 'defining moments'. It is a useful way of looking at the decision-making process. Ethical decision-making is a process of weighing up the different ethical, economic, legal and policy issues that afford choices which have an impact on others. These are quite often long-term issues and affect personal/professional or company reputations with their customers and stakeholders. The balance achieved may be the intentional creation of a particular image, e.g. caring, tough, fair or unintentional, in that there is a limited consideration of all the impacts. Personal ethics may often clash with the company policy. Wesley Cragg (1997)[41] talks about the dimensions of ethics and identifies three outcomes which are important to the operation of decision-making in terms of: (1) actual, rational choices based on ethical principles; (2) ethical awareness and self-knowledge; and (3) personal development which is about fulfilling one's potential. This suggests that the process of decision-making is just as important as the decision and it is hard to impose a single solution even for similar contexts.

## Moral intensity

A descriptive decision-making model tries to look at the priorities. One such model by Crane and Matten (2004),[42] in view of the uncertainty of the outcome of some decisions, measures the moral intensity (impact), by pondering:

- The magnitude of the consequences – are some of the consequences too serious for certain stakeholders?
- The social consensus – is there a strong opposition by some who believe that the situation is unethical?
- Probability of effect – what is the likelihood of an unacceptable ethical outcome?
- Immediacy – are the consequences short- or long-term?
- The proximity (moral distance) of the outcomes – is this affecting my customers or close stakeholders?
- Concentration of effort – is there a small, widespread negative effect or a concentrated one?

This model is mainly a consequential approach and does not hold moral absolutes in order to dilute the impact of a negative effect. This system could be quite helpful in trying to assess the most acceptable decision for a number of stakeholders.

## A business model

The Bagley model[43] is a fairly traditional approach and comprises three questions:

- Is it legal?
- Does it maximise shareholder value?
- Is it ethical?

The model is a consequentialist one, maximising shareholder value and it has two levels for legal and ethical. Shareholder value might be justified on the basis of benefit to other stakeholders, and boundaries will probably be created to link cost to long-term or non-financial benefit. Managers are likely to balance maximising shareholder value with giving benefit to other stakeholders. A company culture will give some indication though culture is strengthened by ethical guidelines to allow some delegation of such decision-making.

The case of Nick Leeson and Barings Bank is illustrative of the way in which an unsupervised policy led to unethical practice for the bank, leading to disastrous results. The bankruptcy of the bank led to loss of jobs, and account holders and customers and others lost money in the crash. The consequences were unplanned, but could be defined as unethical in the process and not the consequence in spite of the high reputation of the bank, based on employee and customer trust previously. The bank collapsed in February 1995, losing £1.4bn after a prolonged trading spree by Far East manager, Nick Leeson, who 'got out of his depth' in the trading of future incomes which failed to materialise in a downward market.

As an alternative to Bagley, a more Kantian approach might be adopted which is based upon 'doing it right' and being consistent in how you approach a problem while ensuring impartiality, integrity, respect, transparency and conscientiousness. This is more of an internalisation of ethical awareness and 'getting it right'. Many organisations would like to communicate this to their clients and customers to build confidence in a range of actions and it is more than a code of conduct which mainly is provided as a basic minimum.

## A professional or virtue ethics model

The virtue ethics model needs to define a community of similar belief and value for the good of society, with a commitment to continuous improvement and building up a culture and social norms. Environment ethics falls into this category, which depends on building a widely acceptable concern to preserve scarce resources for the future of the planet, restrictions on pollution and $CO_2$ emissions to stop climate change and promoting good health. An example of this culture building is Ahrens' (2003)[44] ethics project to educate engineers who had a role to play in forming a consistent European ethical view of engineering education, to improve interaction with partners and clients in terms of cooperation, contracting liability and environmental responsibilities, personal development and the relationship between values. The project to formalise some common beliefs for students of civil engineering came about because many engineers were finding themselves in multi-cultural projects where a culture of respect and commonly understood moral and ethical values could be established to ease conflict over decision-making and professional conduct. In addition, the European Council for Civil Engineers (ECCE) needs to consider the divergent views in awarding mutual recognition of professionals. These views dealt with a common understanding of professional behaviour towards colleagues, employees, clients and the public, and were intended to develop values in the area of functionality, health and safety, economy and prosperity. This project has a sense of creating common language and wider citizenship principles as the moral distance has been shortened by the desire for common trading and a more 'level playing field'.

Blockley (2005)[45] argues that ethics matters in the production of buildings because building professionals are human, and ethics is at the heart of being human and holding a moral responsibility towards each other, and therefore we can define standards of conduct. It also matters because we need to manage differences and relationships with each other and clients, and agree common standards of good and right. More specifically, it matters at the level of defining quality and value in the buildings we produce and the jobs that we do in the built environment, where quality is a 'degree of excellence' and value is 'the worth we give to the purposes we have both collectively and individually'. This is a useful application as it helps to define in terms of quality and value what a 'good' building is. Unethical action will therefore emerge from a superficial definition of quality as designed to specification, and value as functionality, as if all of us thought the same in these areas. Blockley concludes that the latter two levels take us deeper than a professional code of conduct and suggests some soft value scoring. The test that we should apply to our actions, after a responsible consideration of quality and value, is whether we would be able to 'sleep soundly' having taken those actions.

### Reputation and integrity model

Harrison (2005)[46] makes the point that ethical action based on the idea of building up integrity may be an apparent short-term moral action, but the decision to build integrity into the company to gain such an organisational reputation may be based on an egoististical long-term business aim to procure more business and be able to charge more for it. In truth, it is a position of mutual gain and not an entirely altruistic move – 'for the good of society'. Reputation may be attached to a particular decision-maker such as a director or to an organisation as a whole. Buchanan and Badham (1999)[47] define reputation as 'a socially defined asset dependent on one's behaviour and on the observations, interpretations and memories of others'.

Buchanan and Badham are concerned not only with the ethical nature of the decision, but also the credentials of the decision-maker or in their context, the 'change driver', and use the term 'warrant' as a way of assessing the source of authority. A formal warrant is one sanctioned by the organisation's ethical code and a tacit personal warrant relates to a manager's assessment of an appropriate ethical action. They are also concerned with the wider effect on the stakeholders. This leads to a four-part procedure to test the ethics of an action summarised by Harrison (2005):[48]

- Gather the facts.
- The act is ethical if it seeks to optimise stakeholder satisfaction, if it respects individual rights and if it is just.
- If it does not respect all three, then a further inspection needs to test:
  - overwhelming factors which make it reasonable to set aside one or more of the three factors;
  - double facts where an act might have both positive and negative outcomes;
  - incapacitating factors such that the ethical criteria cannot be applied, e.g. lack of information and constraints.

- The final acid test is based on confidence in, and the 'warrant' of, the person (change driver) carrying it out:

  ○ Is the behaviour ethically acceptable?
  ○ Did the change driver have a reasonable warrant?
  ○ Can a plausible account be given of the behaviour applied?
  ○ Has the change driver's reputation changed?

Accounts are the manager's justification of action and reputation is the outcome of the action against social expectation and the observations and memories of others.

Paine (1994)[49] proposed an integrity strategy and saw it as a 'driving force in the enterprise' rather than a set of carefully worded statements for public consumption in the Annual Report. It would then need to influence managers every time they make a decision and give a sense of 'shared accountability'. This type of strategy is connected with training, good communications and frequent feedback and the clear espousal of company values to support an ethical environment. Harrison points out that this type of commitment is equal to any other policy that is important to the company and will also need some financial resources which must be justified to the shareholders.

### A health and safety approach

Patankar *et al.* (2005: 46),[50] in their book on safety ethics, have an approach they have applied to aviation, health and the environment which distinguishes critical health and safety issues. This could be applied to environmental and engineering design as a basis for ethical decisions.

In construction, justification for supporting safer actions and better health in the industry, as an overwhelming decision-maker is supported by the Health and Safety Executive (HSE) who have noticed a continuing high level of health problems and a surprisingly stubborn level of serious accidents in spite of a significant amount of legislation. Building management has also become shoddy in the operation of the business. An ethical approach that prioritises the health and safety of the design, construction and use of the building is quite often a shared objective. For design, it means looking at healthy environments and ways of enhancing the safety of the maintenance team and any production processes. For the contractor, it means reducing accidents and putting the building together in a safe way so that workers are protected, and allowing an adequate bid. For a client, it means having a generous approach to allowing resources and conditions, e.g. not expecting a building to be rushed, so that a building project encourages the resources to be injected into the tender without penalising the safer working contractor. It also involves handing reliable survey data to the design team so that risks can be properly evaluated. For the facilities manager, it means having an effective maintenance regime which also includes training for users to escape safely in the case of danger, but also not to misuse facilities.

## Conclusion

We have seen in this chapter that there is a strong belief that built environment has operated outside an ethical envelope and that it is possible for short periods to operate while ignoring the consequences, on the basis of business as normal, but today there is less tolerance in the industry for a poor reputation following the Egan and Latham Reports. In today's collaborative climate, there is a need to build up a systematic approach which can be identified and shared with others. In identifying dilemmas, we are thinking about actions in context and seeking to deal with specific conflicts of interest as well as our own selfish interest. A designer has to deal with their client's and user's needs as well as the expectations for sustainability and the community-led context. Architects and engineers also need to play a fair role in traditional contracts as arbiter between client and contractor. The property developer and manager have a responsibility to the community as does a commissioning client. These challenge them to follow a wider stakeholder approach with particular interest in community and tenant needs. A planner also has a duty to be fair in the allocation of planning conditions to both consult with the community and to keep to government policy guidance and local expectations. The contractor has many temptations to take short cuts and to take a purely commercial role, forgetting their professional obligations.

A number of different decision-making models and approaches have been presented which show that a single philosophical approach is not sufficient and needs to consider consequences for a range of people while at the same time ensuring the right process. This is the case in competition where the good of the client and the bidder are presented as contradicting each other. An ethical approach requires an ethical process which is transparent and also has a value for money output for the client, but not at the expense of others.

There is room for an interdisciplinary approach here so that developers, designers, contractors and users can come together to design excellent buildings which are good for the environment, have a commercial motivation and give something back to the community.

## Notes

1 Chang, C. M. (2005) *Engineering Management: Challenges for the New Millenium.* Upper Saddle River, NJ: Pearson Educational.
2 Gilbert, J. (2006) 'Beyond Bad apples: The Dynamic of Ethical Erosion', *Leaders Edge Newsletter*, January. Available at: http://www.amanet.org/LeadersEdge/editorial.cfm?Ed=172 (accessed 20 November 2007).
3 Kohlberg, L. (1958) 'The Development of Modes of Thinking and Choices in Years 10 to 16', PhD dissertation, University of Chicago.

4 Piaget, J. (1932) *The Moral Judgment of the Child*. London: Kegan Paul.
5 Kohlberg, L. (1969) *Stages in the Development of Moral Thought*. New York: Holt, Rinehart.
6 Gilligan C. (1982) *In a Different Voice: Women's Conceptions of Self and Morality*. Cambridge, MA: Harvard University Press.
7 Simpson, E. (1974) 'Moral Development Research: A Case Study of Scientific Cultural Bias', *Human Development*, 17: 81–106.
8 INTERCAPE (2006) *The World of Work is Changing*. Available http://www.roehampton.ac.uk/bss/cape/index.asp (accessed 15 August 2007).
9 Patterson, J. and Kim, P. (1992) *The Day America Told the Truth: What People Believe about Everything that Really Matters*. New York: Dutton/Plume.
10 Kelly Services (2005) 'World at Work Survey of More than 19,000 Workers in 12 Countries across Europe', 6 September.
11 Vee, C. and Skitmore, C. M. (2003) 'Professional Ethics in the Construction Industry', *Engineering, Construction and Architectural Management*, 10: 117–27.
12 Gribben, R. (2007) 'Generation Y Talking about a Revolution', *Daily Telegraph*, 30 November.
13 Pew Research Centre for the People and Press (2005). Available at: http://people-press.org/reports/display.php3?ReportID+149 (accessed 15 August 2007).
14 Institute of Business Ethics (IBE) (2005a) 'Survey on Business Ethics: General Trends in Public Perception', included in the MORI omnibus survey of the UK public, September. Available at: http://www.ibe.org.uk/IBE_Briefing_1_Surveys.pdf (accessed 18 August 2007).
15 IBE (2005b) 'Ethics at work – A National Survey', conducted by MORI. 'What Do UK Workers Think of Ethics at their Workplace?' January.
16 Poon, J. (2004a) 'The Study of Ethical Behaviours of Surveyors', paper presented at COBRA Conference, Leeds Metropolitan University, 7–8 September, RICS.
17 Fan, L., Ho, C. and Ng, V. (2001) 'A Study of Quantity Surveyors' Ethical Behaviour', *Construction Management and Economics*, 19: 19–36.
18 Poon, J. (2004b) 'An Investigation of the Differences in Ethical Perceptions among Construction Managers and their Peers', in F. Khosrowshahi, (ed.) *Proceedings 20th Annual ARCOM Conference*, 1–3 September 2004, Heriot Watt University, Association of Researchers in Construction Management, 2: 985–93.
19 ILO Safe Work Codes of Practice Website. Available at: http://www.ilo.ch/public/ english/protection/safework/cops/english/index.htm (accessed 3 December 2007).
20 ILO (1995) *Business Ethics in the Textile, Clothing and Footwear Industries*. Available at: http://www.ilo.org/public/english/dialogue/sector/papers/bzethics/bthics2.htm (accessed 21 November 2007).
21 Forestry Stewardship Council (1996) *Construction and Architects: Claims of Sustainability – What Can You Believe?* FSCUK-FS-306 Fact Sheet. Available at: http://www.fsc.org/en/getting_involved/become_certified
22 ILO (1991) *Conference Board Survey*. Available at: http://www.itcilo.it/actrav/actrav-english/telearn/global/ilo/code/main.htm (accessed 20 November 2007).
23 Fewings, P. (2006) 'The Application of Professional and Ethical Codes in the Construction Industry: a Managerial View', *International Journal of Technology, Knowledge and Society*, 2(7): 141–50.
24 Sokol, D. and Bowman, K. (2005) 'Doctors' Ethical Dilemma: How Much Has the Fear of Litigation Led to an Erosion of Patient Care?', *The Times*, 11 March.
25 Robinson, D. A. (2006) 'Resolving Ethical Dilemmas in Small Business', paper presented at 8th Westlake International Conference on Small Business, Hang Zhou, 15–17 October.

26 NAE (2006) *William LeMessurier: The Fifty-Nine Story Crisis: A Lesson in Professional Behaviour*. Ethics Centre, National Academy of Engineering. Available at: http://www.onlineethics.org/CMS/profpractice/exempindex/lemesindex.aspx (accessed 12 December 2007).

27 Ray, N. (2002) 'Part I, Historical Perspective Introduction', in N. Ray (ed.) *Architecture and its Ethical Dilemmas*. Abingdon: Taylor & Francis.

28 Blockley, D. I. (2005) 'Do Ethics Matter?', *The Structural Engineer*, 5 April, 27–31.

29 Ibid.

30 EU, The Public Procurement Directives 1992–2000. The original ones are implemented in the UK as *The Supply and Works Contract Regulations 1992* (SI 3279), *Public Supply Contracts Regulations 1995* (SI 201), and *Utilities Contracts Regulations 1996* (SI 2009).

31 OFT (2006) *Double Glazing Raw Materials Distributors Fines for Involvement in Cartel*. Available at: http://www.oft.gov.uk/news/press/2006/107-06 (accessed 5 December 2007).

32 BBC News (2002) 'Mont Blanc Tunnel Reopens', 9 March.

33 Tourist leaflet (2006) describing the safety procedures in the tunnel.

34 Bell, S. (2005) 'Mont Blanc Tunnel Fire Trial Finally Begins', *The Scotsman*, 1 February.

35 BBC News, 15 August 2006.

36 BBC News, 11 August 2006.

37 Berenbeim, R. E. (2002), *The Enron Ethics Breakdown*, The Conference Board, Executive Action No 15, February. Available http://www.info-edge.com/samples/CB-EA15free.pdf (accessed 13 October 2007). [The Conference board is established to present issues of interest to companies, established 1916].

38 Considerate Constructor's Scheme (2004) 'A Survey of Labour across 300 Sites', February.

39 Brumwell, P. (2004) 'Migrant Workers are our Lifeline and their Safety must be Protected', *Building*, 8.

40 Badaracco, J. L. Jnr. (1998) 'The Discipline of Building Character', *Harvard Business Review*, March–April.

41 Cragg, W. with Koggel, C. (1997) *Contemporary Moral Issues*, 4th edn. Toronto: McGraw-Hill/Ryerson.

42 Crane, A. and Matten, D. (2004) *Business Ethics: A European Perspective*. Oxford: Oxford University Press.

43 Bagley, C. E. (2003) 'The Ethical Leader's Decision Tree', *Harvard Business Review*, 81(2): 18–19.

44 Ahrens, C. (2003) *Ethics in the Built Environment: A Challenge for European Universities*. The Socrates Project.

45 Blockley, D. I. (2005) 'Do ethics matter?', *The Structural Engineer*, 5 April.

46 Harrison, M. R. (2005) *An Introduction to Business and Management Ethics*. Basingstoke: Palgrave Macmillan.

47 Buchanan, D. and Badham, R. (1999) *Power, Politics and Organisational Change: Winning the Turf Game*. London: Sage.

48 Harrison, *Introduction*, op. cit.

49 Paine L. S. (1994) 'Managing for Organisational Integrity', *Harvard Business Review*, March–April,

50 Patankar, M.S., Brown, J.P. and Treadwell, M.D. (2005) *Aviation, Health Care and Occupational and Environmental Health*. Aldershot: Ashgate Publishing.

# Chapter 3

# Business ethics and corporate social responsibility policy

The ethics of business is a key concern for the captains of industry. There is a need to maintain a good reputation and be open and honest with the wider community, as well as maintaining profit for the shareholders. There are public and political pressures for businesses to meet climate change targets and for construction in particular to 'clean up its act'. Not many businesses have a formal ethical code. The aim of this chapter is to establish the role of business ethics and the associated practice of corporate social responsibility reporting in gaining an ethical approach to business, and to see how this operates in the construction and property industries. The chapter will

- discuss the rationale for a business ethic;
- identify the economic and governance responsibilities of business;
- evaluate the social and environmental responsibilities of business organisations and construction.

A groundswell of public opinion is often the catalyst for the inclusion of a broader view of stakeholder engagement in corporations, often called corporate social responsibility. This is often rationalised by the profit motive – 'sustainable business is profitable business' – and it provides a possible way forward in a capitalistic economy which needs to gain the support of its investors. How sound is this approach? Do companies really mean what they say in the grand reports that they produce? What sort of ethics are they applying in this report, if any? Are they successful in convincing the public and the government that they mean business?

## Business ethics

The US Securities and Exchange Commission (SEC) (2003)[1] defines a code of ethics as:

> written standards that are reasonably designed to deter wrongdoing and that promote honest and ethical conduct . . . ethical handling of conflicts

of interest . . . Full, fair, accurate, timely and understandable reporting . . .
compliance with the law, regulations and rules . . . prompt internal reporting . . . and accountability for adherence to the code.

This provides a reasonable categorised basis for evaluating codes.

The Institute of Business Ethics (IBE)[2] advocates a three-year survey on the
nature of corporate codes of ethical practice. The 2004 survey found that formal ethical codes of practice are admitted by 92 of the FTSE 100 companies.
The other eight are report working on these codes. A further 104 of the FTSE
350 companies, only 46 per cent, also have codes. This is a significant increase on
previous years as 48 per cent of firms in the survey say that they developed these
in the four years prior to 2004. Anecdotal evidence quoted indicates that small
firms are also increasing their interest. In the above survey, only 9 per cent are
construction companies, but a further 27 per cent in the survey are in property,
mining and utilities. Another IBE MORI poll (2004)[3] indicated that 60 per cent
of people do not trust business leaders to tell the truth and the top three areas
of concern are executive pay, discrimination and environmental responsibility.

It is difficult to extract data on construction from the main IBE survey,
but work by Myers (2005),[4] on plc construction and housing development
companies, indicates a fairly poor response in the 2002 financial year to the
CSR agenda, with house builders indicating marginally more enthusiasm. This
initial survey is based on the amount of space given in the 2003 annual report,
the website and any other special reports produced, and inevitably this is
quite an approximate method. Of course, there is great variance, but it does
provide a comparison across the plc housing and construction sectors. What
is rather more surprising in this day and age is the number of construction
and housing companies that actually have no statement at all, given the reputation of the industry for not fulfilling its promises.

In ranking the most ethical and honest professions, Gallup (in *USA Today*
2006),[5] ranked engineers ninth and business eighteenth out of 32 occupations,
with a level of only 20 per cent to be highly trusted for business executives and
61 per cent for engineers, otherwise 56 per cent voted business executives as
average. The trend has been down for business executives and up for engineers.
Teachers and nurses top this list, with construction contracting somewhere in
the middle. Accounting practice has also come under particular scrutiny since
2001 when Enron and other high profile companies went down.

A Mori poll (2005) for the BMA[6] ranked doctors top in the UK, while
business leaders received only 24 per cent of the poll, with high or very high
trust. The challenge of a low rating for business leaders in particular, and
managers and engineers to a lesser extent, is to convince clients and customers
that they really do intend to provide an ethical service. Ethics is important to
business because there is a need to gain a good reputation and to attract the best
customers and employees. Business needs customers to believe what they say
they are going to do.

There are other conflicts of morality which might compromise the manager who wants to provide a flexible and tolerant attitude to personal objectives while needing to enforce and monitor a blunt-ended company policy, such as 'clocking on'. Ethically, both have a strong case. Managers also have a primary role in maximising stakeholder profitability and this conflicts in the short term with some sustainability objectives. One distinction of company policy in this area is whether it is compliance or integrity based. The latter is based on the positive development of an appropriate-based culture.

Corporate bodies have failed us and have been disgraced when the public trust has been broken, as in the case when the fraudulent practices of companies have been exposed such as Enron, Maxwell, Worldcom and Northern Rock. This has been evidenced in corrupt action (Enron and Maxwell, false accounting), or exposure to excessive risk (Northern Rock and other banks' *sub prime* lending), or action which has been considered unethical (Nestlé and the baby milk powder advertising debate).[7] It is also evidenced in health and safety issues such as the excessive lead in paint, resulting in the recall of young children's toys made in China for a well-known toy company,[8] or companies that blatantly ignore environmental standards such as the Third World dyeing factories flushing waste products straight into waterways. The failings of a few companies have had a general effect on business as a whole, but have also led to the public making a differentiation between companies in the same business as well as different types of business. Business has also had to contend with the new regime of terror.

The media, shareholder activists and pressure groups have set up a 'name and shame' campaign to force companies to adopt stricter codes[9] to ensure international suppliers meet internationally recognised labour standards. This has particularly affected the clothing trade where child labour and sweat shops outsourced in developing economies, such as India, are notoriously producing clothes for High Street clothing stores in the UK. In construction, there has been a strong campaign against the poor safety record of the industry, in property, there have been campaigns against inappropriate developments that have priced key workers out of the market and fuelled a buy-to-let culture rather than providing housing for first-time buyers.

### International business ethics

There is often some debate about what action can be taken when dealing with the international trading practices of multinational companies. Many have a global turnover in excess of the individual GNP of the developing countries they trade in, so patently they have a lot of influence. They can:

- Ignore the situation.
- Sign up to a strict international code such as the ILO fair labour conditions.
- Participate in external auditing through subscribing to an ethical index and producing Key Performance Indicators (KPIs).

- Self-impose their own stronger ethical codes and monitor suppliers.
- Innovate to partner with local NGOs to support poverty eradication.
- Avoid trade in countries where there are excessive contraventions of codes.

The current state of public opinion is very sensitive to companies using suppliers in developing countries where international labour standards are contravened, even unintentionally. If the company has followed the cheapest labour trail as a direct labour outsourcing policy and is importing the goods back in, it is under great scrutiny by pressure groups. There is popular unrest when directors are remunerated beyond inflation and golden pay-offs are given to chief executives who have presided over failed strategic policies, or bonuses are given for windfall gains brought about by predatory takeover and asset stripping in the short term. 'Fat cat' pay is not appreciated[10] and has brought in an era of shareholder activism to match pay to real performance.

## Moral responsibility and objections to an ethical code

There are two main objections to the ethical code (Television Educational Network 2004).[11] The first objection is that if an industry is regulated by law, then compliance with the legal requirements is a good enough framework for ethical activity. This leaves the code as a public relations exercise and in practice does little to change ethical behaviour. The second objection is that we have already formed our personal ethical framework during our formative years at school and from our upbringing, and this will determine our behaviour. This will not be greatly influenced by a separate code and detrimentally it might take away our responsibility for making moral judgements. These two objections could be brought together to suggest an alternative framework, i.e. a legal framework that creates a common threshold which individuals will not breach but it will allow them to exercise a degree of moral responsibility inherent in their own environmental conditioning and upbringing which is natural and professionally accountable.

The problem with both of these objections is that they are not focused on the particular activities of the company in question and, as such, do not offer specific guidelines which make use of precedents and experience. If one takes the definition of ethical responsibility as the area beyond legal compliance, and if companies want to build a reputation on practice which exceeds legal requirements, then we will want to make that clear to our employees and also consider in what areas it applies. It is unlikely that standards of human behaviour can be left for the state to decide. The law gives no guidance on conflicts of legitimate interest between different stakeholders and, if it did, it would also be unnecessarily rigid as it would have to be all-embracing.

An individual may also be subject to pressure from others to act against their ethical framework and move into areas where the boundaries are less clear.

Without the benefit of any company guidance as to whether they are supported in this, they may be unable to rationalise resistance or compliance. Objective peer review is an important part of complex ethical decisions where there may be conflicts of interest, and more than one way of looking at the ethical implications, which result in more than one solution. This process confirms that moral responsibility is not abdicated as no guidelines can cover every eventuality. An informal culture where employees make their own interpretations also needs some guidance so that consistency is maintained and so that those who 'toe the line' do not feel a sense of inequality or injustice. Good management needs to justify its decisions ethically, evidencing a sense of fairness.

When making policy, it is quite clear that companies err towards either an input (prescriptive) approach or an output approach, which allows some degree of autonomy with a more senior level of approval regarding the decisions which can make a prolonged or public impact on the company is position. These may lead to evolving guidelines being developed.

## Code framework

Having made a case for more specific corporate code making, it should be possible to identify categories or areas where a statement would be useful. This can be composed uniquely to suit operational needs. In construction contracting or consulting, some of these statements could have common features. Webley (2003)[12] suggests the following structure for codes of ethics:

- Policy preamble helps to identify the objectives of the code, such as maintaining the reputation of the company for fair and transparent dealings, and encouraging all staff to maintain ethical standards.
- Relations with customers need to affirm confidentiality, honesty, bidding and aftercare.
- Relations with shareholders and other investors need to establish a parity of treatment and any special conditions.
- Relations with employees need to cover such things as training, health and safety policy, recruitment and staff retention policies.
- Relations with suppliers need to cover such areas as mutual trust, the giving and receiving of gifts and payments.
- Relations with the government and the local community need to address issues of environmental care and impact, social responsibility, information sharing and local employment.
- Relations with competitors should cover rules of engagement and respect and confidentiality with due regard to the reputation of others.
- Issues relating to international business relate to the recognition of cultural differences and respect for the social and economic development indicated by government policies, pulling back from dealing with repressive regimes and exploitative employment practices of local suppliers.

- Mergers and takeovers recognise the voluntary and compulsory codes of the stock exchange and government concerns for fair trading and outcomes for smaller shareholders.
- Directors and managers will conform to the combined code on corporate governance and provide information to the Board and public meetings which is objective, true and fair.
- Compliance and verification mean that a system should be in place to provide non-biased auditing of procedures and discipline to build confidence in all participants.

Webley (2003) also suggests that there should be several steps to ensure a worthwhile company-wide implementation which critically tries to integrate the code with the processes for running the business. He also emphasises how the code should be clearly distributed and accepted by employees, suppliers, stakeholders and the public so that they know what to expect. For employees, they need to assent, possibly in the employment contract, and training should be arranged taking into consideration the need to connect it to disciplinary procedures where the code is breached. Managers will need to affirm that compliance and arrangements for training exist and that there are regular reviews.

The following are some examples taken from corporate ethical codes.[13]

## Case study 3.1  Corporate codes

We take pride in the services that we provide to communities. We also seek to work in partnership with civic, community and charitable groups where we operate. Through our commitment, these communities will know that they can trust us.

(MMO$_2$ 2004)[14]

We have a clear and unequivocal approach to business integrity and ethics, which underlie the Carillion values of openness, collaboration, mutual dependency, professional delivery, focus on sustainable, profitable growth and innovation.

(Carillion 2005)[15]

Others see a more social commitment 'making charitable donations to financial literacy and money advice' and more personally 'providing opportunities for colleagues to support the development of communities in which they live and work'.

(HSBOS 2004)[16]

## Code credibility

The credibility of the code is more than what it says. The code needs to be effectively implemented and be seen to be fair by the stakeholders who are affected by the operation of the company. This means that the code should have proven relevance to the issues that are important to a wide range of stakeholders, particularly to the employees, customers, suppliers and the community within which the organisation operates. Inevitably, there will be some clashes of objectives between stakeholders which will result in different interpretations of the code.

## A business ethic model

A business ethics model is one that tries to interpret the relationship of a company to its stakeholders. Advice ranges from the rather terse recommendation by Milton Friedman (1967)[17] that the only ethic needed is to keep the shareholders happy within a legally acceptable framework – this is, primarily through managing a fair profit, to a much broader perspective concerning the fair operation of the enterprise to the wider stakeholder community, including customers, employees and the external community. Svenson and Wood (2007)[18] have identified a business ethic model (Figure 3.1), based upon the expectations and perceptions of society which then makes an evaluation of the company outcomes resulting from the codes and cultures that they have implemented. This evaluation of individual company performances makes a reconnection to the expectation for corporate behaviour in general.

There is an iteration in the business ethic model which the authors hope will be an upwards spiral. These expectations are seen through legal and good practice compliance, lobby groups such as Greenpeace and Forum for the Future, media pressure, parity and competition with others, increased education, management aspirations, professional requirements, and international agreements by organisations such as the OECD and the UN. However, in the face of shareholder apathy or defiance, there may be a low level of standards.

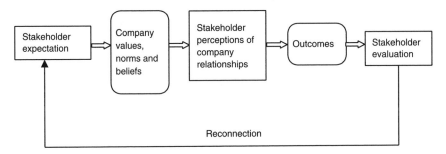

*Figure 3.1* A business ethic model.

*Source:* Adapted from Svenson and Wood (2007).

The idea of a business ethic model is important so a corporation can market itself to its future employees and give some guidance to its existing ones. It may also be relevant to employee retention. Sustainability is a related issue as the commitment to 'futures' is seen as a good barometer of ethical health. It is also true that corporations find it quite difficult to use the language of ethics. Sustainability and CSR have certain undertones that are easier to define. Two elements of CSR, the fair and equal treatment of employees, are covered in Chapter 5 and the ethics of sustainability are covered in Chapter 8. This chapter is mainly concerned with company policy and culture.

## Corporate social responsibility (CSR)

There is a growing movement for leading companies to show corporate responsibility and provide more information to stakeholders through what has become known as the corporate social responsibility (CSR) report. CSR is defined by Jones (1980) as 'the notion that corporations have an obligation to constituent groups in society other than stockholders and beyond that prescribed by law or union contract'.[19]

The principle of CSR provides for a balance sheet with a triple bottom line (Figure 3.2) covering social and environmental as well as economic profit. As discussed in Chapter 1, this is a bid to provide more inclusive reporting in the interests of all a company's stakeholders. Larger organisations now need to employ an environmental manager as well as a company accountant.

The economic *raison d'être* for a company is to make a sustainable profit for its stakeholders which means that long-term continuing investment in a company as a continuing entity. The company will seek to retain a capital base and invest for growth and build in financial contingency. However, the policies of a company will impact on its reputation, both for performance and for investing

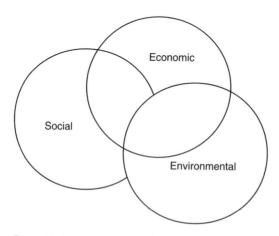

*Figure 3.2* The triple bottom line.

ethically. The company is required to balance its books and to show a profit (if it is profit making), which will pay out its shareholders with an attractive dividend growth (10 per cent where possible), or not make a loss if it is non-profit making). If circumstances conspire to make it loss making, the company is more likely to still try and pay a dividend out of reserves in order to remain attractive to investors. Ethically, the company has a responsibility to provide transparent accounting and, because investors base their opinion on these accounts, it is tempting to hide unattractive financial accounts on the balance sheet. This has led to much legislation to close up loopholes in the accounts because of the catastrophic effect on many people if things go wrong. The victims are mainly innocent, including deceived shareholders, employees losing out on pensions, and facing redundancy.

The *social* responsibility of companies in a more traditional view is to the welfare of employees and the community context in which a company works. In construction, this refers to reducing nuisance and compensating and 'putting something back' into the community. The planning approval system generally recognises this formally in private building developments, by requiring a charge against the development under Section 106 (S106) of the Planning Act, generally called planning gain. The rationale and size of this are negotiated between the local authority and the developer. Developers often recognise this as their social responsibility to provide specific additional social amenity. However, it is controversial and is sometimes seen as a development tax which is off-loaded onto the new house buyer or client, i.e. the value of an equivalent second house would be cheaper because there is no application of S106. In addition, there is a move in construction to set aside resources and time to engage and liaise with the community through what is known as the Considerate Constructors Scheme (CCS). This often includes education, contributions to local causes and employment schemes for local labour. CCS is a voluntary club and contractors hope to gain a reputation as ethical builders by raising the independent score out of 40 as high as they can (see Case study 3.2).

## Case study 3.2 Broadmead Employment and Enterprise Scheme (BEES)[20]

Specific schemes use the 'spin off' of large projects to generate economic opportunities such as employment and business enterprise. BEES proposes to get a further 200 people into employment locally, with 600 in construction jobs and 3,500 jobs in retail tourism and hospitality. This scheme works to bring together various neighbourhood initiatives, local and national businesses and funding opportunities to train locals. This project is in partnership with the developers of the Broadmead city centre expansion and the Bristol City Council Economic Regeneration Office which manages it.

In this model, the developer is in a moral position to consider the local community. These schemes tend not to develop out of an industry code of practice but from meeting the company's CSR commitments and the planning conditions imposed on the project developer. The ethicality of such schemes is heightened the more voluntary they are, but come to fruition where there are dedicated champions who drive the scheme forward.

The *environmental* responsibility of companies is partly statutory and clients have to agree sustainability targets with the planning authority in order to get planning approval. The main requirements for an environmental company approach are usually set out in the CSR report. There are certain standards that companies can sign up to for ethical trading such as FTSE4Good or Business in the Community (BiTC) and the WWF audit. However, there is a rating for standards too in that some are stricter than others and pressure groups accept Social Accountability SA8000 labour standards more readily than the Fair Labour Association because it specifies monitoring subcontractors[21] and not just first tier suppliers, so it is considered more effective. Construction and property companies are also keen to be seen to be subscribing to good practice as approved by these indexes. In the case of a housing developer or a public client, there will be standards which will be considered good practice, such as an EcoHome rating or a BREEAM rating for the sustainability of the house in terms of climate rating, ecology balance, pollution and waste/recycling. These first two apply to projects and are covered in more detail in Chapter 8 on sustainable projects. From a business point of view, there is a need to train people in the use of these indicators and one such training scheme is called 'Natural Step' (Case study 3.3). This is a cultural training applied at the top level of management to help people understand the foundational principles of sustainability, to avoid confusion and contradictions, and to give it a scientific base.

### Case study 3.3  Training in Natural Step

Forum for the Future describes Natural Step as a scientific approach and 'a systematic way of identifying [the] root causes [of environmental degradation and social injustice] and a means of developing solutions that [are] complementary and not contradictory'. Four system conditions are required to stabilise the overuse of our natural environment in four separate ways, given that the environment has a self-righting ability, but that this may be exceeded. These are:

1    Society mines and disperses material faster than it can be re-deposited, e.g. fossil fuels.
2    Society produces substances faster than they can be broken down, e.g. plastics.
3    Society extracts resources faster than they are replenished, e.g. fish stocks.
4    Socio-economic dynamics to balance behaviour with expenditure.

Keeping these in balance by using plant life photosynthesis and natural water and air purification is essential in the long term to human life, and business needs to play its part. This training has been adopted by Carillion & Crest Nicholson.[22]

Figure 3.3 shows an ethical model of how corporate social responsibility can be supported by accountable structures which are transparent. The corporate governance brings the shareholders on board, the environmental audit is verified by an external body, and progress targets are verified by KPIs. The social responsibility can also be measured by KPIs but needs to be backed up by a proper policy approach. Employees in construction are a key asset and there is a need for a programme of development through training and safety improvements and to ensure that incentives are in place to retain staff. Externally in the community is where a firm connects with those around its projects using the Considerate Constructors Scheme.

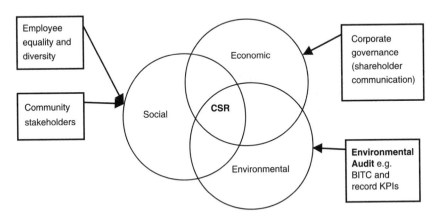

*Figure 3.3* The CSR stakeholder model.

## Reporting and ethical aspects of CSR

Jones' definition fits in well with our basic definition of ethical action to do more than the minimum legal requirements. This raises the question as to what drives a corporation economically, politically or socially to follow a voluntary path. Is it competitor pressure, community relationships and long-term gains or is there a sustainability motive?

There are a number of arguments for CSR and most of them do not advocate a mandatory path as it is felt that the responsibility is undermined and that good (ethical) business is a key factor in boosting the 'bottom line' and shareholder value (Dunnett 2006).[23] On the other hand, some people feel that corporations can never be trusted to look after society's needs and that they depend on market forces which push the price down rather than a concern for human and environmental welfare (Doane 2006).[24] Others feel that it is a missed opportunity for shareholder value not to get the community on your side and this is particularly important for the built environment. Perhaps there is a little of both arguments to win the battle, so that outcome-defined expectations are clearly published by the government, and companies choose how they play their part by publishing evidence of progression and regression against measurable targets. This is the view of the Global Reporting Initiative (GRI) which is an international set of guidelines to try and reach some consistency between reporting so that comparisons can be made. The reporting standards of GRI consist of two parts:

Part 1 – Key impact of the organisation on sustainability and its effects on stakeholders, including the community in general in, for example, pollution, and key actions which the organisation is making to mitigate these impacts and to take advantage of them in terms of the main interests of external stakeholders.

Part 2 – The impact of sustainability trends and risks on the organisation and its long-term prospects and financial performance, more specifically for the shareholders and in financial terms. This will also include a range of financial and non-financial KPIs and the progress made, with a set of targets to be made for next year.

This is an ethical method because it raises specific issues in a transparent and unequivocal way so that consequences are not vague and opportunities are clear, both for the financial health of the company and for the actions taken to reduce risk to others. Terms are expressed tangibly. It also gives a medium-term (three to five years) and a long-term view of the company's performance, which very often is not the case in the annual report. Shareholders may of course object to the social expenditure outlay, but there are commercial justifications for this and the inclusivity of the shareholders is crucial.

There is a 'value for money' (VFM) argument which can be applied to customers in terms of customer after-sales and in terms of the service which would bring returns in similar ways by recommendations to others and by better sales. Ethically, purists say that it should be a quality service without a focus on profits, but in practice there is a strong egoism ethic which works both ways. Customers are particularly attracted by VFM and it is important for developers to create a market differentiation to maintain competitive advantage.

### Shareholder support

Large institutional shareholders are also realising that a long-term investment in an ethical and sustainable business is good for the future profits of the company and ultimately their own investment. The examples of Enron, Worldcom and Maxwell have not proved that there is no profit to be made for some companies from short-term gain, but the sustained shareholder loss in these crashes mobilised shareholders to press for tougher government regulation[25] for governance and sustainability policies.

There are, however, other ethical concerns for CSR reporting concerning trading relationships and their treatment of people. On a global scale, a World Bank survey of 3,600 firms in 69 countries (quoted in Omerad 1997 *et al.*)[26] estimates that 15 per cent of businesses were found to pay bribes and much of this for public work. In 1998, the World Bank estimated this wastage as £80bn (Walsh 1998).[27] Apart from the economic argument, the moral argument by Ades and Di Tella (1997)[28] is that significant corruption 'exacts a cost on additional social efforts to improve economic welfare'. In construction, this would relate to a direct shortfall in the money available for specific social projects such as health and education facilities. Reports, such as the Egan Report (1998) and the Latham (1994) Report in the UK, indicate the client perception that client value for money is particularly poor in construction. This could be one of the reasons for a client drive for a more sustainable solution to their building needs so that lifecycle as well as capital costs are considered. In construction, the publication of reports by Transparency International shine the spotlight on widespread corrupt practices in the industry which exceed even the guns/ defence industry and the oil and gas industries (TI 2003).[29] Chapter 10 deals with corruption.

### Corporate governance and economic responsibility

A company is owned by the shareholders but governed and run by its appointed directors. Corporate governance is the mechanism by which a company shows that it has effective control over its processes and is doing so in accordance with the shareholders. Doing business fairly is important and communicating with the shareholders to understand their wishes is also important. Finally, a company needs to comply with the Companies Act and subsidiary requirements of the stock exchange if it is a public limited company. A limited company means that the shareholders will only be liable for the nominal value of their shares if the company were to go into liquidation, i.e. break up. Debts may be paid to creditors through a company's assets and when these are exhausted, the shareholders will be paid out. Shareholders fall into two types: small private shareholders and shareholders in large institutional holdings.

There is a good deal of pressure on corporations to improve their corporate governance so that they are more accountable to shareholders. By law, corporations must comply with a code of practice which has evolved in the past two decades, since the institution of the Cadbury Report in 1992 for more accountable governance. The Cadbury Report set out clear good practice for independent procedures for auditing, remuneration of directors, nomination of directors and more transparent procedures for shareholder meetings and communications and the introduction of stronger roles for non-executive directors. The non-executive directors' remuneration is not tied to company performance in the same way as an executive and as such they can follow a more impartial line and have less conflict of interest in the receipt of bonuses and the setting of remuneration. The later Greenbury Report (1995) called for more transparency, with director remuneration to be proportionate to performance and for remuneration and share options/bonuses to be published so that shareholders are informed. The Combined Code, which puts together all the previous codes including Turnbull (1999)[30] and Higgs (2003),[31] has incorporated more prescriptive requirements which mean that the role of the chairman and the chief executive should be separate and complementary. The chairman should not preside over more than one other FTSE100 company so that there is no role overload. Non-executive directors should be trained and aware of their duty of care and due diligence, given that they do not have the same amount of time to spare as an executive director. The Code also requires an equal balance of non-executives to executives on the main board with non-executives chairing each of the main committees, i.e. auditing, remuneration and shareholder communications. Companies also have to identify genuinely independent non-executive directors who were not previous employees, major customers or who own significant amounts of shares. These act as a third party for shareholder complaints about directors. Corporate governance is strongly connected with CSR reporting because it naturally leads to a wider accounting for stakeholders other than shareholders.

## Rationale for social and environmental reporting

The rationale for CSR in business is not usually altruistic and companies have to believe that social and environmental welfare gives them added value Toyne (2006)[32] lists the drivers for CSR as: reputation and trust, regulation, company and executive values, preserving the environment such as climate change, reducing pollution and resource use, promoting health and well-being, public awareness and education, affluence and consumer pressure, human rights and diversity, a call for transparency and accountability, better governance and shareholder activism. These are rational reasons for justifying expenditure and also making business gains if a company 'plays its cards right'. Companies do not see a conflict with profit generation if they are legally required to improve their eco performance through the application of a legislative measure such as new insulation standards in new housing. If it also raises an economic opportunity to change behaviour such as increasing recycling to save money on landfill taxes, then it may also be the incentive for more radical changes.

## Sustainability – is it good business?

It will, of course, cost money to introduce special measures, but if these are seen as progressive and sympathetic to public opinion, then companies can also gain custom and charge a premium for more ethical services. The more qualitative arguments are that there is a partnership to be formed with the community and this is done by projects such as educational employment, training and citizenship programmes. These build up partnerships for repeat work.

Contractor consideration for neighbours and the reduction of pollution, noise and dust and good stakeholder communications are other ways of participating. From a developer's point of view, similar things can be done and eco targets can be met to ensure compliance with BREEAM requirements. More examples are given later in the chapter. A lot of these factors are now expected in good contracting, and reputation would be lost if these principles were not carried out, especially on large projects. Winning sustainability prizes for good urban and neighbourhood design builds up a reputation and makes a good marketing prospectus and satisfies opposition pressure groups.

This could apply to a bank making socially, environmental and ethical (SEE) investments which can then provide a savings account with a lower interest rate if it invests in what may cost them slightly more by avoiding, for example, arms companies and tobacco companies. This negative type of investing often associated with underperformance can be enhanced by what is called positive screening, which means choosing between ethical companies for better performance. There is also a climate for shareholder engagement with traditionally excluded companies to help them raise their ethical profile through meetings between the company and ethical investment fund managers and eliciting change through joint shareholder action to table resolutions at the general meetings. This then has a positive reform effect.

Companies using alternative, more expensive non-carbon fuels are finding that fossil fuels are also rising in price and some customers are choosing them because of their policy giving some set-off against the extra cost. Companies may also find that recruitment is easier where they win prizes for a sustainable design and construction and they can command a loyal and more able workforce. Shareholder activism has been possible where institutional shareholders have insisted on company sustainability, KPIs and then looked for improvements year on year, such as the electricity used per person or the amount of waste recycled. These may also be publicised in the annual general report to the company's gain.

Earlier research projects, on balance, have indicated that CSR does pay by providing a premium for company products and the Forum for the Future, in a joint publication called *Sustainability Pays*,[33] identified 47 out 58 studies purporting to give positive correlation. Forum for the Future, however, concludes from later research (McWilliams and Siegel 2000)[34] that CSR benefit is associated with other drivers such as better marketing and more innovation, and when the marketing and R&D costs are taken into account, the CSR effect is neutral. Forum for the Future proposes that CSR is good because it induces change and will be adopted up to a point of neutrality, because of pressures to do so and that non-financial benefits, in terms of reputation, will give a company long-term market leadership and shareholder benefit, and this is the true driver.

There are a number of case studies of organisations who are committed to CSR and these are listed in *Sustainability Pays*.

### Case study 3.4  Pollution prevention pays

3M have instituted a Pollution Prevention Pays programme[35] for 32 years. This involved reformulating products, modifying the process, and recycling and reuse of waste materials. The 3M programme has saved 807,000 tonnes of material.

### Case study 3.5  Stakeholder engagement

A group of 25 Local Authority Pension Funds (LAPFF)[36] following the Forum's idea of adopting a new constitution launched a Shareholder Engagement Campaign to engage retail clothing and other companies in which they invested. This involved these companies being 'named and shamed' until they put in force stronger ethical policies on child and sweat shop labour in their contracts with developing country suppliers.

### Case study 3.6  Pressure groups

The 'cut the cost' campaign by Oxfam and AIDS Healthcare Foundation[37] and the subsequent worldwide patented medicine argument with Glaxo Smith Kline (GSK) to cut the cost to developing countries of high price HIV/AIDS drugs was successful in getting GSK to reduce their price by half on a trade-based argument to 63 countries.

These case studies indicate quite different ways in which companies can be socially responsible and that they often justify their actions to shareholders in specific long-term savings. The actions labelled CSR have to be beyond an existing compliance regime. The next section looks at the particular application of CSR in the construction and property sectors and considers some further case studies

## Sustainable construction

The construction industry provides facilities in the built environment worth about 8 per cent of GDP and is responsible for 32 per cent of energy usage in the ongoing life cycle of those facilities. In response, initiatives to save energy and neutralise or significantly reduce non-renewable energy sources are being designed in. It is also a major creator of net waste and many leading players are anxious to show, quite often in response to growing client pressure, that they are making major inroads into reducing waste in balance to the size of the industry. This affects the efforts of both the design and construction sectors of the industry. TI has suggested that an excessive level of commercial confidentiality has led to a level of secrecy that has sometimes been biased against the public interest. The DTI Sustainable Construction Progress Reports are produced annually and encourage sustainability. Figure 3.4 indicates the expectations for the CSR report.

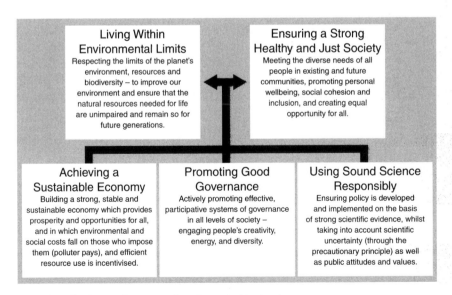

*Figure 3.4* The five guiding principles of sustainable development.

*Source:* DTI Sustainable Construction Strategy 2006, Summary.[38]

These principles cover the area of climate change, waste and materials, costs, water, skills, safety and equity and respect for people. For instance, this sets out zero carbon construction for 100 per cent of new build by 2020 (not just housing which is 2016) and all existing housing by 2030. They also set targets for 100 per cent recycling of construction materials by 2020.

TI (2003)[39] has suggested that an excessive level of commercial confidentiality has led to a level of secrecy that has sometimes been biased against the public interest. This has made it easier for payments to be made to influence the award of contracts or to make conditions which favour a particular contractor or allow them an uncompetitive bid. It also has made it hard to see how construction companies are faring in the CSR agenda and how much they are playing their part to reduce their substantial contribution to environmental ills. A further claim is that the industry too often accepts the status quo by rationalising actions as cultural, too hard to eliminate or of benefit to the client.

## CSR in construction

Government pressures have played their part in more reporting which has resulted in several nationwide and some industry sector-specific publications. These are being strengthened in the built environment by legislation such as more stringent energy-saving requirements in the UK in the Building Regulations,[40] the compulsory Eco-Home ratings for inclusion in the buyer packs for new and eventually all homes, the housing sustainability code, the landfill tax and climate change levy. These are considered as a threshold over and above which CSR starts.

Leading property developers, construction developers, designers and housing developers have made an attempt to apply relevant CSR principles to their businesses in the environmental and social sectors. Some case studies are shown and these include two nationwide schemes developed to work in partnership with the industry. Three of the schemes are taken from the CBI partnership to showcase company case studies.

### Case study 3.7  Development recycling

The use of Natural Step at Bristol Harbourside[41] was a joint project with a local contractor and a regeneration developer. The science of Natural Step led to the reuse of hard core from under the existing road and the recycling of concrete and brick materials by crushing to be used in road sub-base. In addition, recycled plastic-coated wire fencing lined with biodegradable jute mat and filled with dredged river silt was used as part of the retaining structure to create reed beds adjacent to the harbour wall. As the wire netting and jute net degraded, the reed roots would act as the new bank retaining system.

## Case study 3.8  Considerate Constructors Scheme

The Considerate Constructors Scheme (CCS)[42] is a voluntary scheme for contractors enhancing action towards the environment, the workforce and the general public. CCS awards points for contractors of any size involving educational site visits, giving talks and raising funds for local charities and projects through staff and employee release. It has an eight point code of considerate practice: the codes deal with the environment, cleanliness, being a good neighbour, being respectful, safe, responsible and accountable. Registered sites are monitored and visited and a score of three in each category brings compliance. A score of 35 out of 40 brings a gold award for a construction project. Twenty thousand sites had been monitored at the scheme's anniversary in March 2007 and the average score is 30.

## Case study 3.9  Sustainability brief

British Land[43] launched their sustainability brief in November 2004 and it aims to promote continuous improvement in sustainability for all of its projects, by setting targets at the start of each project and reviewing these through concept, detailed design and construction for improvement. It involves sustainability awareness training for all in-house staff, suppliers and contractors in the areas of site and neighbourhood, resource consumption, environmental quality, user satisfaction and stakeholder relations. The British Land manager takes overall responsibility for sustainability improvements. 'Targets are reviewed every 2–3 months and continually refined.' On one project they have introduced a green roof which helps to insulate in winter and keep out summer heat, control carbon dioxide emissions and rainwater run-offs. On all their projects they design in water saving and at the end of 2005 they specified that new projects would have a 15 per cent recycled content for all materials. The cost of this is justified in its enhanced reputation as a socially responsible company and savings passed onto clients.

## Case study 3.10  Company plasterboard recycle scheme

Taylor Woodrow (TW)[44] set up a scheme with British Gypsum (BG) in 2002 for plasterboard waste collection and recycling on condition that BG would supply all of their plasterboard products to its Bryant Housing division. By 2004, TW had moved from recycling no waste plasterboard to recycling 72 per cent (4,600 tonnes). The scheme was environmentally friendly, and saved costly disposal as plasterboard was banned from standard landfill sites and the gypsum was reused by BG, including recycling the paper into fertiliser. The business benefits are obvious for TW who also set up a secure partnership with TW for supply. BG also gained sole trading status.

### Case study 3.11   Basic numeracy and literacy

Serco[45] employs 46,000 people in facility management-type services. They introduced a programme in 2006 called Skills4U which is a basic skills enhancement programme to improve the potential and employability of some of their least skilled workers. It is presented as part of their social responsibility towards their employees and upskilling generally. Employees are assessed and put on basic skills and GCSE programmes within work time to enable them to get over learning difficulties and to improve their levels of learning. It operates within 27 of their contracts and by the end of 2006, 537 workers had been identified with a learning need and had enrolled on programmes. This programme enhances the company's reputation of looking after its workforce as well as allowing it to gain from additional employee skills when tendering for future work.

The above case studies indicate an interesting range of environmental and social concerns to note in the CSR report. Like the previous case studies they are just snapshots of an ethical approach to business that works because it can be justified as a longer-term benefit to offset the cost of such initiatives such as employee retention, more work from clients or a better reputation in the community. They are therefore an egotistic approach to ethics as well as an altruistic approach to helping the environment. Sustainable profit-making means the company can continue. However, it is likely that public pressure and expectation are part of the driver of CSR largesse.

## CSR reporting survey

Pilot research was carried out by Fewings (2006)[46] to test the issues that most mattered to construction corporations in their sustainability reports. Issues of concern for construction companies were drawn from prior interviews with two leading organisations working in the built environment, and after examining the standard codes for reporting sustainability as recommended by the Global Reporting Initiative (GRI) and the World Wildlife Fund (WWF). Both of these codes put an emphasis on the transparency of reporting to make it easier for interested parties to compare information across organisations and also on the production of KPIs on a yearly basis so that progress may be monitored quantitatively or qualitatively with some degree of relativity. It was found that:

- Predefined KPIs were not particularly popular. However, organisations recognised that some hard measurements were needed. Some tended to show targets, but others indicated actual figures obtained.
- Different organisations identify their own most appropriate critical KPIs as leading indicators of their success. These customised KPIs indicate year-on-year improvements effectively as long as the principle of consistency is adopted in measuring declared targets.
- Trust is built up that these targets actually do measure substantial ethical and sustainable behaviour.
- Only three responding companies had a formal ethical policy and no companies proffered a definition for ethical behaviour.
- There was some development of policies on anti-fraud, malpractice, theft, gifts acceptance and whistle-blowing. There was a sense that some were quite recent in response to a perceived need for more transparency and openness.

In order to partially test whether corporations were in tune with stakeholders, a further questionnaire was distributed to undergraduates who were expecting to join the industry and work in similar corporations to the ones which have been canvassed. This group has been classified as potential entrants. The results are used as an indication of both differences and similarities between company policy and the values and expectations of future employees. There are many other stakeholders who may have a more powerful influence on company policy, but dependency on a limited graduate supply was a factor in this choice. Fan *et al.*'s (2003)[47] research also found that sustainable targets are achieved on the integrity of the workforce. It is also more likely that the entrants will have a declared interest to attain their chartered professional status and act conscientiously according to professional conduct rules. It is acknowledged that averaging level of importance, unless showing a single polarised focus, may indicate that entrants and corporations will choose each other according to taste, and that market forces will operate. For this reason, only high scores of over 4 out of 5 are used.

### Survey results

The priority values are recorded in Table 3.1. These had a high average score of greater than 4.5 out of 5 for the corporates and of greater than 4 out of 5 for the potential entrants.

There is a similar level of concern in the environmental areas such as health and safety and waste recycling, which is unsurprising given the current level of

*Table 3.1* Social and environmental values of corporate and potential entrants

| Corporate ethical values (score > 4.5) | Potential entrant ethical values (score > 4.0) |
|---|---|
| Zero accidents, near misses, investigation and improved culture | Getting rid of accidents |
| Cutting down waste and recycling | Reducing waste and hazardous materials |
| Quality (defects free) | Finishing on time |
| Cutting corruption with inflated claims and extra resources | No equivalent |
| Improving motivation and performance | Fair pay |
| Appraisal and feedback | Relationships |
| Training including updates, employee choice, future promotion, personal development | Training and personal development |
| Ethical and professional behaviour | Equal opportunities and rights |
| Integrity and openness | Good relationships – collaboration, integrity, communication and honesty |
| Corporation's greater value on >4.0 | Potential entrants greater value on >4.0 |
| Good scores e.g. Eco-Home | No equivalent |
| Stronger community communications | No equivalent |

*Source:* Adapted from Fewings (2006), Tables 1 and 2.

public opinion. There is also agreement on training and development for personal development and in the case of employers, for promoting employees. There are differences between the employer and employee, in the understanding of the ethical treatment of employees. Corporations put emphasis on ethical and professional behaviour and employees put emphasis on equal opportunities. There is emphasis by entrants on fair pay rather than the employers' emphasis on motivation and incentive, suggesting a much more independent view of motivation. This may be due to the perception of a more mobile workforce who motivate themselves, manoeuvre their position or change company for self-actualisation and achievement – 'a sort of take it or leave it' attitude towards the employer.

### Differences between corporate and stakeholder

In order to compare differences, a more detailed look was taken at the questionnaire. The bottom half of Table 3.1 adds more issues with priority down to 4 out of 5 for both sides. We see that companies value appraisal and feedback more and put a high score on the issues which might serve to boost their reputation such as fighting corruption, developing stronger community relationships and attaining higher environmental scores for their products. As previously noted, the potential entrants value the development of relationships and fair interpersonal practices. Environmentally, they have indicated a greater concern about hazardous materials.

It is difficult to interpret some of these differences, and the issue about equal opportunities for potential entrants was not so clear when it came to assessing equal opportunity policy. Issues such as equalisation of gender employment were not rated highly. There is a consistent male domination in built environment professional jobs which is also reflected in the courses of study. The importance of relationships was taken to mean personal inter-employee relationships as well as promoting external collaboration. The culture of the construction industry has traditionally been adversarial, leading to very competitive behaviour sometimes putting client objectives at risk.

In the area of trading relationships, companies said it was frowned upon to book extra resources that were not used, and to put in inflated claims. There was a strong approval for handing over without significant defects and on time. These issues are not conclusive to an ethical approach as they are leading questions.

Appraisal and feedback had been used to implement policy and to develop sound practices. Only one company did not measure KPIs and the most strongly favoured ones were measuring people performance to improve it, and conformance with client requirements.

## Trading

The issue of ethical trading also affects the way that companies are perceived as employers, and 'bright young professionals' make choices about whom they want to work for. CSRs are building in monitoring systems to provide evidence of their resolve to meet the expectations of their wider band of stakeholders. Clients of the construction industry are waking up to their own responsibilities and demanding life cycle assessments of the new built assets they are commissioning.

In organisations there is also a growing need to revisit the acceptable ethical employment standards. Those who have the privilege of employing expensive labour are realising the benefits of a flexible labour force do not always outweigh the disadvantages of a falling off of loyalty among their employees. With so large a proportion of costs in the labour force and in its recruitment and training, some companies are looking for ways of increasing loyalty and raising their retention figures. Better productivity is being connected with innovative benefits, increasing job satisfaction and security in the job. The latter is patently at risk in a world of acquisition and rationalisation. Young people realise, however, that jobs are not for life and therefore they prepare for mobility. Training and experience are critical to the attractiveness of their CVs and choices for building up skills and experience for later progression are a motivator. It is also possible that a younger generation is more committed to 'ethical companies'.

## Conclusion

The question is whether the new paradigm of corporate self-regulation through corporate social responsibility reporting and action is a sound system for furthering sustainable and ethical development in the built environment. We could also question whether it is a complementary or fairer system than the current dependence on the planning system to guide compliance to a more sustainable built environment. It has been shown that there is some ambiguity about the definition of ethical practices or at least there is a difference in the motivation of young professionals whose aspirations are more to do with justice and equal opportunities, and the corporations. These corporations have a need to maintain their reputation and may well take a utilitarian view of providing the most good for the people who are impacted by the business of development of the built environment while still maintaining sustainable profit growth to satisfy their shareholders. The public pressure for sustainability has become a more important factor in the marketplace and, together with government targets, tax regimes and legal requirements, has led companies to take a broader, much longer-term view. This has needed more innovative practice and progressive companies are predicting moral business will be more profitable.

The case studies suggest, however, different approaches between corporations depending upon the sub-sector in which they operate, leading to a product or a service emphasis in reporting their sustainable and ethical practices. This should give scope for innovation in the development of a responsibly accountable organisation to build its reputation and convince interested parties that they are properly sustainable and may be trusted to deliver.

However, there does seem to be a more independent view of motivation by employees and potential entrants, with the expectation that personal training needs will be met and that there will be equal opportunities for employment and development. The emphasis is on self-sufficiency and mobility rather than loyalty. The targets are likely to be moving 'goalposts' and there is concern that there is some mismatch between the ethical principles of the young professionals and those of their older counterparts.

## Notes

1 United States Securities and Exchange Commission (2003) *Definition of a Code of Ethics.* Quoted in IBE (2004) *Corporate Use of Codes of Ethics (2004) Survey.* January.
2 IBE (2004) *Corporate Use of Codes of Ethics (2004) Survey.*
3 Mori Poll (2004) 'Concerns about Ethical Standards in Business', IBE Press Release, September.
4 Myers, D. (2005) 'A Review of Construction Companies' Attitudes to Sustainability', *Construction Management & Economics*, 23: 781–5, October (quoted on the London Stock Exchange in the 2003 Annual Reports).
5 *USA Today* (2006) 'Honesty and Ethics Ratings', 12 December, Gallup Poll (yearly feature since 1976).

6   Ipsos Mori (2005) 'Doctors Top the Public Opinion Poll on Trustworthy Professions', British Medical Association. Online. Available http://www.mori.com/polls/2005/bma.shtml (accessed 10 March).

7   NUS (2007) *Ethical and Environmental Campaign 'Nestlé'*. Online. Available http://www.nusonline.co.uk/campaigns/environment/11536.aspx.

8   Mattell Recall Notice (2007). Online. Available http://service.mattel.com/us/ (accessed 12 December 2007).

9   Buckingham, L. (2001) 'Curb Child Labour, Stores are Warned', *This is London*, 2 December. Online. Available http://www.thisislondon.co.uk/news/article-884216-details/Curb+child+labour,+stores+are+warned/article.do.

10  Bowie, N. E. and Werhane, P. (2005) *Management Ethics*. Oxford: Blackwell.

11  Television Education Network (2004) *Business Ethics*. London.

12  Webley, S. (2003) *Developing a Code of Ethics*. London: Institute of Business Ethics, October.

13  Svenson, G. and Wood, G. (2007) 'A Model of Business Ethics', *Journal of Business Ethics* 7(3): 303–22.

14  MMO₂ (2004) *Business Principles: The O₂ Guide for How We Work and Conduct Our Business*. Online. Available http://www.o2.com/media_files/bus_prin_ext.pdf, p6 (accessed 6 June 2007).

15  Carillion (2005) *Ethics and Business Integrity Policy*. Online. Available http://www.carillionplc.com (accessed 15 December 2007).

16  HSBOS (2004) *Our Corporate Responsibility Agenda: Our Commitment to the Way We Do Business*. Online. Available http://www.hsbosplc.com/community/corporate_responsibility_home.asp, p. 47.

17  Friedman, M. (1967) Presidential Address to the American Economic Association, cited in Svenson and Wood, op. cit.

18  Svenson and Wood, op. cit.

19  Jones, T. (1980) 'CSR Revisited, Redefined', *California Management Review*, 22(3): 59–67.

20  Bristol Alliance (2006) *Broadmead Employment and Enterprise Scheme (BEES)*. Online. Available, http://www.bristolcitycentre.com/template01.asp?pageid=101 (accessed 15 December 2007).

21  Jackson, R. (2005) *The New Supply Chain Standards: FTSE4Good Enough?*, 3 January. Online. Available http://www.ethicalcorp.com/content.asp?Content ID=3345 (accessed 12 December 2007).

22  Forum for the Future (2005) *Business Futures: Practical Solutions and Positive Thinking on Sustainable Development*. London: FFF.

23  Dunnett, D. (2006) 'Expert Views on Common CSR Arguments', *Corporate Social Responsibility: Your Chance to Make a Difference*, Media Planet, *The Times*. 24 April.

24  Doane, D. (2006) 'Expert Views on Common CSR Arguments', *Corporate Social Responsibility: Your Chance to Make a Difference*, *The Times*, 24 April.

25  Sarbanes–Oxley Act (2002) *Pub. L. No. 107–204, 116 Stat. 745*, also known as the Public Company Accounting Reform and Investor Protection Act of 2002, US Government.

26  Omerad, T., Hurtey, M., Lovgren, S., Palmer, C.C.B. and Cunningham, V. (1997) 'Bye Bye to Bribes: The Industrial World Takes Aim at Official Corruption', *US News and World Report*, 22 December, 22.

27  Walsh, J. (1998) 'A World War on Bribery', *Time Magazine*, 22 June, 16.

28  Ades, A. and Di Tella, R. (1997) National Champions and Corruption: Some Unpleasant Interventionist Arithmetic, paper presented at the University of Pennsylvania, Philadelphia.

29  Transparency International UK (2003) *Anti Corruption Initiative in the Construction and Engineering Industries*. London: TI.
30  Turnbull, N. (1999) *Internal Control: Guidance for Directors on Combined Code*. London: ICAE.
31  Higgs, D. (2003) *Review of the Role and Effectiveness of Non Executive Directors*. London: Department of Trade and Industry.
32  Toyne, P. (2006) 'The Forces Driving CSR', in *Corporate Social Responsibility: Your Chance to Make a Difference. MediaPlanet, The Times*, 24 April.
33  Forum for The Future (2003) 'Is There a Business Case for Socially Responsible Investing?' in *Sustainability Pays*. Manchester: Cooperative Insurance.
34  McWilliams, A. and Siegel, D. (2000) 'Corporate Social Responsibility and Financial Performance: Correlation or Mis-specification?', *Strategic Management Journal*, 21(5): 603–609.
35  3M (2007) *Pollution Prevention Pays*. Online. Available: http://solutions.3m.com/wps/portal/3M/enUS/global/sustainability/management/pollution-prevention-pays/ (accessed 13 December 2007).
36  Local Authority Pension Fund Forum (LAPFF) (2005) *Pioneering Shareholder Engagement: The Work of the LAPFF, 1990–2005*, 24/5. London: LAPFF.
37  BBC News (2003) 'Price of AIDS Drugs Cut by Half', BBC. Online. Available http://news.bbc.co.uk/1/hi/business/2981015.stm (accessed 13 December 2007).
38  Sustainable Construction Strategy Summary (2006). London: Department of Trade and Industry.
39  Transparency International UK (2003) *Anti Corruption Initiative in the Construction and Engineering Industries*. London: TI.
40  UK Government (2006) *The Building Regulations*. London: Office of the Deputy Prime Minister.
41  WRAP (2002) *Bristol Harbourside Regeneration Case Study*. Online. Available http://www.wrap.org.uk/downloads/f_Crest_Nicholson.65ed9e3a.pdf (accessed 13 December 2007).
42  Considerate Constructors Scheme (2007). Online. Available http://www.considerateconstructorsscheme.org.uk/htm-howtobe/index.html (accessed 19 December 2007).
43  Article 13 (2006) British Land CBI CSR Case Studies. Online, Available http://www.article13.com/A13_ContentList.asp?strAction=GetPublication&PNID=1321 (accessed 28 December 2007).
44  Article 13 (2006) Taylor Woodrow CBI CSR Case Studies. Online. Available http://www.article13.com/A13_ContentList.asp?strAction=GetPublication&PNID=1358 (accessed 28 December 2007).
45  Article 13 (2006) Serco CBI CSR Case Studies. Online. Available http://www.article13.com/A13_ContentList.asp?strAction=GetPublication&PNID=1358 (accessed 28 December 2007).
46  Fewings, P. (2006) 'The Application of Professional and Ethical Codes in the Construction Industry: A Managerial View', *International Journal of Technology, Knowledge and Society*, 2(7): 141–50.
47  Fan, L., Ho, C. and Ng, V. (2001) 'A Study of Quantity Surveyors' Ethical Behaviour', *Construction Management and Economics*, 19: 19–36.

# Chapter 4

# The development of professional ethical codes

Professional codes are directed at the responsible behaviour of the professional. It is important not only to identify the nature of the codes, but also to determine the relevance and comprehensiveness of the issues tackled in the codes and the effective ethical delivery of the code by the institution's membership. It is quite possible that there might be a mismatch between the public's perception of an ethical business or profession and the institution's view, and it is also important to determine an evaluation for any mismatch in this area. The objectives of this chapter are to do the following:

- Determine the nature of ethical, corporate and professional codes and their scope.
- Identify the limit and constraints of the codes, particularly in the built environment.
- Define the current use of professional ethical codes in the built environment.
- Compare the codes between the built environment and other professions.
- Evaluate the impact of the codes on the delivery of ethical practice and try to understand the barriers and drivers for success.
- Evaluate the perceptions and expectations of the public.

## Definition

A code of conduct is an expected minimum level of behaviour to protect the good name of the learned society within a body of knowledge. Traditionally, the professions were law, medicine and divinity. These professions have operated independently and mainly autonomously and exclusively on the basis of their status in society. The term professional represents a high standard of education, specialist knowledge and experience. This gives professional practitioners privilege and some degree of monopoly is afforded them by society. They self-govern through a professional body that establishes a code of conduct backed by a disciplinary committee to uphold the good name of the profession and to discipline those members who fall short of the expected standards. Thus an architect who has been negligent or a doctor who has broken patient confidentiality may be censured. A professional who is able to uphold standards is termed competent. A chartered builder who is convicted for anti-competitive behaviour might also lose their chartered status.

The built environment professions of architect, surveyor, planner, chartered builder or engineer have grown up because a body of knowledge has been established which assumes a degree of expertise can be offered in accordance with a level of education and understanding. Professionals also need to update their knowledge in order to offer advice which is up to date and does not restrict client choice.

The status of professionals today is still accepted in society but a more educated population is more discerning towards claims of competence. Competence has a meaning wider than expertise. It also includes ability, skill, fitness, aptitude, know-how and breadth of understanding in context, relevance and application. Hence a client wanting a new school building will seek an architect with experience in school buildings, not just a good architect, though there is some transferability of design skill across applications.

The ethical response to a client's need is to make ability and experience transparent and to only offer advice in given areas of expertise. There is an ethic to delegate or refer clients in the case of other areas of knowledge; hence an architect will not provide a valuation of property, unless they also have valuation experience and local knowledge. A construction manager will not design a structure without sufficient engineering experience. The code by itself is only ethical in that its terms promote a duty of trust towards others to do the service properly, honestly and without concealment. Altruism and public service are expected of most codes. Put together, they promote a modern virtue ethics which depends upon a given standard and an autonomous, balanced and judged approach. It can also assign a moral responsibility to a professional.

### Popularity of professionals

Different professions get different ratings for their trustworthiness and a Gallup poll for *USA Today* (2006)[1] in America has indicated that engineers have a 61 per cent high or very high rating compared with doctors at 69 per cent and contractors at 20 per cent.

A Mori poll (2005) for the BMA[2] (Table 4.1) ranked doctors top in the UK with high or very high trust. Although there are some surprising results with clergy and judges further down the list than you might expect, the medical and teaching professions top the list and the politicians together with the journalists are at the bottom, with business leaders not far in front.

## Professional exclusiveness

One of the problems for professionals in a modern knowledge-based society is that professional institutes are seen to be exclusive. This position needs to be

*Table 4.1* The most trusted professions

| Profession | Tell the truth % | Not tell the truth % | Don't know % |
|---|---|---|---|
| Doctors | 91 | 6 | 4 |
| Teachers | 88 | 8 | 4 |
| Professors | 77 | 10 | 13 |
| Judges | 76 | 16 | 8 |
| Clergyman/Priests | 73 | 18 | 9 |
| Scientists | 70 | 18 | 12 |
| Television news readers | 63 | 25 | 12 |
| The Police | 58 | 32 | 10 |
| The ordinary person in the street | 56 | 31 | 14 |
| Pollsters | 50 | 31 | 19 |
| Civil servants | 44 | 43 | 13 |
| Trade Union officials | 37 | 46 | 16 |
| Business leaders | 24 | 63 | 13 |
| Politicians generally | 20 | 73 | 7 |
| Government Ministers | 20 | 71 | 9 |
| Journalists | 16 | 77 | 8 |

*Source:* Mori poll (2005)

justified if professionals are to retain credibility and relevance. The exclusiveness of a profession lies in:

- its inaccessible body of knowledge, which puts the knowledge in accredited centres away from schools and business and commerce;
- its 'middle class-ness', which makes it a right for this class, but an aspiration for others, imposing class values;
- its confidentiality, which is good, but which excludes some parties and may conflict with 'the good of society' and transparency;
- its ritual – the white coat or the special wig, but more subtly special contracts and 'black box' fee calculations;
- its indeterminacy of its final judgement which also reduces transparency and therefore its ethical nature *per se*.

This is frustrating for the modern questioning mind which demands more access, equality and openness. When customers feel excluded, they may feel patronised, react and extract compensation for outcomes perceived to be detrimental. There is therefore a need for the development of a partnership with a commitment to mutual action by professional and client to solve the problem. This has become a more common feature of modern professionalism. This point is discussed further later.

It has also led to a much wider interpretation of professionalism which refers to objective, efficient service, planned and based on codes of practice. This is a much more universal service and is understood better by 'the person in the street'. The term is also used as a compliment to differentiate good and bad service.

### Professional codes

As a professional, it is important to be seen to be ethical and not to put oneself in a position that might lead to a compromising situation which may be misinterpreted. Professional organisations have favoured a risk conceptual framework approach which encourages professional services to be evaluated on the basis of significant risk assessment for a member's independence. It encourages training in recognition of specific risks that would cause or tend to cause unethical behaviour such as conflicts of interest. This is followed up with the implementation of safeguards which fall into three categories.

- Professional codes with expectations for behaviour, competence and integrity and a disciplinary committee with power to censure.
- Practice-wide controls which prescribe actions relevant to the practice competence and cover things such as training, updating, corporate governance, complaints and disciplinary systems.
- Engagement-specific issues such as a hierarchy of advice when in doubt.

Risk assessment on a case-by-case basis is good practice where operators take independent action and deal directly with clients so that the reputation of the company/profession and confidence of the client are considered. An independent mentor could also be used where areas are complex and an audit trail is required in case of comebacks. The use of a mentor outside the organisation to reflect on practice and to regain perspective is recommended, especially where a professional is working closely with a single client such as in an auditing situation.

For a professional, Kohlberg's stages[3] of moral development are a good health check on their own professionalism. Their codes expect them to be very sensitive to other community interests ahead of their own (stages 3 and 4) because of their privileged position. In addition, those who have a more senior position are expected to make judgements which determine the rights of others (stage 5), such as an architect deciding upon a conflicting client–contractor claim in a traditional contract or an adjudicator making a ruling. These are more a moral judgement than the court's requirement to look at compliance.

## Professional competence

*The Collins English Dictionary* defines ethical as 'in accordance with principles of conduct that is considered correct especially those of a given profession or group'. This provides an essential connection to individual conduct and one of these considerations might be professional or technical competence as indicated in the CIOB, RIBA and RICS Rules and Regulations. The RIBA puts it succinctly in the three principles of integrity, competency and relationships, undergirded by the institutional values of honesty, integrity, competence and concern for others and the environment. The RICS has nine principles of integrity, honesty, open and transparent, act within competence, objectivity, treat others with respect, set good example, courage to make a stand.[4] The CIOB connect the heart of ethical behaviour with professionalism, honesty and regard for the rights of others. There are 16 rules in the Professional Code of Practice which support these basics and give guidance. Each institute will have certain core competencies and other specialist competencies, and the knowledge and practice of those which are declared as upholding these types of generic principles and values are at the core of a Code of Professional Conduct. The specifics of this practice are prescribed through rules and regulations which are relevant to the way business is done in that profession.

### The granting of rights to professionals

Koehn (1994)[5] rightly questions the grounds upon which professionals are able to practise with the degree of autonomy and trust which they would like to claim for their status. If it is simply the offer of expertise with a choice on the professional's part as to whether they would accept a request for help, then as a philosophical approach, there is an element of selfish gain here. A great deal of trust is being placed in the professional competence on the basis of an often broad body of knowledge with a discretionary choice of their particular expertise under the broad umbrella of their profession. For example, a physician who specialises in heart surgery is unlikely to accept a responsibility for healing diabetes even though this might be part of the patient's needs after surgery, thus the wider context of bringing a patient back to health may not be in their power. There will be risks associated with the technique and surgery may trigger other illness-like infections.

In the context of the building professional, misinterpretation may occur when a client perceives architectural design as synonymous with fulfilling their business case, when in fact they have agreed to design a functional building. The basis and terms on which a professional is engaged are very important to their moral and ethical ability to respond to more than just the narrow terms of their contract, which need to be defined in context and with the equal understanding of both sides. Their services are also taken on trust on the basis of reasonable evidence. This would apply not only to the reputation of the professional institution, but also to the actual competence and ability of the individual member. This requires additional evidence based on personal recommendation, regular vetting procedures or self-assessment. The evidence may be taken as comparatively better or worse.

An architect becomes a member of a professional institution on the basis of examinations and a professional experience interview. They are not required to be re-examined on a regular basis, but they are expected to prove that they have carried out continuous professional development (CPD), representing the updating and use of their knowledge. Certificates may also point to the broadening of their initial skills. Elements of trust will exist in regard to the maintained proficiency in the areas important to the client. An award-winning architect does not prove that the building will be watertight and indeed this was the problem identified in the 2006 Sterling Prize speech.

Koehn (1994)[6] speaks of the essence of a legitimate profession and identifies professionals as trustworthy because: (1) they are experts or (2) they provide a service for the client's good. Trust is based on the belief that there will be some goodwill by a professional towards a particular need and this can be undermined by identifying failures in professional expertise or counter-claims for such expertise, e.g. building surveyors and architects both claim to design refurbishments in buildings. A client's trust is violated, for example, when a promised fixed sum contract is in fact exceeded without gaining any extra value. The conditions which she identifies as necessary for clients to trust the professional are:

- aiming at the client's good;
- willingness to act and to sustain that willingness to completion;
- be competent;
- the professional must be able to call on the client's accountability and discipline;
- be able to act with discretion on behalf of the client;
- be able to monitor their own behaviour.

Koehn advocates the need for pledged or covenanted service. She also considers the need for the good of the professional and society for the action to be ethical and for professionals to have discretion in the way they provide a service which matches the pledge they have made. This may be compromised if the client were to insist on action that did not comply with the professional's stated standards of conduct and additionally that contravened the good of others. A client is not entitled to take a passive role or to overreach the pledged service if a desirable and ethical outcome is to be achieved.

Koehn also makes a case for engaging a professional, not on the basis of the explicitness of the contract terms, but on the belief that they care about what you care about and will achieve outcomes in a way in which you approve. The expertise and the fee actually obscure these aims and can weaken the responsibility that the client has towards achieving the final outcome by making them believe they have paid for the choices to be made on their behalf. This state of affairs is unhealthy.

An ethical contract depends on the client and the professional working together to meet mutual objectives within certain standards of conduct. An ethical professional should be able to state that their experience is relevant and that they understand a client's needs. Their professional commitment might, for instance, require them to advise and take action on sustainable and contextually relevant designs. This aspect may not appeal to the client if it means spending extra money, so a compromise may ensue on the basis of legal compliance or on a minimum threshold provision. Is an ethical service simply the advice or is it an attempt to meet the current targets for reduction of fossil-based energy use? Is it important to have a leading solution or to provide client comfort and familiarity? It is of course possible, *in extremis*, to refuse a commission or to report a client.

Is enough choice given to the client or too much choice? These are all important questions to consider and on balance major construction clients have stated that they value open discussion and not just agreement with their own uninformed demands which leads to regret at a later stage. It now is important to see whether current rules justify the privileges which professionals enjoy in client confidentiality and support the service clients expect them to provide.

## Professional rules of conduct in the built environment

The building-related institutions cover a wide range of professional roles which have a tendency to fragment the industry. In the development cycle order, these are:

- The Chartered Institute of Banking, covering financial and funding needs.
- The Royal Town Planning Institution (RTPI), which provides a home for professional planning officers and consultants, and the officers prepare plans and advise on regulatory compliance in a social and environmental context.
- The Royal Institution of British Architects (RIBA), which provides a code of practice for the design sector and supports the requirements of the Architects Registration Board (ARB).
- The Chartered Institute of Building Services Engineering (CIBSE), which covers the design and installation of mechanical, engineering and specialist services. It is also a strong contender for the science and monitoring of the sustainability debate, particularly with regards to energy, heating and cooling.
- The Chartered Institute of Architectural Technologists (CIAT), which covers the supporting technological design and contract administration roles of building design.
- The Royal Institution of Chartered Surveyors (RICS), supporting a range of professional functions mainly connected to real estate, building surveying and quantity surveying in the property and construction sector.
- The Chartered Institute of Building (CIOB), which covers a range of

production and project management professionals connected with building, planning, estimating and maintenance. It also accredits building companies.

- The International Facilities Management Association (IFMA), which provides a broader knowledge of the maintenance and running of buildings.

All the above institutions have international associations and will have links with other country-based institutes. The basis of the chartered codes means that the Privy Council of the UK is responsible for the vetting of their terms and conditions within certain framework guidelines. The Charter allows for the autonomous governance of the institutions and also gives them respectability through certain minimum expectations.

The professional institutions need to be able to articulate some quite specific rules, which act as a precursor for identifying standard behaviour of its members. These rules can be quite prescriptive and detailed or may be based on broad-ranging principles. Some have separate ethical guidelines, such as the CIOB. A member agrees to abide by the standards on election to full membership and also benefits from the kudos and reputation of the institution. In the case of a breach of these rules being reported, a member will be subject to a disciplinary committee, usually internally composed of senior members who will come to a judgement requiring reparation, suspension or exclusion from membership. The credibility of the whole institution rests on its ability to censure incompetence or corruption fairly, justly and promptly so that future malfeasance is seen to be sufficiently deterred. High profile disciplinary investigations in the police force and other professional services have, on occasion, been seen by the public as a whitewash. This is not good for the police service, but is an example of the dangers of making such investigations internal.

A member is normally allowed a right of appeal and this is usually to the senior executive of the institute who will not have sat on the initial committee to allow a further independent hearing.

### Comparison of professional conduct rules

The professional conduct rules can be typified by those required for six of the built environment professions. These cover things such as public interest, competence in the area of advertised expertise, integrity, not accepting work for which you do not have the resources or competence, observing the rules of other countries, fidelity and probity in maintaining client and employer confidences, giving fair and unbiased advice, not accepting gifts or favours that may influence a preferential treatment of the giver, getting professional indemnity insurance or other insurance to cover liability and client impact, not seeking to injure the professional reputation or business of others for gain, and regularly updating their knowledge and professional development, including a working knowledge of health and safety (CIOB 2006).[7]

The RICS (2003)[8] gathers these rules under more general headings and these tend to be in a lot more detail than the CIOB. Table 4.2 represents very similar provisions between a building business and a surveying partnership. The RICS has about 32 pages, the RTPI has about 12 pages and the CIOB six pages.

## Discussion of professional codes of practice

The main differences are in the emphasis in the conduct of business differing primarily in the type of service. A professional builder has more emphasis on the provision of a safe and accurate product or management service. A surveying practice has more emphasis on the integrity of their advice and consultancy. Both have a duty to give high standards, insured services and accurate and reliable advice and information. Both are to be confidential, up to date and to be aware of conflict of interest.

It is interesting that the coverage of environmental and sustainable responsibilities varies from nothing for three institutions' codes to a broad responsibility (ICE).[9] It is also clear that guidelines on environmental responsibility may also appear in other places. The institutions with a more prominent design responsibility are given the most emphasis.

In the personal and professional standards section there is a large measure of agreement in the chartered institutions for integrity, accountability, objectivity, confidentiality, honesty, fairness, lack of bias and upholding an objective and fair reputation for the institution. There is also a well-known requirement for continuing professional development which some have expressed as life-long learning. Only one institution (CIOB) distinguishes between the competence of different classes of member, but the issue of competence between those of different experience is not absolute. It could apply to relevant experience or length of experience and clients do distinguish senior individuals whom they rate as more competent or with whom they can work more easily. Is it ethical to compare individual professional competence or is it provided corporately by an organisation? See below.

In the conduct of professional and business activities section there is a strong emphasis on evaluating and declaring gifts given or taken, or of any third parties who offer financial inducements which indirectly benefit your business/practice. This is most strongly expressed in the RTPI which may be because of its strong connection with public servants administering planning approvals. There are also requirements to declare, or remove conflicts of interest by standing down. There is a hint again that institutions will have a different emphasis on disciplinary action for residual conflict.

Other issues covered by all institutions are only accepting work for which you are competent and can resource, though there is a grey area as to how much you are competent to coordinate any subcontracting of such work and will it be subcontracted to professionals? It is highlighted that others' reputations should not be defamed in the process of making second opinions, or in competing for

Table 4.2 Comparison of professional conduct rules

| Key issues | Comparison | | | | |
|---|---|---|---|---|---|
| | CIOB | BIFM | RTPI | RICS | RIBA |
| Personal and professional standards | Competence, integrity<br><br>Public interest<br><br>Confidentiality<br><br>Compliance standards<br><br>Honesty and fairness<br><br>Interest of others<br><br>Uphold Institution reputation | Professional standards, but not defined<br><br>Keep institution reputation by professional acts, promote and protect and encourage new members | Competence, honesty and integrity<br><br>Confidentiality<br><br>Profession not into disrepute<br><br>Report breaches of others | Integrity, honesty, openness, accountability, objectivity, respect, set an example (core values)<br><br>Courage to make a stand<br><br>Confidentiality<br><br>High and compliant standards<br><br>Keep institution reputation | **Principle 1 Honesty and integrity**<br><br>Not support known untruths<br><br>Respect confidentiality and privacy<br><br>Not offer or take bribes<br><br>(guidance notes 1, 2, 3) |
| Conduct of professional and business activities | Fair and unbiased<br><br>Fidelity and probity | No private gain from membership | Fearless professional judgement (nor subscribe to statements contrary) | Timeliness<br><br>Accurate status<br><br>Reference to the institute | **Principle 2 Competence** Act competently, conscientiously and responsibly |

Not offer gifts for gain

Good practice and regulatory standards

Accept work competent to do or resource or properly delegate

Not offer or accept gifts exerting influence

Health and safety knowledge and risk assessment and clean, tidy and training, planning, records, discipline

Not maliciously injure reputation of others

Due care and diligence

Ensure that services offered are appropriate to need

Ensure work of others for whom responsible is compliant

Disclosure of gift or favour to a third party

Complaints handling procedure

Transparency of fees

Act in public interest

Complaint procedure

Provide knowledge, ability and financial and technical resources for the work

Care and balance of opposing demands including community

Client informed of progress

Meet the clients, agree cost, time and quality requirements

**Principle 3 Relationships (GN7)**

Procedure to deal with disputes (GN 9)

Fair competition – reasonable, transparent and impartial. Or rectify (GN 2)

(Continued overleaf)

Table 4.2 Continued

| RICS Headings | Comparison | | | | |
| --- | --- | --- | --- | --- | --- |
| | CIOB | BIFM[a] | RTPI[b] | RICS | RIBA[c] |
| Environment | | Minimise impact and encourage energy conservation | | | Aware of environmental impact |
| Discrimination, rights and interests of others | Gender, race, ethnic, orientation, marital staus, creed, nationality, disability or age<br><br>Seek to eliminate discrimination by others and promote equal opportunities | Not raised | Race, sex, orientation, creed, religion, disability, age, or equality of opportunity | Never discriminate against others<br><br>Represent correct status of members | Respect relevant rights and interests, beliefs and opinions of others<br><br>Recognise social diversity and treat<br><br>Guidance note 8 |
| Practice details and institution cooperation | Advertising factual and relevant (PCR4)<br><br>Agreed terms of engagement before start – No scale of fees<br><br>Correct use of logo (PCR2) | Not issue public statements on behalf of others without express agreement | Due care and diligence<br>Provide details of business or employment<br><br>Authorised use of logo<br><br>Terms and scope in writing up front | Accurate/fair advertisement<br><br>Agree written terms of engagement before start – No scale of fees<br><br>Provide details of business or employment<br><br>Full and prompt responses to RICS | Advertising (GN 3)<br><br>Clear terms, scope of work, fees calculation and charging, recorded in writing, records (GN 4) |

| Aspect | | | | | |
|---|---|---|---|---|---|
| Conflicts of interest, impartiality and independence | Do not take work conflicting with the client or employer interest. Not injure third party reputation. Act in public interest | | Take steps to avoid all personal, client, family, political, employer, club or business conflicts (comprehensive) | Conflict between business with two different clients. Conflict between member and client interest. Personal interest gain. Business conflict. Public office interest | Not influenced by own or other's interests. Avoid conflicts of interest and declare and/or withdraw. Guidance note 1 |
| Professional Indemnity (PII) and other insurance | Must obtain PII (PCR3). Third party construction insurance as relevant | No mention | PII (add. regs) | Must obtain PII | PII Guidance note 5 |
| Keeping separate trust accounts | No | | No | Yes | No |
| Use of letters | Corporate class (PCR1) | Corporate class | Corporate class | Corporate class | Corporate class |
| Institute activities | | Whenever possible | | | |
| Lifelong learning | Plan CPD up front. Update knowledge – CPD – minimum hours | Keep abreast of new knowledge. Promote training of Facilities Managers | Reasonable steps to maintain professional competence (minimum CPD hours). Encourage others | Update knowledge – CPD minimum hours. Keep records (rule 37) | Guidance note 6 |
| Investigation | Disciplinary regs. Report breach | | Assist investigations. Report breach | Fine for not giving information to RICS | |

Notes: a  BIFM (2006) *Code of Professional Conduct*. Online. Available www.bifm.org.uk/bifm/about/governance/codeofconduct (accessed 30 October 2007).   c  RIBA (2005) *Code of Professional Conduct*. RIBA Council. January.
b  RTPI (2001) *Code of Professional Conduct*, RTPI Council, 17 January.

the same work. The public interest is only mentioned explicitly by four institutions, although the others imply this. Obvious statements are made about staying within the law and declaring criminal convictions, bankruptcy, etc. Only three of the institutions require members to explicitly report the wrongdoing of other members, though this must surely be the main way that contravention of the code is discovered. Discrimination is presented in terms of respect and equality of opportunity as underlying business practice, but does not exclude Koehn's proposal that professionals should not choose who they provide services for.

There is agreement on business issues such as transparent fees, written specification of work covered, terms of engagement declared before work starts and fair advertisement. None of the institutions declares a scale of fees. These indicate that competition has now become more important and it is hoped that the dangers of hiring professional services for the lowest fees are understood by the client.

These rules are generally backed up by disciplinary and competence procedures to give a fair investigation of alleged misconduct or incompetence. The effectiveness of these rules in assuring ethical behaviour of registered professionals is that they are policed and that misconduct complaints are seen to be fairly investigated and miscreants receive due censure (Vee and Skitmore 2003).[10] If it is perceived that there is internal protection of members then the status of the institutional membership is at stake.

## Built environment and the rules

Construction has a particularly complex supply chain which creates opportunities for the code to break down. Transparency, integrity, trust and fairness are common aspirations written down in ethical codes. These values will become meaningless if they are not backed up by a real concern to ensure their efficacy. There also needs to be a trusting relationship between the professional or business who claims compliance with the code and those that commission services from that professional.

This puts an emphasis not just on membership of the organisation, but on the *reputation* of that business or institution. This reputation is built up over time and is known by the collective reports of others. Good reputation can equally be lost by a single non-complying (unethical) act. Tolerance of non-compliance is dependent on the degree of broken trust perceived by the other party and the sympathy they have for the nature of the contravention. For example, if there is discrimination against the use of ethnic minorities, this is a contravention against human rights in most codes, but in the collective conscience is likely to clash with a degree of nationalism which would tacitly approve of such an action.

Credibility then is a relative value which is crucial to the benefit of a code, but may also be gained for a less than fully compliant standard. Some tolerance on the basis of outcomes which meet expectations can be found in the utilitarian model that maximises good and minimises harm, but there is less tolerance under the Kantian model of compliant process in spite of income which is the basis on which the codes are set up in order to protect the public level of service.

A public organisation has to satisfy a tougher credibility standard because of the more direct accountability of the public official to their elected position and to the taxpayers' expectations of fair play. An example of this is the tendering process. The planning application process will be discussed in its own chapter.

## Transparency in construction trading relationships

Relationships will be viewed differently between suppliers and contractors, clients and the design team. They will each have their own view of the meaning of fairness, transparency trust and integrity. Suppliers will see fairness as being paid on time, as agreed for each piece of work they have carried out, plus compensation for any disruption experienced because of others. The contractor will only want to pay for work which is completed with precision, highest quality and timeliness which has not caused them any delays or extra costs charged for deficient work by follow-on contractors. They may also want to impose penalties for delays caused by suppliers over and above costs and hold retention on whole pieces of work of which only half is defective. Quantity surveyors have a key ethical role to play in this. A collaborative relationship should be seen as a more transparent one, where payments are honoured with a desire to respect the abilities of both parties. Latham[11] pointed out the lack of trust in the building industry between contractor-client and contractor-specialist.

### Case study 4.1 Late payment – an ethical requirement

The Latham Report has particularly challenged the poor treatment of suppliers by larger main contractors imposing punitive damages for risks that have been passed down the line and delaying whole payments for partial non-completion and large and long-standing retentions of payments to cover for defect. Clients have often exercised their dominant position to delay or withhold payments for delays and changes of their own making especially to designers and tenderers who have invested up front to provide evidence for feasibility. This has caused a vicious circle which Latham tried to break in the Construction Housing and Regeneration Act 1997. The Egan Report has supported a regime of managing value and attempting and identifying risks in the planning stage of the contract and this is discussed in a later chapter.

An industry in general may also get away with a norm which is patently below ethical action such as the late payment of subcontractors by a main contractor who has not received that payment from the client. Although this has also been disallowed in the UK in The Housing Grants and Construction Act, there is a sympathy that paid when paid is a logical position and is unlikely to lose that MC's reputation, except among those subcontractors who may refuse to do work with the contractor. Because of their weaker position, however, they may price a contingency to deal with such a situation. Those making an ethical stand will therefore be in the invidious position of losing commercial benefit through prompt payment of their subcontractors and later payment from their client. Here only an ethical partnership between client and contractor and the supply chain can hope to regain the higher ground.

The transparency between an inexperienced client and the contractor will be depend upon the architect's explanation of the standard conditions which prevail under the chosen type of procurement and the assumptions made by the client for appropriate contractor behaviour. For example, in the case of changing the brief, they may expect to pay *pro rata* and not appreciate abortive and disruptive costs. They may feel if they alter the design during the scheme drawings, they are justified in causing redesign work for no extra cost if they have not yet received planning permission. In this case, the ethical action of ensuring transparency and trust is tied up in fully briefing the client on the cultural assumptions that are normally made by various parties beyond the letter of the contract.

The contractual fragmentation between design and contractor means that clients often become involved in extra cost because of a contractor claim for insufficient, late or inaccurate information received from the design team. In this case the design team is not usually required to pay the cost even if the contractor becomes liable for liquidated damages from the client for late completion. This is due to the direct contractual responsibility of the contractor to the client when the problem has occurred in the indirect communication link through the designer with whom they have no contract. In this case, a client strictly would have to sue the designer on behalf of the hurt the contractor has received with plenty of opportunities for 'opaque' evidence and counter-claim and breakdown in trust. In any case, it does not make it easy to have integrity in settling matters out of court when there are all sorts of conflicts of interest being implied by the architect who is also the arbiter in the contract. The public have rightly questioned the professional impartiality of the architect in these situations.

## Ethical partnering codes

At the heart of the ethical code which applies to trading is the need to properly define the procurement route with its advantages and disadvantages, making specific references to the needs of the client and good practice as normally understood by the supply side. This should enable an informed choice *by the client*, based on their own values and their willingness to conform to best practice. This should not preclude the introduction of innovative approaches, but these are best introduced in the context of early involvement of key contractors where a degree of mutual negotiation and understanding can be developed.

This can be done on a long-term (strategic) partnering basis where the partnering agreement covers a specific period such as a five-year framework agreement covering a share of a specific client's future projects. Alternatively, it could be a single project or programme of projects where there is the commitment to a partnering charter to further trust and transparency between the client and the principal contractors.

Projects are particularly vulnerable to misunderstandings as each project is a prototype design and has a markedly different context and project team where familiarity, loyalty and trust need to be built up. Partnering is a way of recognising the benefits of keeping teams together, usually with the same client. It can build upon the knowledge gained of a client's value system and project type, standardisation of project design or components, or from continuous improvement of a project delivery process. These are not as easily gained in open competitive tendering.

Partnering could use a prime contract[12] or it might involve a standard voluntary framework such as the Joint Contract Tribunal (JCT) Framework Agreement, or a contract such as the New Engineering Contract (NEC). The JCT non-binding agreement has a number of sections covering good practice procedures such as tendering, risk and value management and health and safety management, in order to develop a culture of collaboration, early contractor involvement and management. It sits upon the contractual conditions which already exist. This partnering agreement is more complex than a simple partnering charter which agrees to basic principles. It would, however, be ineffective unless it was well understood by *all* members of the project team. It does have the benefit of still being voluntary. The benefit of a binding contract is seen by small organisations as protecting against a costly one-sided compliance by them with no recourse to recovery in the case of non-compliance. This raises the question as to whether it is operated in a spirit of cooperation, which is the basis of partnering. In Chapter 13, a case study has been prepared on the use of the JCT Constructing Excellence contract.

Ethical behaviour can thrive in an atmosphere of partnering, but may have a 'shelf life' due to the belief of a dominant partner that they are able to control the conditions of continuous improvement with maximum benefit to themselves and reduced incentive to others. For example, a supermarket that wishes to partner in order to continuously reduce the programme time for the completion of new supermarkets will identify genuine savings to its business where demand is proven by gaining weeks of profitable takings. Initially, similar savings to the supplier may ensue in reduced overheads in supervision and temporary accommodation, etc. In the third or fourth round of programme reductions, however, the supplier may encounter reduced productivity gains due to the congestion of activity on the site – inefficient levels of resources for short periods of time disproportionately make the mobilisation costs escalate. These can be offset by perhaps a new phase of off-site manufacturing of larger components, reducing the labour force. This will be a new research and development cost (a new prototype of design) and will also require more lead time to arrange component delivery. If a client ignores these issues, incentives for the extra risk and effort by the supplier will be constrained. If a client ignores the strategic planning needed for earlier lead times, then programme times will be capped or promises will be less sustainable. The partnership has arrived at a step change which will initially cost more or need to be reformed.

The cycle shown in Figure 4.1 is hypothetical, based on driving out waste in the system. Ethically, effectiveness is as important as efficiency and is the ultimate aim, as some benefits are intangible and align themselves to the client's value and culture. It is not strictly indicated in an ethical code, but is implied that a service will be provided that gives best value to the client.

The contra argument that partnering is open to abuse is implicit in the publication of a code of practice. Not all partnering arrangements need to be long term and whether long or short term, they can refer to the development of mutual objectives, working to agreed decision-making processes and actively searching for improvements in productivity. A partnering charter or pledge, however, puts a greater emphasis on trust between the two parties and this trust has to be maintained over a period of time, theoretically making the contract obsolete or more practically giving a fallback position in the case of breakdown of the charter and goodwill.

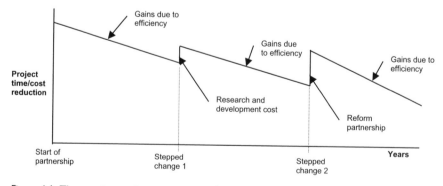

Figure 4.1 The continuous improvement cycle.

### Small building works codes

As the industry also has easy entry, then there is also scope for ignorance and the construction industry has a cowboy image at the less regulated end of the size spectrum. Myers (2005)[13] points out that 93 per cent of the sector consists of small organisations employing less than eight people which is a disproportionate percentage compared with other sectors and will depend much more upon individual values of the proprietor and be less subject to public pressure except in substantial legal compliance IBE (2005).[14]

The Trustmark scheme[15] is an example of a government/CIOB scheme that has been instituted for the purpose of registering companies which agree to certain trading and conduct standards and is run by the DCLG. The core aim is of consumer protection by providing accreditation and monitoring services by other organisations accredited to check competence, standards and customer service. A similar scheme is run by the Federation of Master Builders (FMB)[16] on a membership basis which is open to those who agree to provide certain standards of service, which will be renewable annually on the basis of a code of practice which they sign up to as members. This code covers a formal contract, an agreed programme of work, a warranty backed by the FMB and a respect for health and safety and the customer's property. It is a basis for good practice and ethical trading considerations and includes for most a vetting procedure through the Trustmark scheme. The benefit of these schemes is that there is a standard contract with encouragement for good practice for the builder and the client and a warranty back-up for things that go wrong, even if the builder goes out of business. It even builds in a dispute resolution mechanism.

Ethical trading also depends upon two sides communicating and trusting each other's judgement and values even when unexpected things occur. Customers are encouraged to obtain more than one quotation and to agree reasonable payment terms, doing some checks for themselves as a basis of trust. Builders are encouraged to check the ability of customers to pay and to advise them in advance of the cost impact of too many and late changes. They are also responsible for ensuring compliance with planning and Building Regulations requirements, many of which may have changed since the customer last commissioned building work.

### Community/societal codes

There are a number of codes that have been drawn up internationally that apply to multinational corporations. Lovell (2005)[17] cites the Social Accountability Code SA8000 which was policed by an international accreditation agency for ethical trading for the commitment not to use suppliers who use 'sweat shop' and child labour. They give guidance to ethical activity to set a standard and have produced country-wide codes or sector codes. These tackle particular issues such as the conditions of manufacture.

## Professionals and sustainability

Other aspects of professional codes deal with the need to provide a service which does not impact unreasonably on the community or society as a whole. These issues relate to the need to provide a sustainable environment which is not adversely affected by the construction work itself, or the ultimate use and operation of the building in the long term. Many issues such as energy savings and climate change, as well as the impact of the building on the space and views of the community, are well documented but difficult to measure. Although there are many pressure groups, progress is slow in reducing the use of scarce resources or creating pollution which cumulatively has a global effect, such as climate change, or a local effect, such as noise and water pollution. The general consideration of environmental ethics is considered in Chapter 8. However, most corporate social codes and some professional codes will also include reference to the responsibilities which they perceive to provide a better quality of life for their neighbours and contribute to a securer long-term future.

The UN Global Compact (2000)[18] introduced by Kofi Anan in 1999 gives a high profile to sustainability. This is discussed in more detail in Chapter 8. Figure 4.2 shows the sustainability principles enshrined.

There is room here for built environment professionals to support this through their own codes and this would overcome Roth's (2000)[19] three-fold criticism of the compact – that it is not legally enforceable, there is no monitoring and enforcement, and there is a lack of clarity.

A sustainability reference has more recently been included in codes such as the new Royal Academy of Engineering (RAE 2007).[20] These, under a heading 'Respect for law and the public good', cover such issues as 'Minimise and justify any adverse effect on society or on the natural environment for their own or succeeding generations', and 'Take due account of the limited availability of natural and human resources'. These represent the growing and specific importance of the sustainability agenda in professional codes and will be underpinned in each of the engineering disciplines the Academy represents.

There is sometimes a conflict of interest between the code of competence which asks professionals to 'be aware of the environmental impact of their

---

Principle 7: Support a precautionary approach to environmental challenges.

Principle 8: Undertake initiatives to promote environmental responsibility.

Principle 9: Encourage the development and diffusion of environmentally friendly technologies.

Principle 10: Businesses should work against corruption in all its forms, including extortion and bribery.

---

*Figure 4.2* UN sustainability and anti-corruption principles.

work'[21] (RIBA Principle 3.2) and the need to balance the needs of different stakeholders with the community needs (RIBA Principle 2.1). These conflicts of interest are often very subtle, but may even lead to a contractors not taking on a commission where they feel unable to meet their client's requirements and aspirations. An example of this is the refusal of a client to recognise the need to comply with basic energy saving compliance as indicated in the regulations. Many of these issues may work themselves out through the planning requirements to provide some indications of the sustainability of the building, in order to make a case for its suitability for future needs. This is often a case of balancing different aspects of sustainability to give an overall 'eco score' which is acceptable.

For professional practices there is often a phrase in the ethical code which acknowledges the responsibilities that they have in a community – maybe in terms of trust as the following extracts clearly show.

## Case study 4.2 Typical ethical statements taken from corporate literature

Independence and high ethical standards are our core values ... Ethical behaviour in this role is paramount and Integrity is therefore the first of our three core values, alongside Innovation and Collaboration.

(Davis Langdon 2007)[22]

Ethics can also be connected with excellence.

Ethics must play a major part in driving our financial decisions and our quest for excellence.

(Atkins 2006)[23]

However, society will primarily judge organisations on their actions. The perceived motivations will differ if the outcomes of policy do not produce an ethical result. For example, the withdrawal of support by a bank for a charity which subsequently collapses and affects the lives of primarily poor people, may be seen by the public as corporate greed, especially if the bank is making record profits that year with shrinking service provisions. In fact, it might have been poor money management by the charity.

### Professional practice and corruption

In construction, the publication of reports focusing on widespread corrupt practices in the industry exceed even the guns/defence industry and the oil and gas industries (Transparency International (TI) 2005).[24] TI has suggested that an excessive level of commercial confidentiality has led to a level of secrecy that has sometimes been biased against the public interest. This has made it easier for payments to be made to influence the award of contracts or to make conditions which favour a particular contractor or allow them an uncompetitive bid. A further claim is that the industry too often accepts the status quo by rationalising actions as cultural, or too hard to change without eliminating benefits to the client. The UN Global Compact since 2004[25] also includes a principle and responsibility for the private sector to share a responsibility to 'eliminate' corruption and makes an ethical and a business case and a connection with sustainability.

Stansbury (2005)[26] recommends that there should be a specific rule of conduct outlawing bribery, with training and awareness of bribery and corruption in the UK construction industry (which he shows to be rife), and disciplinary action and the need to report these misdemeanours to the authorities. For corporations projects, he recommends that there should be a sector-wide non-bribery pact which may only be practicable in the case of a high cost entry industry. However, Stanbury goes further to introduce a project-specific Integrity Pact which would bind signatories of all project organisations to eliminate bribery and automatically report offending organisations and individuals to the authorities. This may not be so easy in an international situation and it may also not cover the client.

This report has been taken seriously by professional institutions and prompted the CIOB to commission a survey in UK (Construction Manager 2006),[27] taking some of the TI-defined corruption issues such as cover price, falsely inflated claims, favoured treatment of a tenderer, hospitality for clients and exclusive negotiations with the planning officer for Section 106 agreements. All of these practices can potentially be prosecuted with a prison penalty of up to seven years. Respondents were not unanamiously in agreement as to any one of these practices as corrupt, but professionals have a unique opportunity to influence clients in either direction. More is discussed in Chapter 10.

## Ethical leadership

Leadership is often interpreted in terms of power from the top with direction from the leader to the follower, but it can be influenced in a more subtle way. Guillen and Gonzalez[28] sum up a wider definition as 'a dynamic free interpersonal relationship of influence, beyond formal power, and based on the technical (*skills*), psycho-emotive (*follower motivation and satisfaction*) and moral (*ethical behaviour*) dimensions of trust' (italics added). Leadership in ethical matters means leadership in the grey areas which are not covered by regulations or codes of practice, but are covered in the area of best practice. Clearly the proper interpretation of regulatory codes in their spirit as well as the letter is important, but regulations may create an area of blindness to normal practice such as in the issues discussed above. This is not chiefly an audit activity, but one of ethical leadership.

Professional leadership is an important part of the credibility of ethical codes. Codes need an internal credibility so that staff will take them seriously, because corporations and professional bodies depend upon their members for the sensible and consistent implementation of the policy. They will not do so if there is no commitment by the leadership in terms of a consistent application and enforcement across all areas of trading and consultancy, and also if there is no back-up in terms of financial resources and authority to make the tough long-term decisions, which may not be so profitable in the short term. The commitment should show in terms of continuous professional development and the feedback which is honestly made available by senior members. There needs to be a no blame environment for employees and senior managers/partners to express their ethical concerns without their career progression being affected. These issues can be summed up in two key areas which are

- openness so that there is a willingness to discuss a wide range of issues which are connected with the way of doing business with a willingness to learn;
- courage where there is a willingness to 'stand up and be counted' and it is believed that the company/practice has to make an ethical stand against existing practice or corrupt practice which has become acceptable through common use and competitor pressures. This might also be defined by the commitment of a company/practice to an ethical principle or practice.

The first can be applied as a commitment to, at its minimum, a greater dialogue with the stakeholders as well as clients, with a genuine option for change to take place as a result. So, if a water company is looking for ways to clean up its environmental act it could honestly approach its customers for a rise in water prices for a set period, in return for a more than proportionately higher investment in water leaks, over and above the OFWAT minimum. This would involve a joint investment by customer and company to save waste and to reduce future costs, which otherwise might be unobtainable. Unethical communication consists of making false promises about leakage reduction targets which are missed and lead to maximum band water price rises and no future reduction.

Courage is required by a professional to turn down a commission on the basis of a client's refusal to put any investment into sustainable features in the design. An ethical proactive approach would be to offer alternatives to the client, based on research into their stakeholder community in order to make the proposal more attractive. The life cycle and operating costs should also be used as part of the argument. However, these may need to be attractive to potential buyers or tenants if the client is a developer or is only planning to use the building in the short term.

In some countries, professionals may be asked to get involved in bribes in order to get building permission or permits or to take gifts from suppliers to specify their materials. This tests the integrity of the professional and there needs to be a clear policy as to the influencing nature of these payments. To maintain integrity, there should be a clear policy on whether any payments are necessary and, if some are taken, that they are declared in the accounts and that they are proportional to any service received. It is easier not to accept gifts and hospitality, but in countries where this is social practice they need to be proportional, declared and not for personal gain. Payments or gifts received that are conditional explicitly or implicitly to the benefit of the giver are clearly unethical. Actions that distort competition are clearly unethical.

## Case study 4.3  Gift giving and taking

Human interactions are complex and the size of the gift will differ in its impact dependent upon the person, the way it is done and the expectations in the context of the gift. It is easier where the culture is familiar and the parties are known. Culturally, Europeans and North Americans are happy with no gift or small gifts. In Japan, gifts are traditional and expected in order to seal future relationships. In an unfamiliar context, expectations have to be judged and it is easy to misjudge this or to contravene your company or professional code. The receiver may interpret the size as insulting if it is below expectations or a way of bribing the receiver if it is too large or even thought necessary. For these reasons, it is useful to have a model. The model in Figure 4.3 has been adapted by the International Society of Accountants IFAC (2003). The code allows for reasonable reflection, but needs to be answered honestly to be effective.

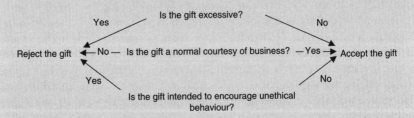

*Figure 4.3* When is a gift acceptable?

*Source: Proposed Code of Professional Accountants* (IFAC 2003).

Courage is illustrated by standing up against a hidden cartel, which is operating to the advantage of a group of oligopolistic companies, by reporting it. In the short term, it would be seen as a way of hurting the industry as a whole and might incur serious fines, but in the longer term it could be argued that waste is stripped from the system in order to give value to client and company profits. In the UK, the Office of Fair Trading (OFT) actually has a leniency scheme to encourage the reporting of cartel activity by allowing an amnesty to the first reporting cartel members, as will be discussed in Chapter 10.

The IBE (2004)[29] survey on ethics indicates that 42 per cent of people in the institutional investor community mentioned honesty as an issue in their judgement to invest and, when not prompted, these numbers fall to 6 per cent compared with 10 per cent in 2002. This indicates a tenuous significance for shareholder/stakeholders and the pressure for honesty is mainly from other stakeholders. An ethical minority representing £4.5–4.7Bn (2 per cent of UK funds in 2004) are ethically invested. NGOs also pressure change in ethical behaviour and this is a growing concern which has influenced company behaviour against maximum short-term returns. Clearly, there is a long way to go, but the need to have a viable CSR are having some effect.

## The Considerate Constructors Scheme

This scheme has been adopted by many contractors as a way of implementing, on a local basis, the ethical issue of working as harmoniously as possible with the local community. Building sites, especially in built-up areas, have an impact on a local activity and may be noisy, dusty, dangerous, dirty and inconvenient. The aim is to be good neighbours, avoid or reduce this impact on them and to underplay expectations of the inconvenience caused. Sound planning and resources allocated to maintain good relationships are essential to the success of such a scheme which is also monitored by experienced volunteers who make third party reports by scoring the measures implemented. This score is converted into poor, satisfactory, good and very good. A good score can also be used in promotional literature to promote image and reputation, and is a motivation for top management commitment to provide resources. A well-organised site should also have pay-offs in productivity and a better sense of fulfilment for all.

The UK scheme covers the eight areas shown in Table 4.3. Measures in each section should be creative and applied.

The problems may be that sites are often selected as a flagship and shown off for maximum effect and other more difficult sites which have done more are indistinguishable in scores from the 'easy' ones. Claims for widespread participation in company literature are vague and only cover up patchy or poor compliance. It may be perceived that a building development will always be a rough experience for neighbours and that measures to mitigate impact are only cosmetic.

The UK scheme is a voluntary one, which also specifically encourages environmental compliance and creativity. The scoring of the UK system[30] is from 1–5 for each section and 3 is a satisfactory score. A satisfactory site in all areas would get a score of 24. The average in the scheme (20,000 sites up to beginning of 2007) is 29.9 with 600 visits a month. However, the scoring is

*Table 4.3* Measures for Considerate Constructors Scheme

| Categories for action | Possible measures |
|---|---|
| Considerate | Restrict delivery times, inform scope of works and consult, clear access, visitor arrangements, operative car parking, disability access, personal touch and providing hoardings for visibility |
| Environment | Reduce waste, respect flora/fauna, spillage bunds, energy saving, stop pollution, segregate and recycle, reduce noise, directional lights, delivery control, reduce cars |
| Cleanliness | Clean and tidy site, wheel cleaner, clean roads, damp dust, clean toilets and showers, food hygiene, no litter, efficient material access, paint hoardings |
| Good neighbour | Complaints courtesy and action, 'hotline' information, reduce security alarms, clear and limited hours of work, non-obtrusive lighting and deliveries, no radios |
| Respectful | Screen toilets from public, dress sense especially for local shopping/contact, conduct rules, exclude offensive material, reduce noise, good welfare, viewing |
| Safe | Warning signs, effective protective barriers, lighting, walkways and segregation, record accidents/near misses, escape routes and rendezvous, safe material storage, protection of public |
| Responsible | Community participation, be aware of other activities and adjust work to suit, first aid help, information to public services |
| Accountable | Records of schemes in place, information to the public, operative training and induction, photographs, etc. |

designed to remedy all sections equally and any site with any section scoring below 3 is asked to remedy it, and may be pulled out of the scheme if it does not reach a minimum two compliance in all sections. There are prizes for high scoring sites. In 2006, this amounted to 298 prizes from a pool of 3,475 eligible sites.

These schemes require ethical inclusiveness and refer to some of the wider principles of sustainability, but are also recognised as a code of practice which operates at a local level. The voluntary nature of such schemes is considered to promote the creativity in which construction projects enhance worker experience, reduce nuisance and provide a community service.

## Conclusion

The professions can be seen as exclusive clubs to protect privileged membership and remuneration. The modernising process requires professionals to justify their existence and privileges in a more sceptical world, particularly if they are to act as ethical gatekeepers. The concept of professionalism is far more than the keeping of a professional code of conduct although this might be a good starting place. Contrary to some expectations, Koehn[31] shows there is need for a social contract of mutual trust and action between the professional and their client, with the professional providing an equal service to clients. This recognises that a service can only be carried out properly if there is commitment on both sides. This may be more of a challenge to the less experienced client who would simply like to hand over to an expert, but the ethical decision framework needs the support of the client. The codes themselves do not avoid the need to make ethical decisions.

It is possible to break down the codes in the built environment institutions into an integrity section, a competence section and a relationship section. The professional codes themselves show a strong ethical bias to a Kantian duty to perform for the 'good of society' but there is also a pressure for the professional to be narrowly client-focused. Societal aims have been paid lip service to and it is increasingly realised that the professional needs to work out their own solutions in practice. This conflict of interest needs to be resolved to maintain the trust of the community.

There is a growing trend in the codes to be more specific about building futures through more explicit comments about the conservation of resources and a more sustainable approach to design for the present as well as the future (RAE, 2007),[32] (CABE, 2007).[33] This trend is being pushed by government policy which recognises the property and construction sector as a major producer of carbon and of waste. Built environment professionals have a unique opportunity to influence the future of society through the development control and design process and by using construction materials more efficiently to reduce waste. There is also some opportunity for property professionals, designers and developers to exceed government targets for sustainability by working together.

Transparency in construction relationships has not traditionally been strong often due to the contentious nature of construction contract conditions. There *is* room for more ethical behaviour in collaborative contracts, where parties are encouraged to speak to each other more plainly and bring the contractors 'on board' earlier so that risks can be allocated effectively. The concept of good faith is beginning to be recognised more as a duty in the courts (see Chapter 11) and, if this becomes so, it gives material advantage to collaboration in the case of dispute.

The recent reports that have dealt with widespread corruption are a professional as well as a business problem as they may often be accountant-led. Professionals may have an ethical conflict of interest with the commercial interest of their employers, but this should be resolved by the opening up of channels to discuss ethical concerns with management at a higher level without fear of reprisal and to make all payments transparent. New standards of ethical practice in the area of competitive practice are expected.

## Notes

1  *US Today*/Gallup Poll (2006) *Honesty and Ethics Poll* (poll shows trends since 1976).
2  Ipsos Mori/BMA (2005) 'Doctors Top the Public Opinion Poll on Trustworthy Professions'. Online. Available http://www.mori.com/polls/2005/bma.shtml 10 March (accessed 11 June 2007).
3  Kohlberg, L. (1969) *Stages in the Development of Moral Thought*. New York: Holt, Rinehart.
4  RICS (2006) *Professional Ethics and Core Values*. Online. Available http://www.rics.org/management/businessmanagement/businessethics/ethics.htm (accessed 3 January 2006).
5  Koehn, D. (1994) *The Ground of Professional Ethics*. London: Routledge.
6  Ibid.
7  CIOB (2006) *Rules and Regulations of Professional Competence and Conduct*. Online. Available www.ciob.org.uk (accessed 30 May 2007).
8  Royal Institution of Chartered Surveyors (2003) *Professional Regulation and Consumer Protection Rules of Conduct*. Online. Available http://www.rics.org/NR/rdonlyres/A6554ADC-403C-4C76-8D30-CAFD3C75FE83/0/rules_of_conduct_02.pdf (accessed 26 October 2006).
9  ICE (2004) Code of Professional Conduct Rule 4.
10  Vee, C. and Skitmore, C. M. (2003) 'Professional Ethics in the Construction Industry', *Engineering, Construction and Architectural Management*, 10: 2 April, 117–27.
11  Latham, M. (1993) *Trust and Money: Interim Report on the Construction Industry*. London: HMSO.
12  A prime contract is a contract developed by the MOD to encourage more upfront contractor involvement and is generally let with a close design and build liaison and equal partnership of designers and contractors under the prime contractor or project manager. It encourages the rewarding of innovation and the sharing of value improvements. There is a JCT Prime contract form being developed.
13  Myers, D. (2005) 'A Review of Construction Companies' Attitudes to Sustainability', *Construction Management and Economics*, 23: 781–5.
14  Institute of Business Ethics (2005) *Corporate Use of Codes of Ethics, 2004 Survey*. London: IBE.
15  Trustmark (2005) Online. Available http://www.trustmark.org.uk/default.asp DCLG (accessed 24 November (2007).
16  Federation of Master Builders (2006) Online. Available http://www.findabuilder.co.uk/why/practice.asp, FMB (accessed 6 June 2007).
17  Lovell, A. (2005) *Ethics in Business: A Literature Review*. Edinburgh: Institute of Chartered Accountants of Scotland.
18  UN Global Compact (2000) *Environment*. Online. Available http://www.unglobalcompact.org/AboutTheGC/TheTenPrinciples/environment.html
19  Roth, K. (2000) *UN Global Compact*. Online. Available http://hrw.org/corporations/un-compact-new.htm, Human Rights Watch.
20  Royal Academy of Engineering (2007) *Statement of Ethical Principles*. London: RAE, in collaboration with Engineering Council and leading engineering institutions.
21  RIBA (2005) *Code of Professional Conduct for Members of the RIBA*. London: RIBA.
22  Davis Langdon (2007) *Corporate Responsibiliy*. Online. Available http://www.davislangdon.com/ANZ/OurBusiness/1726/Ethics-and-Integrity/
23  Atkins, J. (2006) *Corporate Responsibility Report*.
24  Transparency International (2005) *Global Corruption Report: Corruption in Construction and Post Conflict Reconstruction*, 16 March.
25  UN Global Compact (2004) *Transparency and Anti Corruption*. Online. Available

http://www.unglobalcompact.org/AboutTheGC/TheTenPrinciples/anti-corruption. html (accessed 16 December 2007).

26  Stansbury, N. (2005) *Anti Corruption Initiative in the Construction and Engineering Industries*. London: Transparency International UK.

27  Smith, K. (2006) 'Ring Any Bells?', *Construction Manager*, CIOB, October.

28  Guillen, M. and Gonzalez, T. F. (2001) 'The Ethical Dimension of Managerial Leadership: Two Illustrative Case Studies in TQM', *Journal of Business Ethics*, 34: 175–89. Quoted in Foster-Back, P. (2006) 'Setting the Tone', *Ethical Business Leadership*. IBE.

29  Webley, S. and Le Jeune, M. (2005) *4th Corporate Use of Codes Survey*. London: IBE.

30  Considerate Constructors Scheme (2006) Online. Available http://www.ccscheme. org.uk/ (accessed 10 October 2007).

31  Koehn (1994) op. cit.

32  RAE (2007) *Code of Ethics*, op. cit.

33  CABE Briefing (2007) *Sustainable Design, Climate Change and the Built Environment*. London: CABE.

# Chapter 5

# Discrimination and human resource ethics in the built environment

Employment is a privilege in that an employer is recognizing the worth of employee skills, abilities and knowledge. On the other hand, for most employers, human capital is their most valuable asset and to lose the respect of the workforce is a serious issue. Ethically, it is also important to refer to diversity and equality in a built environment context. Discrimination of certain groups on the grounds of gender, age, race, disability or religion is not permissible and yet there is a lack of diversity in built environment occupations. There are also some allegations of unequal treatment, where differences have been discriminated against because of particular working conditions such as long hours, heavy manual work and poor welfare. The objectives examined in this chapter are:

- to consider the impact of treating workforces fairly in accordance with human rights, starting from Cadbury and including reference to culture, employability, less discrimination, conditions of employment, motivation, incentives, moral climate, training, career development and progress, and the psychological contract;
- to look at the ethics of legislation and best practice such as ethical leadership, professional discretion, fuller employment, recruitment, trade union roles, public disclosure, fair conditions in outsourcing, harmonious workforces, and better community engagement.

## Diversity and equality

Another ethical approach to employment is equality and diversity. This is reflected specifically in construction by the fact that only 9.2 per cent of the workforce are women, compared with 34 per cent as a whole, and with 1 per cent in manual trades and only 4 per cent in membership of professional institutions connected to construction and consultancy. There is 1.9 per cent of black and ethnic minorities in construction compared with 6.4 per cent in the economy as a whole. There is no compelling reason for this except that the industry has traditionally had a male image and is still in the process of change, but attempts to right this imbalance have been in place for 15 years (Gale and Davidson

2006).[1] Gale and Davidson also make the point that diversity and equality is a more positive concept and can be valued as a benefit to better change the culture, while discrimination is associated with a failure to comply. From an ethical point of view, equality of opportunity is a human rights issue for freedom of choice, and constructive exclusion of certain categories of labour is something that should be eliminated if this is not to cause frustration. Diversity is more easily discussed as a Kantian approach where it is seen to be a duty to reflect the proportionality of the population. Equality is a justice ethic where it is seen as a human rights issue to be treated equally with others of equal ability. There is a tradition since the 1960s in the UK to apply discrimination legislation to try and enforce an equal approach on employers in particular. This has some effect, but does not tackle the underlying attitudes of the workforce where discriminatory behaviour between employees is ignored. Managers have to be seen to be complying.

## Human rights

Work is a right enshrined in the UN Universal Declaration of Human Rights (1948)[2] and it is important that employers recognise their responsibility to provide fair treatment, pay and conditions which are legally compliant. It is also ethical so there is a duty of care to engender a sense of dignity, equality and pride in the process of work and respect for all levels of the workforce. It is hard to formalise this in rules and regulations and there are many contexts unique to cultures, type and level of work and management style. The responsibility for ethical leadership applies at all levels of management. Historically, employers have managed to keep the workforce compliant by limiting educational opportunities. In return, the workers received limited reward in addition to working in second-class conditions. Expectations were low, with a line drawn between the professional and working classes.

Trade associations developed towards the end of the nineteenth century and latterly trade unions had more muscle as workers grouped together to secure better working conditions and rates of pay increased the gap between worker and employer expectations. Pay had become the predominant form of remuneration for the modern worker and other rights were subjugated. In recent times, there has been a worker-level movement for quality of life and equality, which has been reflected in fuller market economies, and some employees have felt a greater need to find intrinsic satisfaction. The human rights movement has also become more widely interpreted in the western workplace. Globalisation of many manufacturing activities and the transfer of labour to cheaper emerging economies have occurred in recent years. Campaigns rage to modernise the conditions in developing countries.

The following is a quotation from the first three paragraphs of the Declaration for Human Rights preamble giving a rationale for the ethical and basic treatment for the workforce.

Whereas recognition of the inherent dignity and of the equal and inalienable rights of all members of the human family is the foundation of freedom, justice and peace in the world.

... Whereas disregard and contempt for human rights have resulted in barbarous acts which have outraged the conscience of mankind, and the advent of a world in which human beings shall enjoy freedom of speech and belief and freedom from fear and want has been proclaimed as the highest aspiration of the common people.

... Whereas it is essential, if man is not to be compelled to have recourse, as a last resort, to rebellion against tyranny and oppression, that human rights should be protected by the rule of law.

It goes on to list a full 30 articles. These articles deal with such weighty issues as dignity, *equality*, freedom of movement, thought, and from discrimination, right to life, recognition and fair trial, treatment and asylum to escape persecution. Articles 23 and 24 refer to fair working conditions and Article 25 refers to adequate living standards and special considerations to motherhood and children (Figure 5.1).

### Article 23

Everyone has the right to work, to free choice of employment, to just and favourable conditions of work and to protection against unemployment.

Everyone, without any discrimination, has the right to equal pay for equal work.

Everyone who works has the right to just and favourable remuneration ensuring for himself and his family an existence worthy of human dignity, and supplemented, if necessary, by other means of social protection.

Everyone has the right to form and to join trade unions for the protection of his interests.

### Article 24

Everyone has the right to rest and leisure, including reasonable limitation of working hours and periodic holidays with pay.

### Article 25

Everyone has the right to a standard of living adequate for the health and well-being of himself and of his family, including food, clothing, housing and medical care and necessary social services, and the right to security in the event of unemployment, sickness, disability, widowhood, old age or other lack of livelihood in circumstances beyond his control. Motherhood and childhood are entitled to special care and assistance. All children, whether born in or out of wedlock, shall enjoy the same social protection.

*Figure 5.1* Declaration of Human Rights (clauses relevant to employment).

These articles represent nothing more than the basic expectations of a worker and a company should expect to exceed the basics and to do more. In most cases, governments will have legislated to cover issues like a minimum wage, minimum holiday entitlements, working hour limits, maternity (and paternity) leave, unemployment and sickness benefit, equal opportunities and provisions for special needs such as disability. Broader applications such as loss of dignity are harder to protect against, and there will be a varying treatment of migrant workers. The UK situation will be considered as a typical developed economy where equal pay, employment rights, anti-discrimination and trade union laws are used to try to standardise conditions.

*Article 23* is about fair and ethical employment with adequate remuneration. This also refers to the conditions of work and to the right to have equal pay for equal work. Most of these conditions have been in place for some time in developed economies. For example, in the UK a minimum wage in 1997 was enshrined in law. There are, however, situations where people fall through these safety nets, as in the case of immigrant and home workers who have no choice but to be on piecework which may not be proportionately remunerated to achieve minimum wages. It is also not a guaranteed source of work.

Dignity is associated with choice and the fallbacks that are available through a social security network. Social security means that people have some choice of work to claim dignity rather than a totally forced employment. There is also a right to join a trade union which can provide collective support and representation. In construction, many workers are self-employed and this is a choice also, but there are some situations where the status of workers is being shown as self-employed, but because of the dependency on a single source employer, they are effectively employed by them, but without the built-in benefits and security. This means that some people lose their basic employment conditions – notice, minimum pay, holidays and redundancy and pension entitlement for a slightly increased pay rate. Some employers have tried to get rid of TU representatives and have an agreed no-union situation. This has to be agreed by 100 per cent of the workforce

*Article 24* is an attempt to give rights for a balanced life. The working time regulations refer to a working week not exceeding an average of 45 hours, with allowance for rests. Many workers in construction choose or are pressured to opt out of this privilege in order to earn more. The opt-out indicates that many depend on a low wage and also that many on a salary actually do very long hours for no extra pay.

*Article 25* is about an adequate minimum wage and other basic benefits for living. The industry is not very good at flexibility because of the amount of self-employment and also because there is a need for supervision of others who do long hours. This means that it is harder to recruit females and mothers though there are enlightened ways to introduce part-time, job sharing and other flexible working arrangements for managerial and professional staff.

The development of an ethical framework for employment of personnel and the elimination of either direct or indirect discrimination is based on this contention of human rights that were confirmed in the UN 1999 ethical code (see Chapter 4). There is also a UK Human Rights Act 1998 to ratify the rights enshrined in the UN Declaration of Rights.

The aim of such legislation is to ensure basic rights, but also to provide some liberty by employers and other authorities to police these rights so that they can exercise normal powers such as law and order and the restriction of activities which inequitably affect the freedoms of others. The Cadbury case study considers the historical development of these rights.

## Case study 5.1  The Cadbury Brothers[3]

When George and Richard Cadbury moved their chocolate-making factory premises to Bournville in Birmingham in 1847 they were acting upon a corporate vision to improve the working and living conditions of their workers. The Victorian industrial scene was harsh and Birmingham was no exception. One quarter of the population of Birmingham (220,000 people) lived in 2,000 streets in back-to-back housing. The streets were undrained and flooded regularly with sewage. Poor quality water was sourced from wells or brought on carts which were sold at extortionate prices and sewage was often into open channels. The back-to-back houses consisted of 120–130 sq ft of living/kitchen space on the ground floor with a bedroom above and an attic room. The terraces faced out onto the squalid street scene with another row of houses directly attached behind, facing out onto a parallel street or built into small courtyards. This arrangement left little communal space and the cramped conditions cut out the daylight. Houses were also mixed in with private slaughter houses which, together with the flooding and sewerage, created seriously unhealthy conditions. Workers' lives were foreshortened and families had high rates of infant mortality. The factories nearby that employed these people were losing productivity through sickness, extensive absence from work and early loss of skilled workers.

John and Richard Cadbury were aware of these conditions through their work with adult education which involved visits to worker areas to improve the lot of the workers. It made them ask: Why should an industrial area be squalid and depressing?, Why should not the industrial worker enjoy country air and occupations without being separated from his work?, and If the country is a good place to live in, why not to work in?

Their vision for their new factory was a 'factory in a garden'. In order to stop the factory being hemmed in by a demeaning street pattern as described above, they bought land around it. They set about creating a village worker community called Bournville on the River Bourn which they rented out, or sold to workers. This was designed to improve the lot of the workers on the basis that a healthy worker was not only a happy worker, but a productive and loyal one. The village was open to a mixed community. It was not tied worker housing and provided affordable housing to a range of workers as states the Bournville Village Trust's Statement of Purpose:

> To provide high quality housing developments, distinctive in architecture, landscape and environment, in socially mixed communities, using best management practices to promote ways to improve the quality of life for those living in such communities.
>
> (Cadbury 2007)[4]

They go on to describe the garden village urban and architectural design principles as:

- Cottages grouped in pairs, threes or sometimes fours.
- Groups were set back from tree-lined roads, each house with its own front garden and vegetable garden with fruit trees at the back.
- All cottages were well built with light airy rooms and good sanitation.
- A typical cottage would consist of a parlour, living room and kitchen downstairs and three bedrooms upstairs. Some early houses lacked a bathroom (easily added later).
- Houses should cost at least £150 to build: they were to house 'honest, sober, thrifty workmen, rather than the destitute or very poor'.
- Building was restricted on each plot to prevent gardens being overshadowed and to retain the rural aspect.
- The first houses were sold on leases of 999 years to control the rural appearance of the district: mortgages were available for would-be purchasers.

Special almshouses were also built for pensioners, under the village scheme, and these still exist today, and a pension fund was first set up in 1906. In addition, recreational facilities were provided and holidays and shorter (5½ day) working weeks were introduced for worker welfare. Young workers were encouraged to leave an hour early to go to night school. Workers Councils were set up to discuss and feed back to the employer on such issues as worker safety, social conditions, education and training. The Selly Oak colleges were set up to encourage worker education. The estate also campaigned for greenbelt status for the nearby Lickey Hills and purchased land to ensure continued parkland areas nearby. Staff dining rooms were established and discount fares to transport workers from central Birmingham were introduced as the factory grew bigger. The Cadburys themselves were also involved in wider social reforms, such as outlawing child labour in chimney sweeping and animal welfare reform. The Cadburys also purchased the much older J. S. Fry chocolate business which set up a factory in Keynsham, Bristol. This factory was subject to similar reforms and had a similar Quaker tradition of fairness.

The Cadburys' successful experiment informed the Garden City movement and gave it impetus with Letchworth Garden City beginning in 1902. This concern for worker welfare was in opposition to conventional scientific management which was to treat workers as a factor of production, plentiful and exploitable for the maximisation of profit. The revolutionary philosophy was copied by others as a more socially aware treatment of the worker community which was believed to bring in commercial as well as an ethical dividend. It laid the groundwork for modern personnel management principles and protective laws for the workforce to protect their rights and gained a wider compliance in the 'cut throat' world of the Industrial Revolution and the development of a modern industry.

There are a lot of discrimination laws which protect basic rights and the 'comfortableness' of the workplace, but these cover technical and emotional conditions and do not cover a wider *intellectual discrimination* associated with professional discretion. The built environment requires an intervention which must consider the human rights of others and the individual obligation of the professional to meet these in design, safety and health.

### Professional discretion and human rights

Ethically, employers are accountable in professional practice not to affect the fundamental rights of their employees to freedom of thought and the interactions with and between their clients and the public. Acting honestly and openly is an underlying principle for ethical action, but it also ensures that the workplace itself is not hiding information or blocking people's liberty and security for freedom of thought. This can affect their ability to make decisions for their own good and the good of the organisation and society. For example, an engineer who is told to design a new coal-fired power station and does not know why there are no other options in a culture of climate change could be unfairly restricted. Opinions are expressed widely by professionals (Uff 2005; O'Neil 2005)[5] about the need for individual discretion in complex professional and ethical decisions. This is generally put down to the paralysis caused by complex bureaucratic procedures. It is expressed well by the difference between a consent form for an operation in 1954 which said, 'I agree to leave the nature and extent of this operation to the discretion of the surgeon' and one now which consists of a multi-page form to cover all aspects of the process of consent. The bureaucracy is understandable, but it is in danger of upsetting the informed and open relationship between doctor and patient, because it now confuses the patient (Giddings 2005).[6]

Public positions have their own strong codes of practice which are often ethically focused. However, the imposition of a conflicting range of arbitrary KPIs, which is based on performance, places the professional in a binding position intellectually. For example, the management requirement for criteria-driven housing waiting lists can destroy the human rights of community support and family life by forcing emergency allocations from different areas away from their families. The criteria can also ensure that in resource scarcity there are extremely lengthy waiting lists for non-priority tenants who feel that their needs should be listened to as a matter of right. Public officials should be given discretion by their employer to act with some freedom of thought and solve local problems fairly otherwise they lose their humanity. In short, there is a call for ethical flexibility and moral imagination by our employers.

### Ethical leadership

When ethics is looked at in the context of employability, there are some strong crossovers with building up character and identity, virtues and related employment skills, which pick out people as unique operators, but also equip them to work with others (Robinson 2005).[7] Ethical leadership has an important role in imparting and developing these skills.

Most companies will seek to comply with the basic legal standards, but may have a culture of minimalism, which is sub-ethical and seeks to comply with the letter rather than the spirit of the law. Ethical leadership defines actions above the legal minimum in the responsible development of employees and the reasonable treatment of customers so that they feel looked after and feel they can come back again.

Employment law deals with such areas as minimum wage, equal pay and non-discriminatory treatment in race, religion, gender, age, orientation, parenthood and disability. There are also minimum requirements for health, safety and welfare, employment contracts, fair dismissal, sick pay, redundancy and trade union activity. Part-timers also require equal treatment. Employment law is less able to deal with staff development, informal communications, team spirit, integrity, empathy, conflicts between basic employee culture, values and belief systems and capacity for critical reflection. One of the key areas is the recognition of diversity and tolerance. Robinson (2005: 6)[8] rightly recognises the move away from absolute ethical standards in what he calls a 'postmodern,

postmodern' world to a more relative approach with no grounds for challenging personal ethical views other than harm to others. In a collective situation, i.e. a business team, this puts pressure on individuals sitting down together and working out a common, normative approach in what he calls 'ethical autonomy'. As stated previously, this needs to be managed and not imposed.

Traditionally, the teleological and deontological approaches have polarised a business approach to ethics. This essentially refers to an emphasis on the *means* or an emphasis on the *ends* when assessing the ethics. In this context, the teleological ethic would stress widespread, better remuneration and happiness of employees. The deontological ethic would stress intrinsic satisfaction for employees in carrying out business in the right way, among themselves and their customers. This might mean some subordination of the short-term profit motive. Both ethics need to pay attention to customers and wealth creation, but the underlying motivation and driver are subtly different and lead to different ways of achieving ethical leadership in employment issues.

### Full employment and ethics

There is also a populist view that low benefit levels, a light touch with employment legislation, tight employment conditions and low levels of collective bargaining are the only path to full employment (OECD 1994).[9] A report by the Work Place Foundation (2006)[10] disagrees and suggests that governments do not have to use a single model to achieve workplace efficiency.

Full employment was really a 1970s concept and the levels of unemployment vary in developed countries between 3 per cent in Japan, 5 per cent in the USA and 10 per cent in some European countries. The latter has increased from 10m in the OECD countries in 1970 to 36m in 1990 with a further 15m who are in the employment bracket, but not actively looking for work. This gives a current average of 7 per cent unemployment in these countries. There are generally much higher rates of unemployment among young people, except in Germany and Austria and this is the case for women, except in the UK. However, the concept of full employment must be considered as a relative term because many are employed in marginal jobs outside their full economic potential (OECD 1994; Britton 1997).[11]

The concept of whether non-economic employment is ethical depends upon the desires of the individual and the fair allocation and equality of pay for the actual work carried out. In developing countries, unemployment is often as high as 80 per cent, but this rate is under-estimated by concealed employment, which operates in the black economy and therefore does not appear in the statistics. This proportion of formal/informal employment has an effect on the working attitudes of the workforce and represents more entrepreneurial, but less ethical conditions where employers circumvent a weak legislative control. The lower employment of women and young people may not be a healthy situation.

# Employment and the psychological contract

## Conditions of contract

The minimum conditions of the employment contract in the UK are summarised in the Employment Rights Act 1996 which requires companies to give an employee a written statement of the conditions of work, such as hours, paid holiday, redundancy, pay, discipline, grievances and appeals procedures. This acts as a basic communication between the two parties based upon the formal job roles. The basics expanded give a contractual basis for work administration and in the case of any dispute:

- right to an itemised pay statement including tax and deductions;
- place of work, especially if need to travel around;
- date when employment began (and when it will finish if temporary), including any periods of previous employment that count towards qualifications for certain rights e.g. pensions;
- scale of payment as well as the rate and how often it is paid and how it is calculated;
- conditions relating to qualification for redundancy and pension schemes or employer/employee contributions;
- terms and conditions relating to hours of work and entitlements for holidays and public holidays;
- any collective agreements (not necessarily trade union-based) that apply to the post;
- conditions for disciplinary procedures, dismissal and appeal for unfair dismissal so that third parties may be involved.

Details of the pension scheme and employee rights, the grievance scheme, disability, non-smoking, and discrimination policies that are specific to the employment, must be accessible and any changes in the details, like change of employer, must be passed on personally to the employee. Many other pieces of legislation may be involved to cover specialist areas. There are often company documents which cover more than minimum entitlements, such as a right to receive non-paid leave without leaving post. In the case where there are no written contracts, then the legal minimum is assumed to be the default if any cases are brought to employment tribunals. The key rights in UK employment law, as summarised by Lott (2007)[12] are quite comprehensive:

- not to be unfairly dismissed and to be accompanied in the case of grievance, disciplinary or dismissal hearings;
- the rights to redundancy payment and for employee's representatives to be consulted before redundancies are implemented;

- not to be treated less favourably on the grounds of sex, race, age, disability, sexual orientation, religion or belief;
- to receive equal pay compared to a person of the opposite sex in the same organisation;
- limitation of working time, minimum rest breaks, daily and weekly breaks and annual leave;
- no unlawful deductions from pay and to receive more than the minimum wage;
- time off work (not necessarily paid) for specified duties such as worker representation, maternity/paternity leave, up to three months' unpaid leave for parental care for a child or for an emergency;
- request with reasonable consideration for flexible working to care for a child under six or a disabled person under 18;
- protection for detriment in the case of public interest disclosure as in the case of whistle-blowing, provided they are made to an appropriate person (the media is permissible in certain defined circumstances).

These types of protection are more common across European states and Western society. Traditionally, there is a greater welfare protection in Europe. There are notable exceptions in terms of employment protection even though employers 'walk quite a narrow tightrope' in not contravening some of the rights for their employees above. For example, the consideration of bullying and harassment at work is enshrined in the UK in the context of the various discrimination acts so, for instance, a single person virgin, a pacifist or an activist who feels they are being prejudiced against because of their position does not have a ready recourse to law unless they find a pretext such as sexual orientation or religious belief. A person under 18 working more productively than a person over 18 does not have a right to equal pay (young people have a lower minimum pay scale).

### Compliance

Employers that flout the legislation need to defend themselves through an employment tribunal, which has a tradition of awarding moderate amounts that reflect recognition of short-term damages rather than punitive compensation. Employers have, in most cases, to prove that they carried out fair procedures in making decisions as well as not contravening minimum rights.

The laws are often judged on the basis of a technical misfeasance, giving benefit of the doubt to the employee, which ethically gives strength to the weaker party. In the case of a smaller employer, this may make it difficult to remove a troublemaker or an employee patently unsuitable for the post. A misfit employee may affect the rights of other workers and companies need to take a strong principled line to ensure a just solution. It is reckoned in a recent survey that fear of being taken to court is the main driver in introducing diversity and not an ethical imperative.[13]

There is recognition for providing conditions for more flexible working in the EU which seeks to redress equality of rights of part-time workers with full-time workers' rights and where possible to get employers to consider job share and unpaid leave. These issues are to do with quality of life and choice of the worker who, in a situation of adequate supply, may find their choices for work patterns limited. Retirement is still connected to a certain age, but some workers, for intrinsic or economic reasons, have a right to be considered for work after 65, as this helps Western economies, who are short of labour. It also provides a better balance of tax payments towards publicly supported pensions where the pensioned population is putting a strain on the tax levels of the working population.

## Case study 5.2 Flexible working ethics

According to a survey among personnel workers[14] in UK companies, there was an almost universal opportunity to work part-time. Flexitime and job sharing were not far behind in the firms surveyed, but were not offered to the same extent in private companies. Options like term-time working were only significantly offered by organisations composed of at least 50 per cent of women. The greater tendency for public organisations to offer more flexible options was more than double. Flexible working was perceived to be a positive benefit in most organisations for the employees, where improved staff morale and motivation and better staff retention were quoted, and it was considered a right by most employees. It was considered that a prevalent 'long hours working culture' was counter-productive, though not many organisations thought they should discourage long hours working, and employees thought they had a right to opt out of 48-hour limits. However, it appears that managers are not comfortable with flexible working and management training was suggested as a way of overcoming the gap and helping flexible working maximise its benefits.

In construction, flexible working has not been given a high priority and is generally connected with women working, of which there are fewer in construction. There is a more flexible working practice in the design professions, but job share and flexitime may not fit with the supervisory and management posts on site. However, there is a legal requirement to sympathetically consider flexible working for care purposes and to give clear evidence if refusing requests, when traditional practice does not count. It may even be considered as discrimination by a tribunal. The benefits of more flexible working for construction might be to draw in more female employment and a wider range of personnel who do not care for the long hours or who have care responsibilities.

However, there is a range of unspoken perceptions and expectations which properly 'oil the relationship' between the employee and his/her boss. These are sometimes referred to as the psychological contract.

### Respect for people

Rethinking Construction (2003)[15] brought out a toolkit called *Respect for People: A Framework for Action* which was trialled across the construction industry and benefits were assessed for the introduction of people-friendly measures for the business and employees. Overall, it was assessed that a 2 per cent increase in profit would be gained by the companies that adopted the toolkit, through

- improved communication and relationships with the client;
- improved communications and involvement with the workforce;
- improved use of best practice through the requirement to measure so that improvement was noted;
- improved image and quality;
- directly improved standards.

If the ethical treatment (respect) of the company's workforce and supply chain is proven to produce economic benefits, then this could be an additional driver to sustain company interest and improvements. Mansell have published a case study,[16] where they have claimed major economic benefits over a programme of partnering with its clients and bottom up involvement of its employees. Rethinking Construction claim that client leadership is critical in insisting on evidence from contractors of the quality of their workforces and their ability to retain quality staff.

In 2001, only 9.2 per cent of the broader construction workforce was female in the broader construction category compared with 34.6 per cent in industry as a whole (Rethinking Construction, 2000)[17] and the average working week was 43.9 hours. A large proportion of companies (90 per cent) employ less than seven people. In the *Respect for People: A Framework for Action* there is a suite of toolkits which are:

- Diversity and equality
- Working environment and conditions
- Health and safety
- Career development and lifelong learning.

These toolkits are investigative to pinpoint problems with existing systems.

**Case study 5.3  Working environment**

Ove Arup used the RFP toolkit on the working environment because they were aware of problems with the working environment. As a result of the question-naire they realised that their induction procedures were poor (CE 2003).[18] They have substantially increased the information they give to new employees and have assigned mentors.

## The psychological contract

The psychological contract is defined as 'the informal beliefs of each of the parties as to their obligations within the employment relationship' (Herriot 1998).[19] These less formal aspects of the employment relationship refer to the unwritten and sometimes unexpressed perceptions and expectations of the employer and employee. The contract recognises the two-way process in employment relations. It replaces a more mechanistic view that the employee seeks work in return for a reward. It also means that employer and employee are able to redefine their relationship as the job evolves and both sides are probably looking for something more as the employee grows into the job.

The expectations, particularly in the construction industry, are much greater, and younger employees will expect to get intrinsic satisfaction in return for longer hours of work and transient workplaces, with sometimes sub-standard office space for project-based staff. Promises made of compensation for parti-cularly remote contracts and pressurised time schedules are balanced by flexibil-ity for work hours, time off in lieu and time off when in between projects for management staff. Office-based staff have more travelling to do to visit sites, but have comparatively consistent hours and standard facilities. An imbalance in the informal contract of management staff occurs when work overload is caused by staff working longer and harder with fewer resources for the same or a small increment in pay (Snell 2002),[20] when there is a lack of control over work decisions which directly impact on that workload or on the intrinsic satisfaction of the job, where there is a poor parity with other similar work-places in pay or conditions, there is a breakdown in community and social interaction due to poor teamwork or because of frequent changes in job loca-tions or teams, where there is an absence of fairness and trust is lost in the employer or where there are conflicting values (Loosemore *et al.* 2003: 81–82).[21] Breaches in expectations result in the deterioration of employee relations. People are used to poor conditions and long hours, but managing the balance and compensations for that is important to aid productivity and to stop the worsening of relations.

## Discrimination and diversity legislation

This section covers the equality and diversity legislation that has been introduced in Europe since the 1970s. In the UK, the main threshold Acts are the Equal Pay Act 1970, which prohibits different pay for the same job, the Sex Discrimination Act 1975, which requires equal treatment in employment for men and women, and the Race Relations Act 1976, which prohibits race discrimination for employment, education and services on the grounds of race and colour. The legislation has expanded broadly to cover other areas of equality and protection for sexual orientation, religion or belief, racial and religious hatred, harassment, parity for civil partnerships, disability, gender recognition, ex-offenders and age. Various amendment Acts have been necessary to emphasise particular issues that have become more prevalent, and regulations have been set up to clearly outline good practice. There are three Equality Commissions for disability, race and sex, which deal with awareness, education and compliance. These are now merged under the Equality Act 2006 into a single commission called the Equality and Human Rights Commission. Employment Tribunals deal specifically with the resolution of claims and disputes to deal with incidents that occur in employment.

There are two types of discrimination – direct and indirect. Direct discrimination is automatically unlawful and occurs when a person is treated less favourably than another employee because of any of the reasons above. This also applies to recruitment, with one exception which is that there is a genuine occupational requirement (which is rare) and that it is proportionate. This would not apply in excluding applications from men in the recruitment of nurses, but it would be relevant if the nurses were being recruited to work in a safe hostel for women subject to abuse by their husbands or partners.

Indirect discrimination arises out of a provision, criterion or practice being applied equally to all, but a person is put at a disadvantage because of their particular group and the law acts to protect against that group being discriminated against.

### Case study 5.4 Indirect discrimination

For example, a person who is asked to work all day, every Sunday, might be upset if they were a Christian, but not if they were a Moslem or a secular person. It would be discrimination if it was impossible to attend a Sunday service at all, if the employer could organise and plan other ways of *sharing* Sunday duties with other employees. However, this case is only valid under the discrimination against religious belief. There are no protected groups for family life in this situation and a religion or belief would need to be an organised one. An atheist in a general sense would not be able to have time off on a Wednesday just because they liked a mid-week break.

In another situation, a height requirement might exclude women as it used to in the police force or men as it still does for tall men who apply to work as aircraft cabin staff. Some airlines exclude staff who might be a health and safety risk.

Diversity legislation refers to the recognition of differences and the equality of each individual. Sometimes there is a case for a degree of positive discrimination for those who are under-represented in the context of the corresponding potential employment of those groups in the areas where an organisation works. This is a broader ethical issue, with the complementary interest and competence of under-represented groups an essential prerequisite and the acceptance of reverse discrimination of better qualified majority groups. It is mainly monitored through a company's diversity policy which should include the collection and review of statistics to feed into strategic policy.

Companies may choose to develop local workforces as part of their social contribution. This could be established by establishing quotas. In some carefully conceived cases, a seriously under-represented group may be positively discriminated against, but this should be as broad as possible and rationalised for a short period only on the basis of proportionality.

## Case study 5.5 Diversity

In 1999 there seemed to be institutional racism in the Metropolitan Police.[22] In order to put this right and also to deal with social problems arising from representation of ethnic minorities, a positive drive for coloured and black ethnic minorities was encouraged. In this case it was proportional and could be rationalised.

### Disability discrimination

This is often interpreted in construction as the implementation of access and other arrangements into the design of a building. As we are dealing with employment ethics, we must consider the other aspects of disability which are to do with how disabled persons are treated in recruitment and in their place of work. The key concept is proportionate adjustment in the workplace. According to Lott (1997)[23] there are five areas of unlawful discrimination which are direct discrimination:

- Treating people less favourably because of their disability, e.g. paying them less.
- Treating people less favourably because of a related reason, e.g. not allowing them time off for treatment (this is different from sickness) unless the employer can show that it is not appropriate, e.g. the time off is not proportionate to the treatment.
- Reasonable adjustments are not made in order to make it possible for a disabled person to do their job, e.g. the provision of a higher chair and adapted computer workstation.
- Harassment which is behaviour towards them that is bullying or attacks their dignity as a person, e.g. sustained or acute name calling.
- Victimisation, where a disabled person is treated differently because they have brought a claim against their employer, e.g. they have given evidence against them at a tribunal.

Some of these five areas seek to ensure that disabled people are treated differently in order to give them equality with other employees. However, there is a close interface with patronising behaviour and most disabled people value inclusive treatment with conditions – access, reasonable adjustments, etc. – so that they can carry out their role with equal skill and dexterity as able-bodied employees.

## Case study 5.6 Disability

A deaf person employed to be a call centre agent can overcome the disability by the use of ICT technology to give an instant read-out of customer calls. In this case, it requires the customer to dial an additional two numbers to use the amanuensis provided. A dedicated phone number and space for the agent will also be required. This would be a reasonable adjustment which would cost the employer money and require some planning if the work required 'hot desking'. The customers are entitled to complain about any delay in the phone service, but this would not affect their core agreement unless the agent was on the trading floor of the Stock Exchange.

In order to ensure that disabled people are simply not overlooked in recruitment, a company needs to show that they have diversity in their workforce. This will include shortlists where those who are disabled are interviewed without interviewers initially knowing about their disability. In addition, employers have to look at flexible work arrangements.

A construction site or a manufacturing facility is a particularly vulnerable place for the deployment of some types of disability, but the health and safety argument is not always a relevant one. It would be reasonable to presume that a blind person could be a danger to themselves and other if they needed to work at heights alone, as they may unwittingly kick materials off the scaffold. They also would not carry out visual inspections or read from drawings. However, if they are accompanied, they may still be able to carry out relevant data collection to record progress and compile programmes, if the relevant computer technology is available, and they have the relevant technical knowledge. Access arrangements for wheelchair users would need to be proportionate to the effort involved. A large site with lifts in place would have no excuse not to keep walkways clear, or access to all floors. A small site with ladder access to the roof would be inaccessible.

### The ACAS model[24]

The ACAS model has been developed to establish systems which are fair, clearly understood and do not lead to a blame culture so that people can learn from their mistakes and have a culture of learning and development. Employees need to feel confident and equal partners in the workplace. The ACAS conditions are:

- Formal procedures for dealing with disciplinary matters, grievances and disputes that managers and employees know about and use fairly.
- Ambitions, goals and plans that employees know about and understand.
- Managers who genuinely listen to and consider their employees' views so everyone is actively involved in making important decisions.
- A pay and reward system that is clear, fair and consistent.
- A safe and healthy place to work.
- People to feel valued so they can talk confidently about their work and learn from both successes and mistakes.

This model is ethical in the sense that it is a way of making sure an employer has some guidance to provide a proportionate model which is also a way of understanding employee concerns and working with them to create an enabling rather than a censorious approach based on minimum and grudging facilities that help employees to enjoy and motivate them in their jobs.

## Recruitment, selection and retention

Because of the dynamic and fragmented nature of the building industry, with many small subcontractors, many workers are self-employed or agency workers. Project-based technical staff move around because of the short-term nature of projects, so staff are not retained to the same extent, but become 'project hoppers'. So, although employers do employ core staff, a lot of people move around to retain work and improve themselves.

### Equal opportunity (EO) recruitment

This refers to the concept and the process of giving equal opportunity in the recruitment process. There are four stages of recruitment:

- job specification;
- job advertisement;
- selection process including shortlisting and techniques for choice;
- induction and training to settle the candidate into the system.

Job specifications relate to the personality and experience of the previous employee and also to the actual working conditions, culture and workload of the organisation. Thus a job specification has the potential to be adapted to the unique skills of the recruitee so that benefit can be obtained from these skills and training, in turn, will be offered that will bridge the gap between current and required ability for the key tasks as illustrated in Figure 5.2.

EO is not only matching the job to the candidate, but it is developing the job with the changing direction of the company and the aspirations of the employee. Pay can be enhanced on the understanding that workers grow into greater responsibility as an incentive to do so. As they become proficient, so they are groomed for promotion. This development is in itself motivating and part of employee expectations. Promotion is also a part of the expectations for developing responsibility. This is a way of showing trust in the employee and helps build up good ethical relationships.

Young, less experienced employees or new professionals in the industry are also looking for the opportunity for personal development to improve their CVs. An attitude by the employer to restrict access to training in order not to lose employees is retrogressive, unethical and unlikely to work. Staff retention is more likely to be improved by affording developmental opportunities within the organisation, in offering continuing, competitive rates, good security and holding a philosophical attitude to some who will inevitably leave. It is recognised that you 'lose some and gain some', because younger staff will move around to get a wider experience. The recruitment process is helped by making attractive training opportunities available. Public organisations may play an unequal role in training as they are unlikely to offer attractive rates, but are committed to training. This may mean they lose 'young blood'

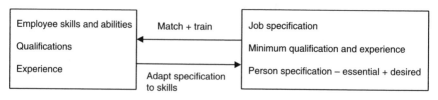

*Figure 5.2* Matching the candidate to the position.

who are glad to accept the training, but will move on quickly as they gain confidence.

Consultant and contractor retention rates are also quite low because they move around for more lucrative jobs in other organisations which have won contracts. This requires these organisations to expand quickly and so they make attractive offers to staff who feel insecure at the end of project-based jobs, or for whom the lure of better conditions entices them to change employers. This is not a healthy ethic and some companies are seeking ways to retain staff more effectively so that they can retain hard-won experience and knowledge.

### Employability, skills and ethics

Employability is seen as more than qualifications and experience and refers to certain deeper qualities for being effective in employment, often identified as interpersonal skills and the ability to learn and develop within the position to which you have been recruited, but still with reference to one's own beliefs and values. These skills could be self-confidence, negotiation, presentation of ideas, self-awareness, capacity for networking, team working, taking responsibility and the capacity to make decisions (Robinson 2005: 8).[25] Capability is allied to competence and, although it includes continually learning, it is more to do with taking appropriate and effective action and working with others to achieve this (Stephenson 1998).[26] Robinson says that employability involves reflectivity, responsibility, connectivity and innovation so that there is not just a sense of being employable, but an intrinsic good that leads to good outcomes in employment. It is also a holistic way of doing a job in context with the environment and community values, and generating ongoing innovation and improvement. There is also a sense of ownership of the decisions that are made because of confidence and self-belief in actions taken. The latter ideal in practice may lead to some conflicts of interest between employer, community, client and personal values and require an additional skill to integrate requirements fairly. This is where an external ethical framework could play a part in delivering beneficial outcomes.

The particular ethical values that Robinson picks out of Carter's (1985)[27] taxonomy to be developed in professional employability are valuing or respecting others, integrity, which includes trustworthiness, and high standards and empathy in seeing things from the point of view of others. These indicate particular qualities that might be sought out by an ethical employer. These are unlikely to be expressed convincingly on a CV and they may be made evident in a face-to-face interview or an assessment day, so that candidates are scrutinised in an applied and practical situation and possibly under some degree of pressure.

An ethical candidate/employee will be evidenced by the ability to make holistic decisions that consider other people. They will recognise an open and transparent interaction with others without giving away client confidentiality.

They will recognise some balance of decision between commercial and social factors and they will consider a wider range of stakeholders. Finally, they will evidence an ability to make a spirited defence of relevant, clearly held ethical values even if held against the majority. As no one is up to a perfect standard of practice, it helps to see intent to reflect on practice to learn and improve and not just to accept existing moral norms.

## Developing economies

Developing economies have significantly different conditions and many multi-nationals have taken advantage of the global economy to use cheaper labour which might not be rewarded with fair conditions. The use of global out-sourcing on home workforces can have a demotivating and demeaning effect, affecting job security, pay and promotion.

In construction and property, there is an emerging market in Russia, China, India and South Africa and Brazil because many goods and labour are imported and work is outsourced from Europe. Issues of interest revolve around employment protection, social security, social protection and employment migration.[28] There is a need to recognise difficulties of high unemployment levels, and the need to provide social security to fight poverty through benefit systems, the poor and insecure employment conditions surrounding many jobs with long hours and low pay, and under-age labour and the black market which avoids the payment of tax to provide benefits. Many employees have no minimum wage and there is a problem with the prompt payment of public wages because of difficulties in cash flow and in the banking system. There is also an issue of labour migration abroad, where some of the best educated workforce is disappearing from the domestic economy to better paid jobs abroad. There is often little state pension so only the well-off can afford private schemes out of reach of the majority of the population. The lack of maturity in the labour market means that unions are weak and the health and safety of working conditions leave a lot to be desired. As unions have the power to stop employers cutting corners, workers are at less risk of injury and death. Welfare provision is also reduced to save money and there is a gap between reward and the subsistence level employee pay because of the shortage of available jobs.

### Case study 5.7 Developed, emerging economies comparison[29]

The disparity of costs across different economies is startling and can be demonstrated by comparing the 1996 spinning and weaving labour costs globally. The UK is noted by the ILO as having a rate of about $12/hour with $22/hour in Germany and $27/hour in Switzerland. Many developing countries are paying about $0.5/hour. When taking into account the greater social contribution in wage costs in poorer countries, this mirrors a 1 in 80 difference in $GDP/head with a 1 in 70 difference in $wage rate. This gap is not as bad as it looks in terms of purchasing power as prices are much lower, workers are younger and education is less, worker for worker in a developing country, but there is still *substantially less* purchasing power, forcing more of the family to work for longer hours.

## Ethical leadership in the developing countries

Ethical leadership is the exception rather than the rule in the developing countries. This means that many employers are able to ride roughshod over any employment protection and workers work to survive and not for the ideal of dignity and freedom of speech and choice of employment, since they get what they can get and not what they would choose if they had that choice. Employers do not do more than they can get away with.

In construction, many emerging economies are more labour-intensive, at least in some of their work. This leads to the skills required for supervising and controlling a large workforce. The intimate nature of a small team is replaced by an appropriate technology to use an under-employed workforce. Attempts to carry out offsite manufacture and import highly processed goods are not appropriate economically either because of the poor balance of payments, the resulting delays and high embodied energy represented by the heavy use of carbons in transportation and manufacture. 'Labour-intensive' as an appropriate technology should help with job creation and help to move labour out of the informal sector into the formal sector of construction. It is often used on infrastructure projects such as unpaved roads, water supply, roads and is particularly appropriate for self-help projects and for where heavy machinery is not easily accessible to remote areas. It has also suited local authorities who can motivate the labour force by offering either a 'food for work' programme or a mixture of cash and food remuneration. The injection of cash can move communities away from food dependency to using money to develop the local economy, if well managed.[30]

## Case study 5.8 Expanded public works programme, Durban, RSA

Other schemes have been urban-based, such as the Durban-based labour cooperatives scheme which consisted of a positive discrimination programme for local black South Africans to bid for building, infrastructure and maintenance schemes called the Expanded Public Works Programme (EPWP), which was launched on a province-by-province basis in 2004. Its objectives were to creat jobs among the marginalised community, improve the skills base and provide social services through infrastructure and environmental management. As part of the deal, an EPWP Training Committee was set up to use Department of Labour funding for 'on the job' training for those recruited. This was to be enabled through the tender document conditions.

One such project completed under this scheme in 1997 was the International Convention Centre in Durban. This building, a large ambitious design with plenary hall, seating 1,800 people and seats that can be lifted into the roof, restricted the use of plant in order to use labour-intensive carrying and lifting. The extension of the convention centre in 2004, which will make it the only centre in South Africa to house 10,000 people, is being carried out by a Masinya Consortium, a 51 per cent owned black empowerment group with a handicapped foreman and many female tradespersons. The project is worth R396m and was completed by the end of 2006.

## Learning organisations, training and development

Training and development is generally considered to be beneficial to the internal growth of a company. Briefly, training is defined as imparting information to benefit the current job, and development can be distinguished as training and preparation to deal with a changed role or more responsible role in the organisation. There are, however, a number of benefits to training identified by McNamara (2007):[31]

- increased job satisfaction and morale among employees;
- increased employee motivation;
- increased efficiencies in processes, resulting in financial gain;
- increased capacity to adopt new technologies and methods;
- increased innovation in strategies and products;
- reduced employee turnover;
- enhanced company image;
- risk management, e.g. training on sexual harassment, ethics and other more technical areas.

These benefits are tilted towards the business, but training also has benefits for the employee, chiefly to increase their personal competence and marketability and to help them feel comfortable in the role they have been assigned. This is a key ethical rationale for training.

Training is ethical in that it ensures that employees are competent to do the job they are employed to do and understand it properly to do it safely. It consists of basic training for new responsibilities, updates and health and safety.

Development is the grooming of an employee for a new role, promotion or for change in the organisation. It will often incorporate an element of 'off the job' training in order to incentivise innovation and creativity. There is a competitive need to keep up with the developing norms of the industry and keep 'ahead of the game'. This adds an edge to the importance and success of the training and there may be some resistance to change if the reason for it is not obvious. A company may need to win employees over to motivate them out of their comfort zone and make an attractive business case which also includes benefit for them.

Ethically, there is a responsibility on the employer to look after the welfare of their employees and development, which affords some element of choice and empowerment, is likely to be more effective. The cost of training and development is the employer's responsibility and some have also sponsored personal non-direct development as a powerful tool to ensure more autonomy in their employees. The introduction of a new culture is much more difficult and requires the changing of attitudes.

Loosemore et al. (2005: 97)[32] conclude that the learning organisation is one where the individuals in the organisation learn to learn; this is differentiated from training where essentially individuals learn things which help them in their jobs at a relevant time and place. They also identify a more corporate learning process where individuals learn together in an organisation so that a critical mass of understanding is generated in order to enable cultural progress. There is also a sense in which information is captured centrally on an intranet or a database and it makes information accessible on demand. This is a process that is relevant to the application of expert knowledge. This corporate learning focus is a powerful concept and requires the employer to organise the learning system systematically so that the workforce and management ability progress in a coordinated way and gain synergy. Dainty quotes Druker et al. (1996)[33] as saying that 'construction companies are far from learning organisations with training schemes deducing with no evidence that the same is not true for projects'. In property development, the relationship may be a lot closer between the outcomes of a disastrous project and the financial well-being of the directors who are keen to learn what they can about the conditions for a successful development.

Because of the integrated nature of the built environment, this concept needs to be transferred to the learning project where many organisations come together and their skills and training are paramount to the successful outcome of the project. With so many organisations involved, it is harder to ensure comprehensive training and a weak link in the chain will cause a breakdown. Many workers complain about their remoteness from the centre of information and lack of control of the flow of information to them, such as a contractor who is unable to get design information, and the steelwork subcontractor who is starved of architectural information. Learning projects have the potential to work through their extranets to bring equality of information and to provide a wider view.

## Absenteeism and presenteeism

The issue of absenteeism is huge for some companies as it costs UK business £3.8bn for lost working days. Thirty-three per cent of the lost working days are due to long-term illness which accounts for only 5 per cent of cases. Many companies monitor multiple short-term absences and are keen to identify causes of such absences. Workers have rights for sickness pay, but in return must fill self-assessed sickness notes for absences up to a week and supply doctor's notes (which they may have to pay for) for periods over a week off.

Presenteeism is the phenomenon of workers feeling under pressure to come to work when they should be home sick or have to or feel pressured to work very long hours. They have a tendency to spread germs to others and are also likely to be less productive in their workplace. They may also be a safety hazard to others. An issue is raised here about the culture which applies these pressures. Colleagues are particularly unsympathetic about those who have regular time off and those who have children often have no choice but to attend to them if they are sick, and this and any other sick absence can be disruptive to others who cover at the last minute.

In construction, there are less absence days due to minor sickness because of a large self-employed workforce who will lose money, and also it will have an impact on their insurance premiums. However, any employer needs to consider ways of controlling the disruptive aspect of sickness, much of which may affect project progress and cost. In teaching and nursing, supply networks exist to cover teacher absence to avoid halting education and stretching healthcare standards.

Employers may consider that absences are because of a lack of commitment to the employer or to the project. They should consider the effect of their own actions and consider flexible working, incentives and more attractive ways of working. For example, the number of employees who wish to take a career break is increasing right across the age range and flexibility in covering for this may be a fairer system than simply sticking to compulsory maternal and paternal rights. The perception of fairness and parity is connected not just to the employment contract signed, but to what other companies are offering. Productivity may drop where a hard line contract is enforced, especially where the company is making a lot of profit which is not reflected in bonuses, commission or profit sharing.

## The labour force

In construction and property, the labour force is only partly unionised because of the mobility and the flexibility of labour. Many workers are agency workers or self-employed and follow a more independent path. The main operating unions are Union of Construction and Allied Trades and Technicians (UCATT) and UNITE which includes the mechanical and electrical unions and the former TGWU. The transport and mechanical and electrical unions merged to gain more influence in negotiation with employers, because of a decreasing membership. They have traditionally played a role on the really large projects like Heathrow T5 and have gained significantly higher benefits for their members (see case study in Part II).

In the main, though, unions in the UK construction industry believe that major and subcontractors have been unethically using temporary, agency and self-employment labour to opt out of their employer responsibilities (see case study in part 2) and are convinced that more direct labour will improve long-term security in the industry.

*Temporary workers* are part of the culture for the construction industry because projects are short-lived and different trades concentrate their effort over very short periods of time (except on the large projects) and need to move from project to project, and contractors will have a wide variation of demand for any one trade. This is accentuated because of the tendency to flood sites with resources to meet tight timescales. Workers are therefore made redundant regularly without qualifying payment from large and small projects when they complete. The industry has tended to outsource extensively even the most basic trades, which has led to a 'package mentality' rather than the management of labour logistics to move direct labour around, so that most labour is employed by small contracting organisations and many supervisors are employed through agencies. Self-employment and agency work are common in a series of letting out labour down the supply chain. Control of workforce labour and supervision are more remote for the contractor. Even direct workers leaving very large projects end up with few options.

*Self-employment* means that individuals are responsible for the provision of of their own national insurance, health and safety, personal protective equipment redundancy, pension payments, holiday pay and sickness pay, but they *are* able to deal with their own tax deductions and negotiations with the Inland Revenue. Although they are paid more than direct labour, these payments are not lucrative and it needs discipline to put aside specific payments for long-term needs and for a 'rainy day'. In addition, they see that the self-employed, who are mainly dependent upon one contractor for their work, have less of a choice and they may be forced to work in poor conditions and take a drop in pay because of their dependence upon a dominant contractor.

The tax office in the UK is seeking to tackle pseudo self-employment and has required employers to make deductions at source, including making national insurance contributions for all workers who work for a single contractor and are confined by their hours of work. They will, in this category, also be directly responsible for their personal health and safety, for making provision for paid holidays, and rest breaks under the Working Time Regulations 1998, and, under further regulations, will need to provide all the other benefits such as pension provisions and redundancy protection of directly employed workers. This is a major additional responsibility for employers and there has been a sort of running battle between the unions and employers (Goodman 2007).[34]

*Agency* employ workers direct on behalf of a range of contractors. Some of these agencies supply labour and others specialise in certain skills. Agency workers are sometimes not local and have to travel long distances with multiple pick-ups. The contractors have a fluctuating workload for different trades because of the short-term and varying nature of construction work. Agencies can juggle labour forces to fit the industry workload and workers get variety, but not security. This type of arrangement can militate against training and quality as workers move around and the agency training is not apprentice-based, so worker development is minimal. The long-term prospects for quality workers are poor where apprenticeships by the Construction Skills Agency are not attached to long-term jobs. Health and safety provisions for PPE are the responsibility of the agency and long-term provisions for pensions and redundancy are haphazard as agency workers are likely to move around agencies to get what they want. The concept of loyalty and labour retention is lost.

*Migrant* workers are at risk in that they are often seen as a cheap alternative to local labour and may be recruited either for their willingness to do more menial tasks or for the acute shortages in skilled and quality labour which may have been partly caused for the reasons above. Migrant labour fills a gap and in some cases is favoured because in the short term they are more productive. However, the path is filled with obstacles in that there may be health and safety vulnerability because of a language problem, workers may be illegal and supplied by agencies that are unethical in practice, creating more risks and low pay for their workers. Employers are required to check out credentials, but if there is no information, they may not bother. Migrant workers also may not qualify for training schemes. Qualified trades people may only stay with an intention to earn some 'good money' and return home after two to three years. However, they are an essential source due to some of the labour resource shortages in the industry and they can increase diversity and learning on the sites, especially where they are taking on supervisory positions. The essential health and safety training supervisors taken on for these positions are adding value.

Ethically, the labour supply needs to be legal which in some cases it barely is as the responsibility for employment is pushed more and more at 'arm's length' and benefits and conditions are not always in place, against the intention of the legislation. This leaves workers more open to discrimination and abuse as they are subject to the vagaries of their temporary and mobile site conditions. Diversity is almost impossible to manage in the case of indirect workers. Huge efforts are being made to shift some of the responsibilities to the project – mainly in the critical area of health and safety. The project manager has very little personnel management responsibility and people management has been the diplomatic skill to get work completed on time. The building up of ethical teams becomes an important role and needs to be supported by the senior management of the main contractor and consultants.

*Consultants* are used broadly for the design team and provide a service to the client, the contractor or development company in return for fees. In many procurement routes, they are little different from contract workers and are depended upon for their professionalism. The nature and development of professional responsibility in a project management matrix organisation are key to ethical teamwork. Each professional is responsible for their own behaviour and attitudes to their institution and a project culture of professionalism needs to emerge and be planned. The concept of an integrated team including the client and professional team has been proposed by the Forum for Construction 2007[35] and their target is for 20 per cent integrated teams by the end of 2007.

## Whistle-blowing

Whistle-blowing policies are beginning to appear in company policies and guidelines, partly because of the protection offered by the Public Information Disclosure Act 1998 (PIDA). The Act is designed to give employees a safe way of reporting unethical and unfair practices *which impact on the public interest*, without fear of victimisation or harassment. In practice, this remedy is most often used following dismissal which is considered unfair or because of discrimination following reporting of events. The American system has some protection, but the overwhelming common law in the USA is 'employment at will' which affords less protection (Agapiou 2005).[36] The personal consequences of unprotected whistle-blowing are severe and an American study[37] showed that 90 per cent lost their jobs or were demoted and many others suffered from health or family problems, financial consequences in law suits or even committed suicide, so some amount of protection and confidentiality is required for those who are still in employment not to keep silent in the face of serious ethical violation.

Ethically, an organisation needs to stamp out illegal or sub-standard practices and provide protection to those who report it so they can encourage whistle-blowing and are alerted to put the problem right. If and when a reported shady practice becomes a criminal activity, it will need to be reported outside and dealt with by the police and dealt with through the courts.

The internal solution is preferable for all sorts of reasons, but ethically it will open up a more transparent company culture. However, the perception of their employees regarding fairness is dependent on the culture that is engendered within the company by the leadership. A culture of fear may mean dominant or even corrupt managers hiding the problem and the morale of the workforce falls. A culture of confidential disclosure within the company organisation may help to resolve issues and maintain a productive climate. A key issue in reporting taking place is the protection of the whistle-blower from victimisation by the perpetrator or the company. This means identifying a third party in the organisation who will protect their privacy from the person being reported. If this cannot be arranged in small organisations, then confidential outlets to government bodies such as the HSE may be appropriate.

It is still a difficult step to take for an employee to 'squeal' on their colleagues. A few may be less inhibited and go straight to the media, with the promise of financial reward, but they are only protected in media disclosure in certain very limited circumstances. The UK probably has the most comprehensive whistle-blowing legislation in the world which makes it easier to 'blow the whistle' than in the USA. An organisation called Public Concern at Work (PcaW) has been set up since 1993 to help organisations set up policies and deal with complaints that are received fairly. PCaW aims to promote a culture where people can question dangers and risks in the workplace in a constructive way.

## Case study 5.9  Whistle-blowing[38]

A plumber who had refused to work in a particular house due to the unreasonable behaviour of the tenant was sacked by the maintenance company because of breach of contract. The plumber wrote to a senior director claiming that he had heard of under-handed payments by this manager for personal gain. His claim for unfair dismissal was upheld and the claim investigated.

### Public disclosure

In the UK, under the Public Disclosure Act 1998 (PIDA),[39] whistle-blowers are protected, where the disclosure is not malicious, to encourage responsible citizenship. In the first three years of the Act, 1200 claims were made, with two-thirds withdrawing or settled before a full hearing. Tribunals have been set up to resolve issues of public disclosure. Of the tribunal hearings, 54 per cent won their hearing and an average award of £107,000 was made in restitution for fraudulent activity.[40] Many of these cases concern whether an employee has been unfairly dismissed actually or constructively because they have addressed concerns about working practices.

Typical issues arising relevant to the built environment are danger to health and safety or the environment, stealing, planning bribery and corruption, cover ups and miscarriage of justice. It is not meant to cover issues of personal grievance, which have other mechanisms built into employment law. It is popular these days for the media to 'name and shame' cowboy builders who are caught in the act of extortionist practices. These apply in a general sense.

The UK has been encouraged to legislate in the belief that major catastrophes have been caused by employees noticing contributing causes, but not reporting them for fear of 'rocking the boat'.[41] Examples are given of the Piper Alpha disaster when dangers were generally known by the workers which could have caused the explosion, but were *not* reported because they feared losing their jobs if they made a fuss. There are also situations such as the Maxwell Pension Fund where an autocratic management precludes a no blame culture which is necessary for transparent review of problems in an organisation. In this situation, an employee is free to report directly outside an organisation if they genuinely believe that there is a serious breach that affects people's livelihoods and that it will not be taken seriously if reported internally.

Policies for whistle-blowing need to make staff feel confident that they are encouraged to raise genuine concerns and that they will not be victimised and to make the channels for confidential disclosure clear. The following guidelines appear on the PcaW leaflet:[42]

- The organisation takes malpractice seriously, giving examples of the type of concerns to be raised, so distinguishing a whistle-blowing concern from a grievance.
- Staff have the option to raise concerns outside of line management.
- Staff are enabled to access confidential advice from an independent body.
- The organisation will, when requested, respect the confidentiality of a member of staff raising a concern.
- When and how concerns may properly be raised outside the organisation (e.g. with a regulator).
- It is a disciplinary matter both to victimise a 'bona fide' whistle-blower and for someone to maliciously make a false allegation.

Getting the right balance is important. However, there may be a number of concerns that are raised by employers such as the use of lightly based accusations in hindsight under the protection of PIDA to get compensation for what is genuine dismissal or fair redundancy. Companies also require protection from employees who make vicious claims that have an unfair effect on the company reputation and are untrue or unsubstantiated. Company confidentiality may be breached unfairly in these cases and an employee can be disciplined for breaking their employment contract. They may also be prosecuted.

PIDA does not require a policy, but it is good practice and also favourable to an employer in the case of claims for the employer to have one. It also induces a better, more transparent working culture.

## Conclusion

There is a vast amount of legislation which has been connected directly with the rights of an individual in the workplace, but there is also a concern for the responsible employee so that the employer and the employee can be in partnership to promote ethical and best practice. The need for a mutual blame-free culture to encourage transparency is essential to reward integrity and to promote trust. In itself, this is not a matter of legislation although protection from exploitative employers, and in some cases employees, is necessary under legislation.

Labour retention is difficult in the industry because of the short-term nature of projects. Young people are likely to search around for the best conditions as they look for widening experience to improve their employability and their value to employers. Employers find that their workload varies and they need to hire and fire to suit supply and so they give the impression of poor security and thus lose loyalty. Ethically, it would be good to improve labour retention and provide attractive long-term conditions which do not falsely escalate the price of labour. This has also led to a wider use of agency, self-employment and short-termism by the manual and white-collar workforce.

Discrimination and lack of diversity in the industry are a problem and there is potential for innovative practice to make the industry more attractive and in many ways increase its ethical profile. The temporary nature of project employment and the fragmented nature of the industry into lots of small organisations has always made it difficult to ensure proper training for future skills, but training is the ethical core of the industry. There is no excuse for abandoning the current efforts. This could be developed between the unions, employers and training organisations.

## Notes

1 Gale, A. and Davidson, M. (2006) *Managing Diversity and Equality in Construction*. Abingdon: Taylor & Francis, pp. 2–3.
2 Universal Declaration of Human Rights (1948) *GA Resolution 217A (III)*. Online. Available http://www.un.org/Overview/rights.html (accessed 19 December 2007).
3 Cadbury website on History. Online. Available http://www.cadbury.co.uk/EN/CTB 2003/about_chocolate/history_cadbury/social_pioneers/bournville_development.htm (accessed 28 February 2007).
4 Ibid.
5 Uff, J. (2005) 'On Medicine, O'Neil on Philosophy', *Proceedings RAE Conference Ethics and the Engineer: Embedding Ethics in the Engineering Community*, 13 October at The British Library, London. Online. Available http://www.raeng.org.uk/ news/publications/list/reports/Ethics_transcripts.pdf (accessed 13 November 2007).
6 Giddings, A. (2005) 'Presentation on Medicine', *Proceedings RAE Conference Ethics and the Engineer: Embedding Ethics in the Engineering Community*, 13 October at The British Library, London. Online. Available http://www.raeng.org.uk/news/ publications/list/reports/Ethics_transcripts.pdf (accessed 13 November 2007).
7 Robinson, S. (2005) 'Ethics and Employability', Learning and Employability Series, 2, *Higher Education Academy*, UK, September.
8 Ibid.
9 OECD (1994) *Jobs Study: Facts, Analyses and Job Studies*. Paris: OECD.
10 Coats, D. (2006) *Who's Afraid of Market Place Flexibility?* Workplace Foundation. June.
11 OECD, op. cit.; Britton, A. (1997) 'Full Employment in the Industrialised Countries', *International Labour Review*, 36.

12 Lott, P. (2007) 'Employment Law in a Nutshell', one-day course for HE Managers, 24 October 24. UWE.

13 Paton, N. (2006) 'Fear is the Main Driver for Workplace Diversity', online in *Management Issues* July. http://www.management-issues.com/2006/8/24/research/fear-is-the-main-driver-for-workplace-diversity.asp (accessed 10 December). Survey was carried out by the CIPD.

14 Equal Opportunities Commission/IFC Research Ltd (2005) *Flexible Working Practices*, Research Report, April (238 respondents). London: EOC.

15 Rethinking Construction (2003) *Respect for People: A Framework for Action*, Report of the Respect for People Working Group, Constructing Excellence.

16 Rethinking Construction (2001) *Mansell Case Study 274 November*. Online. Available at: http://www.constructingexcellence.org.uk/pdf/case_studies/mansell_kpis_274.pdf (accessed 22 Dec. 2007).

17 Rethinking Construction (2000) *A Commitment to People: Our Biggest Asset*. Online. Available at http://www.constructingexcellence.org.uk.

18 Constructing Excellence (2003) 'Arup: Re-engineering the Induction', in *Respect for People Case Studies*. Online. Available http://www.constructingexcellence.org.uk/pdf/rfp/rfp_casestudies_201004.pdf (accessed 22 December 2007).

19 Herriot, P. (1998) 'The Role of the HR Function in Building a New Proposition for Staff', in P. Sparrow and M. Marchington (eds) *Human Resource Management: The New Agenda*. London: Pitman.

20 Snell, R. (2002) 'The Learning Organisation, Sense Giving and Psychological Contracts: A Hong Kong Case study', *Organisation Studies*, July–August.

21 Loosemore, M., Dainty, A. and Lingaard, H. (2003) *Human Resource Management in Construction Projects*. London: Spon Press.

22 Guardian Unlimited (2003) 'Metropolitan Police Still Institutionally Racist'. Online. Available http://www.guardian.co.uk/lawrence/Story/0,,941167,00.html (accessed 22 April 2007).

23 Lott, op. cit.

24 ACAS (2005) *The ACAS Model Workplace*. Online. Available http://www.acas.org.uk/media/pdf/g/e/Model_Workplace.pdf (accessed 28 February 2007).

25 Robinson, op. cit.

26 Stephenson, J. (1998) 'The Concept of Capability and Importance in Higher Education', in J. Stephenson and M. Yorke (eds) *Capability and Quality in Higher Education*. London: Kogan Page, pp. 1–13.

27 Carter, R. (1985) 'A Taxonomy of Objectives for Professional Education', *Studies in Higher Education*, 10(2): 216–25.

28 ILO (2006) *Programme of Co-operation between the International Labour Organisation and the Russian Federation 2006–2009*. June. Online. Available http://www.ilo.ru/countries.htm (accessed 15 November 2007).

29 Lindbaek, J. (1997) 'Emerging Economies: How Long Will the Wage Gap Last?', Background paper for Meeting of International Finance Corporation. 3 October. Online. Available http://www.itcilo.it/actrav/actrav-english/telearn/global/ilo/seura/ifcemer.htm#notes. (accessed 13 December 2007) A Mexican–US study indicated that out of the original difference of $11.30 to $1.30/hour these allowances would reduce the difference by $4.70 giving $6.60 to $1.20 a factor of 5.5.

30 Miller, S. (1992) 'Remuneration Systems for Labour Intensive Investments: Lessons for Equity and Growth', *International Labour Review*, 131(1).

31 McNamara, C. (2007) *Employee Training and Development: Reasons and Benefits*. Online. Available http://www.managementhelp.org/trng_dev/basics/reasons.htm (accessed 19 Dec. 2007).

32 Loosemore *et al.*, op. cit.

33  Drucker, J., White, G., Hegewish, A. and Mayne, L. (1996), 'Between Hard and Soft HRM: Human Resource Management in the Construction Industry', *Construction Management and Economics*, 14: 405–16.

34  Goodman, E. (2007) 'Union Demands Talks with Carillion over Agency Labour', *Building*, 20 November.

35  Strategic Forum for Construction (2007) 'Headline targets'. Online. Available http://www.strategicforum.org.uk/targets.shtml (accessed 23 December 2007).

36  Agapiou, P. (2005), 'Whistle Blowing Protection Provisions for Construction Industry Employees', *Construction Information Quarterly*, 7(4).

37  Dempster, 1997, cited in Agapiou, op. cit.

38  PCaW (2003) Case Summaries. Online Available http://www.pcaw.co.uk/policy_pub/case_summaries.html (accessed 21 Nov. 2007).

39  UK Government, Public Disclosure Act (1998).

40  PCaW (2003) 'Whistleblowing Case Summaries', ibid. http://www.pcaw.co.uk/policy_pub/case_summaries.html (accessed 21 Nov. 2007).

41  AIG Insurance (2000) *Overview of the Public Interest Disclosure Act and the Consequences for Businesses.* Online. Available http://www.pcaw.co.uk/pdfs/aig_brief.pdf#view=FitH (accessed 9 September 2007).

42  Audit Scotland and PCAW (2003) 'Don't Turn a Blind Eye'. October.

# Chapter 6

# The ethics of construction quality, safety, health and welfare

Ethically health and safety means the reduction of harmful accidents or dangerous incidents which will cause immediate or future health problems. Conventionally this can be done by ensuring compliance with regulations, significant induction and follow-up training of staff and workers in safe working procedures, risk assessment and management, and methods of working, investigating accidents or near misses and using statistical data to know where to prioritise preventive action. This can apply during building or facility use, during construction and during maintenance procedures. This chapter makes a case for health and safety, indicating a wider duty of care that is about trust, reliability and effective long-term relationships. When integrated with a total quality management approach (TQM), it is tied up with the achievement of a focused objective working together to achieve a positive win–win outcome. It is, however, not only focused within the objectives and time frame of a project although there is a need to develop an appropriate collaborative culture for each project. It starts more in the development of longer-term supply chain habits, based on multiple organisational commitments to health, safety and zero defects.

Quality in this chapter is taken to mean optimally fit for purpose and include health and safety considerations and the benefits of a collaborative culture. The quality (fit for purpose) aspects of health and safety will be looked at as the driver of a moral and ethical process.

## Fit for purpose

Purpose is a principle, rationale, function, use or intention.[1] Fit for purpose generally means providing a service which is well suited, competent and appropriate. If quality is fit for purpose, it will also infer a safe and healthy product in its finished form, in its context and in the process selected to construct the product. For example, to be fit for purpose,

- a wall should be built to specification in accordance with its finished use;
- laying bricks for that wall should be as safe as the finished brick wall;

- the use of the wall needs to be safe, so a load-bearing wall should be able to bear the dead and live loads that will be applied;
- where the wall is needs to be safe – wall ties and concrete for building walls by the sea need to be stronger than those that are not.

In most of these cases, quality is synonymous with safety and there is a sense that safe work has been designed carefully and works well. If a wall in a dwelling or working place is to be healthy,

- it should not emit hazardous fumes from the insulation;
- it may need to be fire proofed so as not to give off excessive fumes in the case of a fire or collapse;
- the design should be sympathetic so it does not cause stress.

Other health risk examples in building are the use of asbestos, contact with chemicals in contaminated ground, dust and chemicals causing contact dermatitis from wet concrete and various fine dusts or causing acute respiratory arrest or skin burn.

In addition, there is a need to consider the environmental aspects of health in fit for purpose work such as the cumulative or acute effects of pollution and emissions which cause later health problems such as chronic respiratory problems, asthma, skin and other allergies.

Usability is a term which indicates the user's point of view and refers to efficient to use, easier to learn or more satisfying to use.[2] Any of these could be applied to the building design or location. ISO 9241–11 (1998) *Guidance on Usability*[3] defines usability as 'the extent to which a product can be used by specified users to achieve specified goals with effectiveness, efficiency and satisfaction in a specified context of use'. Flexibility may also be important to quality if it is needed to adapt for different users or last for a long time.

## Learning from the statistics

In the UK, the Health and Safety Executive keep statistics[4] on the number of reported accidents and the type, trade[5] and rate of reported accident. The number of fatal accidents is 77 in 2006–7[5] up from a low of 60 in 2005–6. This is 31 per cent of all fatal accidents in all industry sectors. There are about 2,200,000 people employed in the industry. The number of serious and over three-day accidents per 100,000 workers was 314 and 630 respectively in 2005–6. This is about three times the average of all industries. This figure has been steadily dropping over the last 10 years by about 20 per cent and 30 per cent respectively. The most common type of *serious* accident is slips and trips and falls on the same level followed by falls from a height and handling and back injuries. About a third of all accidents are falls from a height, but the slips and trips and manual handling injuries are on the rise. There are about

400 convictions per year mainly involving a fine. About a 30 per cent of the fatal accidents occur to the self-employed.

Interestingly, the rate of accidents per 100,000 by trade is negligible for brick-layers, electrical workers and wood trades and over 6 times higher for painters. As is to be expected, those working at height such as steelworkers, crane drivers, roofers and cladders and scaffolders have the highest rates of accidents. Because there are a large number of bricklayers and carpenters there are still, signifi-cantly, nearly 500 accidents for each category, with painters at 380.

Morally it seems that there is still a lot to be done in construction to make construction sites a safer place to work. Compared with other industries which have a fixed and non-changing layout for work with much more passive activ-ities, they are inherently more dangerous, but can you develop safer ways of working which will eliminate much of the increased risk? Working towards a zero accident regime is the only real moral position, but often viewed as unrealistic. It is actually a cultural change in the industry and pulls everybody on board. It has the benefit of taking away the perception of greater danger perceived by site workers.

One method of trying to systematically tie down the risks is through the influence network HSE (2003).[6] This is a way of understanding the key human influences at site, design, organisation and external levels to work towards a culture of eliminating the risk and evaluating the potential effectiveness of improvements. Benchmarking was carried out in workshops to separate best practice from the norm and to identify effective risk control factors. These were adapted for key activities such as falls from height.

## Occurrence of accidents, dangerous and health incidents

Most significant accidents that happen will be the subject of reporting and an investigation. This is unlikely to be so clear in the case of an incident that has been hazardous to health over a longer term though more of these issues are being defined for reporting. Dangerous incidents that do not cause injury or health problems are a learning opportunity. Major incidents are reportable under the Reporting of Injury, Disease and Dangerous Occurrences Regula-tions (RIDDOR 1995)[7] and may be investigated externally. There is also a need for internal examination to see why the danger occurred to change procedure and improve practice.

Accidents can take place during construction, occupation, maintenance or refurbishment (e.g. space reorganisation, updating, etc.). There is a responsibility for clients commissioning building work, designers, contractors, co-ordinators, safety representatives and building managers to identify risks and implement any measures that reduce risks and manage the residual risk that remain.

In the UK an investigation by the HSE would be looking for negligence in any of these responsibility areas and would also be concerned about co-ordination of activity where more than one organisation needs to work safely together. As knowledge and experience are gained, so there are fewer excuses for ignorance and the bar needs to be raised as time goes by for preventative actions that are assumed reasonable and practicable. One of the identified key risks on construction sites and multiple occupancy buildings is the management procedures to design, sequence and co-ordinate the work in such a way that accidents do not occur at the interfaces of work and overlaps/gaps in responsibilities and risk assessments do not in themselves cause dangerous circumstances. For example, the laying of floor tiles with highly volatile fumes being done at the same time as welding work produces the danger of explosion while by themselves the tasks are individually safe. The protection of the public from accidents means deterring entry onto the site and having a control system for entry and induction where access is required by visitors and new workers or maintenance staff. Construction sites are inherently more dangerous as they are in transition from a design to a finished building. They will be carrying out activities at height where materials and people can fall and there is mechanical and manual lifting on a routine basis.

The health aspect of health and safety is much wider and the effects of actions for future harm are not so clear. Construction health risks also affect the most number of people through incidents like asbestosis (4000 cases/year), stress, asthma and allergies, back pain, eye sight deterioration, dermatological complaints, radiation sickness, legionella disease, building and radiation sickness, hearing loss and white finger. Compensation is often offered but it is often more chronic and persistent than injuries from which, though more acute, you may recover.

## Ethical approach to health and safety

Health and safety management is a Kantian ethic where there is a duty to protect the employee and to act by absolute standards of safe working for all parties to the best of your ability, where the means does not justify the ends and playing fast and loose with people's well-being, health and safety to gain a commercial or any other outcome is immoral. However, more risk is likely to ensue from construction even after a comprehensive risk mitigation exercise is completed. Even if this is perceived safe, then worker and public behaviour may render it unsafe by not complying with a duty to consider the safety of others or by refusing to use safety equipment or abide by safety rules. The morality of health and safety is often tempered by the commercial imperatives to be able to build within budget and to get it ready for market as quickly as possible. Society appears to have a value placed on life through the court compensation awards against employers who have been prosecuted. However, they may have insured the lives of their employees for more or less. The average fine in the UK is £4,000–8,000[8] for causing a major accident and a compensation award to the injured party.

## Value of life

The value of a life is hard to tell as it depends on how you calculate it. On the basis of compensation pay-outs by insurance companies to the relatives, it is quite low at about £40,000. This is a lot less than the money paid out on average per death of military pilots which was £2.5m/death in 1960, while a calculation on the basis of the yearly hazard pay paid to miners in the US was £17,000 at the same time. However, this may be an unfair reflection of the fact that you can buy private insurance at whatever level you like. Campbell and Brown (2003)[9] indicate that it might be acceptable to take less compensation for less risk of injury or death, i.e. if you accepted £x for working in a risky industry, you might compare risk by accepting £x/p where p is the probability of harm so that you accept more for the rising risk, but they suggest that it may not be a linear relationship and there may be a limit at which risk becomes reckless.

## Ethics of risk

Where an employer is offering a safe working place, as they must do under the UK Health and Safety at Work Act (HASW) 1974, then they are obliged to take relevant measures to ensure freedom from harm. However, they will have calculated a limit to the cost of preventative measures depending upon their budget and the trust they put against their management supervision and workforce. In short, a financial assessment of significant risk. Measures for preventing harm may well be recklessly ignored by a minority of the workforce and this may put others at risk in spite of adequate measures and in spite of legal recourse against these employees. It is therefore morally right to have additional measures in place to counter reckless employee and contractor behaviour, such as investing authority and resources in site staff to monitor and suspend employees who do not respond to warnings. This requires extra training and resources for supervisors.

Some of the ethical questions are: Does an action to reduce or eliminate one risk cause a greater risk elsewhere? Is there a limit to the resources for prevention? How do contractors remain competitive with more negligent rivals? Will clients select contractors without due recourse to their ability to be safe? Is it possible to have zero accidents?

Eliminating one risk in order to employ untested new technology creates a further risk and the relative size of the risk should be compared as life itself is a risk and the ease of managing the risk is part of the mitigation of the risk.

Some contractor's judgements are commercial, based on the probability and cost of a risk occurring – which may also turn out to be insufficient to cover the cost of incidents. Some contractor's judgements are strategic believing that long-term planning has a built-in efficiency for good health and safety management which reduces costs. Others may dispute the efficiency factor and have less prevention and respond reactively to incidents, putting themselves in a less moral position if more incidents do cause more personal distress.

It is generally considered that a smaller organisation cannot easily absorb the start-up costs of a health and safety regime and therefore is stuck with a more risky commercial choice but still needs to spend the money on training. However, they may have statistically more chance of having their trust fulfilled with a workforce which is more close-knit, but supervision may be proportionately more stretched. Accidents statistics in the UK have shown that more fatalities have occurred on small contractors' sites than on large contractors' sites so the choices that small contractors are making are more risky as the value of work covered by both is similar, but the impact is much more stretched in the case of a huge number more sites. Organisationally, then, serious health and safety improvement is much harder to communicate to smaller contractors because they have fewer dedicated staff and the geographical/organisational spread and the impact of accident and injury dilutes the seriousness to a few and limits reactions to a change culture.

## Responsibilities for health and safety in the construction life cycles

### The client

The client has a key responsibility to set the standard for and to ensure a quality environment by allowing the resources to construct safely and by insisting on a building of appropriate quality, including safe and healthy use. In many cases there will be strict guidance for them in minimum standards of quality and sustainability as indicated in building regulations and development control conditions and other legislation such as the Construction Product Regulations. In respect to resources, they have an ethical/moral responsibility to make sure that they do not select suppliers who are cheaper, but under-resourced or incompetent to the task. This passes cost cutting down the supply chain to the subcontractors and the designers. Under the UK CDM Regulations[10] Section 10, the client also has a duty of care in terms of providing information about hazardous materials, conditions or contamination and in ensuring that a suitable health and safety plan is in place before work proceeds. They have a vicarious, but back-to-back liability with the contractors and design consultants to ensure safe planning and a final health and safety file is received for the safe use of the building.

### Ethics

The ethical concern for the client is to provide information regarding the safe delivery of the building. There are major unknowns in the area of contamination and asbestos, but there are also issues about the social conscience of a client in putting money upfront to meet realistic targets for health and safety. They are also required to ensure they have appointed a competent health and safety co-ordinator.

An improvement in the sustainability of new buildings is a requirement of the planning regulations in the UK. This advocates a healthy building. The client has an ethical choice to increase energy efficiency and reduce its carbon-related emissions more than the current targets and in a way that is transparent. Toxic emissions are a health hazard for their neighbours. There is also an ethical responsibility to the building user in the degree of basic comfort that the clients offer and the general quality standard to increase the building's efficiency in use.

## The designer

There is an unavoidable duty on the designer under the CDM Regulations section 11 (2007) to provide a design which is safe in construction and in use. The designer is required to assess hazardous materials and site conditions in the light of investigations and information received from the client or other registers. From this they must design a safe structure avoiding using, or eliminating the harm of, any hazardous materials or conditions and providing a design that supports safe maintenance and encourages safe use. All this is in the context of fit for purpose and managing the other values and client requirements for the design. They will need to advise the client and contractors of any residual risks in their proposals and to be innovative in the way they ensure safe ways of achieving client business needs. They also have a duty of care towards contractors that the design does not encourage dangerous exposure to other hazards.

This duty is relative and it is recognised that the contractor also has a duty to organise and plan their construction process safely. This will depend upon the designer passing on information to inform them of residual hazards such as radon or chemical contamination in the ground, the correct use of new materials and working with the contractor at an early stage to ensure safe buildability to ease sequences and lifting operations and fall hazards. There is a principle of proportionality rather than elimination. For example, heavy components are not eliminated because of a manual handling hazard, but the sequence of their construction is possible if there are lifting eyes on large structural members and that repetitive tasks are limited to components of manageable weight. Tall buildings are permissible, but their sequence must limit work near edges and afford safe access to fixings.

The designer has a clear responsibility to design a safe building in use, such as compliant escape routes which protect people and provide minimum escape times with information, immediate fire fighting and signage which is clear and understood by all. They also have a responsibility for those who maintain the building and have to regularly put life and limb at risk balancing on ladders, crawling above ceilings or accessing roof edge details, working with live electricity, cleaning windows of tall buildings, changing light bulbs in shopping malls or working near very hot services.

Interestingly in action over non-compliance on the CDM Regulations in the UK over the five years 1999/2000 to 2004/2005, there have been more convictions for clients than principal contractors (60) and very few designer convictions (10).[11] This suggests that there is a major problem with the client ensuring they are able to take their responsibilities seriously and since then there have been more onerous duties on clients under the CDM Regulations (2007) as well as a new stronger role for a health and safety co-ordinator to support the client.

### Ethics

If a building or tunnel collapses or creates a major health problem, there is a design duty of care, making it unacceptable if precautions were not taken for foreseeable risk. Ethically it is a betrayal of the trust that people have in the design. The designers recognise that they have professional liability and they need to take special care not to be negligent in their research. In the case of using asbestos in a new building, it is professionally irresponsible, as it is banned. In the case of ignoring guidelines, it is ethically wrong unless proper research has been done to guarantee alternatives, and in the case of discovering a hazard after a building is handed over, the client must be informed of the dangers.

### The health and safety co-ordinator

Health and safety co-ordinator is a term in the CDM Regulations (2007) for a special agent role to reduce risk. The health and safety co-ordinator is appointed at the inception stage by the client to co-ordinate the pre-construction and design stages and has a strong client advisory role, although does not become the client's agent in client responsibilities. They have a responsibility to ensure the suitability of the contractor's health and safety plan and to produce a file of 'as built' design drawings and maintenance manuals to be handed onto the facilities management team for the building in use.

Over the 13 years since CDM (1994), there has been some difficulty in binding together the ethical approach of the designers, client and contractors. In CDM (2007), there is a greater emphasis on the clients' influence and ethical responsibility to ensure the resources and information passed to designers and further passed to contractors and users are seamless. It becomes the prime responsibility of the health and safety co-ordinator to ensure the due regulatory framework is in place and is working. This was resisted very strongly on the basis of lack of knowledge of the industry and their desire to be able to hand over delivery completely to their chosen specialists.

## Ethics

Ethically it might be unrealistic to ask an inexperienced client to carry out strict competency checks without the help of a professional, and this role strengthens the client's hand though they are still accountable. If they feel they have the experience, then they may adopt this role in house, but in both cases the close association of the client and co-ordinator responsibilities strengthens the compliance regime. It would be impossible to do the job without constant communication as the client issues the key instructions or censures.

## Principal contractors and contractor supply chain

The principal contractor (PC) is the principal co-ordinator in the mobilisation, planning and construction stages of the construction cycle. They produce and are responsible for the construction safety plan. As the second co-ordinator, their main role is to co-ordinate and control the wide range of contractors working together from different organisations in the supply chain on site and to be a link with the designers. They also liaise with the client/building team in ensuring preparation of as built drawings and the maintenance and safety manuals for use.

This legal duty and their legal control of the site afford some authority to make site rules for the safe conduct of the contractors. This is an opportunity to sequence and control the different activities and protect dangerous areas, policing the common access routes and escape procedures and ensuring that there is good interface control between contractors so that their risk assessments don't clash. In the case of change, it would be the PC responsibility to ensure that risks were re-assessed and the information relevant to others is communicated. It is a complex, but critical task to ensure that risk assessments and documents are regenerated for change and received by all parties affected. Otherwise it would be too easy to lose control of changing risks. The co-ordinator's role is to assert control and to effectively ensure compliance.

The health and safety risk register needs to identify risks, obtain method statements and assign responsibilities for management. Activity-related risk assessment and method statement is the role of the individual contractor who is also responsible for managing their employees' safety health and welfare and not impacting on others. This needs a back-to-back agreement with their subcontractors. Subcontractors are the weak link as there is difficulty in ensuring compliance with a given method statement.

The provision of personal protective equipment in terms of hard hats, boots and fluorescent jackets and safety harnesses is important, but is secondary to reducing or eliminating the risk in the first place and providing protection for a wider corporate body. For example, the provision of ear muffs to an individual on the jack hammer is unhelpful to those working nearby and it would be better to look into quieter ways of doing it when others are not around.

Typical site-wide concerns are separating pedestrians and site vehicles, screening off dangerous areas physically and co-ordinating safe vertical lifting, access platforms, deliveries and forklifts so that there are planned movements around the site. Other areas are the provision of sufficient welfare and the opportunity for workers not to be frustrated or stressed by the workplace by sequencing and resourcing activities efficiently and fairly.

The public who pass or work near construction sites are also in danger. The public has to be completely protected as they have no knowledge of identified risks, are offered no PPE and temporary structures may collapse or things may fall off them when they are adjacent to the boundaries of the site. There may also be more emerging traffic caused by deliveries and waste/muck removal and noise, dust and mud and altered routes are a nuisance and unfamiliar. Access should be highly regulated with banks, people at site entrances, regular scaffold and crane checks, information notices, hoarding and protected walkways structures for shoring up need proper checks to save excavation and building collapse. However, accidents still seem to happen on a regular basis that affect the public and cause significant nuisance.

There are many unknowns in the ground and the collapse of buildings or explosions through hitting gas pipes or electrical cables can be dangerous to both workers and public alike. Maintenance of scaffolding is a key area of interface with the public above busy streets and additional scaffold sheeting and fans back up failure of systems. Crane collapse seems to be at epidemic proportions and they can crash into surrounding buildings and people.

Road works are adjacent to fast-moving traffic and create noise, mud and dust on the road, causing dangerous conditions for workmen and public alike. They also cause narrowing of access and diversions for traffic. Works within buildings are the responsibility of the owner of the building who needs to work safely with their employees or tenants to create safe working areas and provide warnings and information.

## Ethics

The number of accidents occurring on sites is mainly due to the lack of planning, supervision and co-ordination on the site. It is therefore unethical not to work towards a programme of continuous improvement and to implement the plan with the full authority given by the regulations. It is no excuse to say that resources are not available, but there is a skill in knowing how to identify priorities and keep ahead of the game. This comes with experience, so training is not enough by itself.

## Building managers

Building managers need to train staff in safe maintenance procedures and instil a safety-conscious maintenance team who are trained to use work equipment

safely and healthily. Escape routes should be maintained and safety evacuation procedures practised and instructions clear for all including visitors
to follow. Access to ceiling spaces, boiler rooms, window cleaning and roof
and gutter maintenance is typically hazardous, together with the handling
of certain materials. One of the biggest dangers for a building manager is
evacuating the workforce in the case of fire or any other site-wide threat.

Ethically it is important to make users aware of the dangers to their safety,
health and welfare. Their own responsibility is to comply with given instructions and not to put others in danger by their actions. Buildings can be used
safely when users know how to work plant and machinery and use the proper
protective equipment. Healthwise the buildings need to be maintained against
legionnaires' disease, leaking gas appliances and effective extraction systems.
Hygienic food preparation should be instinctive with separate hand washing
and food washing and cleanable surfaces which do not host germs and keep out
pests and vermin. A poorly designed building makes hygiene dependent on good
practice alone. The access to maintenance activities needs safe procedures if
working at height, in confined spaces and near dangerous equipment.

### Occupied older and refurbished buildings

There are always vulnerable people in buildings and different types of disability including sight, mobility, or deafness need particular facilities and access.
Making use of automatic opening doors for getting to safe havens in wheel-
chairs, providing signage that can be read by blind or partially sited users
and educating people on evacuation procedures may be more expensive, in
older buildings, but must not be less comprehensive and escape plans should
be in place with this in mind. Access to ceiling spaces, lights and windows
is varied but is likely to be at greater heights with more complex access externally and internally and with more fragile materials. Some components may
be without the benefit of instructions. Materials may be more hazardous
or worn out and preventative or reactive routines may mean more frequent
access.

In refurbishment of buildings then operation and maintenance manuals and
health and safety files need to be accessed for new owners so that procedures
and maintenance routines are understood and particular hazards assessed.
Refurbishment often means the uncovering of unknown structural arrangements which should be amended with care and design specification consulted if
possible.

Ethically greater care is required in older buildings and a dedicated manager
is needed who has time to 'learn' the building. There is no excuse for not
spending more on a building which has more expensive components and
repair costs.

## Risk assessment and moral hot air (virtual morality)

Health and safety assessment is often seen as assessing a uniform harm which is confused with identifying significant risk and it is important morally to distinguish the difference between the overall situation in which a harm could occur and significant risk where there is a real danger of someone getting hurt. There is a lot of virtual risk assessment which lands up on bits of paper which do not get into the hands of the frontline worker. This sense of 'going through the actions' is a travesty of justice and allows workers to do jobs unguided and unsupported with unknown safety and health risks and the effect this has on other workers. Situations can also get out of control so that if there is a lack of co-ordination between two relatively harmless activities a situation of panic or excitement is created which makes the ordinary safe behaviour of people become unsafe because of a mass perception of harm (e.g. the mass stampede that took place among pilgrims in Mecca).

Risk assessment is measured on the basis of impact/severity of an incident *and* of probability (Figure 6.1). This means that for a frequent small harm incident, such as slips and trips, there is just as much moral obligation to take preventative action as there is trying to stop a person slipping off a high platform where injury is likely to be very serious, but for which there may be very little frequency, i.e. in an office building. This context matters, as on a building site, falling from heights is more frequent and is statistically the most likely cause of major accidents. Here, measures to limit falls to below average danger would be morally justified as the impact and harm are great.

Cut hands, slips and trips on a building site are more frequent than falls from a height, but the harm they do in most cases is not perceived to keep an employee off work (the HSE method of assessing seriousness of impact is off work more than three days), which if you look at the statistics is untrue. It can also reduce productivity and in some cases, like slippery staircases or muddy ramps or unprotected edges near excavations, will build up a perception of danger in almost everyone. It is wrong for an employer to say 'It doesn't matter if we have a few cuts and bruises each day because we don't have to report them.' This culture of moral silence on building sites has been common for smaller injuries, but is often an indicator of poor planning.

It is commonly said that a clean and tidy site is a safe site. This is mainly

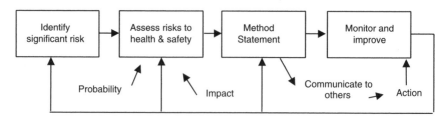

*Figure 6.1* Risk assessment process.

because of the impression of orderliness and pre-planning, which encourages excellence in small improvements. It is also more likely to cut down on slips and trips which have now become the most commonly reported accident on site and CIOB members have been asked to report sites they visit which are untidy.[12] However, improvements should be targeted, e.g. supplying gloves to *all* employees and visitors as a protective agent sends multiple messages. One message is to remind you that cuts and bruises are common and this reminds you to be careful. However, negative and paternalistic messages may be psychologically weaker because, in making it compulsory, it stops people thinking about the particular risks of their unique situation and there is an wastage of effort in putting gloves on when in fact the greater danger is, say, to use wet gloves near electrical equipment. Well-thought through training is more effective and respect for an employee to learn about relative risks and in some cases by their own mistakes is also ethical.

There is also a danger that zealousness in the over-protection of a workforce will produce an inverse reaction to the need for that protection. So for an employee to be told they must be careful not to fall over the edge of a desk or slip in the corridor when they walk around the office can produce a disdainful or scornful reaction which could lead to less attention to other instructions such as the use of hazardous chemicals which are not so obvious. In the former case an employee is only likely to hurt themselves slightly and so is likely to learn about obvious hazards quickly and painfully. If there are dangerous sharp desks or slippery corridors, they should be a temporary hazard to be removed by the employer. In the latter, if hazardous chemicals or trailing leads are left around carelessly and unexpectedly by employee neglect, then others are affected adversely and a regime of planned use, tidiness and order is needed. Correct presentation and management of health and safety are a shared ethical responsibility for a safe and healthy output.

On a building site there are more opportunities for a 'careless obstacle' to escalate a safe workplace to an unsafe one, because of the temporary nature of the layout and storage regime. Lots of materials do not have a risk-assessed storage place and become obstacles which get walked around in unsafe diversions. Roofs leak and access gets wet and slippery. Workplaces exposed to the elements are harder to standardise and are temporary for different work teams who follow on from each other. Excavation or work overhead is in juxtaposition to access routes and slips, trips and falls are more to blame on the lack of care of others. Temporary access points for vehicles are no longer segregated from pedestrians, making access unexpectedly dangerous. It is therefore a moral responsibility to institute strict protocols for ongoing risk assessments and briefing to workers each day. Unfortunately unintegrated and laborious site-based risk and method statements don't efficiently deal with this dynamic situation and EU research[13] has shown that 66 per cent of accidents are due to management and design shortcomings. They are not inherently unsafe in task execution, but in transition, co-ordination of information and unpredicted circumstances such as overlaps with other work or poor supervision of design awareness.

### Ethics

There is moral responsibility for the contractors to provide a safe working place for their employees and for others working near them or passing by. There is a resistance to heavy health and safety risk assessment because people feel controlled and often don't know the reason why.

- Often risk assessments are done by those not in the workplace and workers are not encouraged to participate in the planning of their own safety when they can see better ways of doing things and they are not informed of the danger caused by others. This lack of consultation and communication is unethical and encourages people to cut corners without knowing the whole system.
- Health and safety is not given relevance to the *individual* and a culture of shifting the problem to others 'who don't know what they are doing' is built up.
- Risk assessment procedures for individual tasks conceal the overall planning and co-ordination of the design and construction process in keeping an organised and tidy site with good communications to identify risks caused by overlapping factors.
- Priority is put on time and cost and *quality* is *disregarded* as a priority. If quality in the sense of 'fit for purpose' was put first, there would be a greater control over the workplace by the worker about existing or developing hazards. This does not necessarily have to take more resources or time, as accidents and aborted or reworked defects are time-consuming, demotivating and wasteful in time and cost and materials.
- The cost of accidents is also immediate, giving rise to direct delays and demotivation which can lead to cutting corners.

There is often a desire to make health and safety a technical rather than a moral or ethical problem. The overall picture of killing and maiming many more people in the construction industry, three times the number killed and seriously injured in manufacturing in the UK, is a moral problem for the employer not to provide a safe workplace. Through CDM, this responsibility also transfers to the designer, the client and the co-ordinators.

## Quality and moral imagination

The development of a health and safety culture is not in isolation and is better carried out in association with a quality regime. A quality regime is one which encourages excellence and doing things right first time and better on subsequent times. People build up a sense of pride in the value and quality of their work in the intrinsic motivation theories and it is a more positive and satisfying culture to encourage excellence and not to simply push absence from accidents. A negative objective like that is seen to cost money and resources and only gains a productivity which should not have been lost in the first place – a retrieval rather than a step forward.

Pratley (1995, quoted in Harrison)[14] has indicated some reasons for a moral silence because the language of morality is seen by managers as a threat to harmony, efficiency and the manager's image of power and efficiency. This has meant that achieving health and safety and quality has been presented as a technical achievement in terms of standardisation which has blurred the issue of trust, transparency and fit for purpose in the delivery of quality as a commitment to the client and *the failure to achieve effectiveness is not seen as breaking a trust.* Morally it is wrong to force a handover of a sub-standard building, but it accords with the specification. This creates conflict with users that sully the otherwise excellent relationship. It is equally immoral to *reject* a building which is fit for purpose, but fails to reach an arbitrary standard, but there is no recourse in law. A precise (contractual) product that does not work, is correct, but needs mending. It also demotivates workers who believe the client is not getting a good deal, because they have been told to build to a flawed standard. A building that works well, but is rejected on the basis of non-technical compliance needs to be reviewed and the specification amended.

Total quality management (TQM) could be expressed in moral, democratic and collaborative terms to involve the ethical conscience of all involved. It is a bottom-up democratic continuous improvement regime which operates on consultation and feedback from all involved to achieve fit for purpose and to fine tune it. It tries to iron out the unethical anomalies above and encourage a more dynamic and improved standard. Ethically, improvement could be in terms of wider client values such as health, quality of life and environmental long-term aims and not just in terms of immediate efficiency. There is a strong correlation between TQM and zero accidents in that the aim is to prevent quality defects as well as improve the product. Preventing defects is tantamount to preventing accidents. The principles of TQM according to Hanson[15] are

- The quality standard is defect-free.
- Quality can and must be managed.
- Everyone has a customer and is a supplier.
- Processes and not people are a problem.
- Every employee is responsible for quality.
- Problems must be prevented not fixed.
- Quality must be measured.
- Quality improvements must be continuous.
- Goals are based on requirements.
- Life cycle costs and not front end costs.

- Management must be involved and lead.
- Plan and organise for quality improvement.

The points above indicate the way in which every worker is involved in quality including management and how the system promotes prevention and improvement. The business rationale for such a system is that it improves productivity by improving quality (Smith 2002),[16] but he notes the need for a longer upfront investment in information systems and training prior to payback. Smith reckons a 10 per cent improvement cycle.

These points could be equally applied in bringing about a zero accident culture which could be run alongside TQM where workers are drawn into an improvement regime with incentives and encouraged to provide expertise bottom up. Managers would have an obligation to measure progress against targets, provide training and to plan and organise the system so that it ran smoothly and identified the risks created by overlapping activities and general safety management of joint welfare induction, access and sequencing. Their equal involvement in monitoring and improvements is illustrated in the case study indicated in Part 2.

### Quality and distortion

This affects some of the public sector regimes for quality is usually expressed in numerous KPIs and other types of targets. In the health service, KPIs for reducing waiting lists have often produced distortions for the quality of treatment for those treated, with additional risks such as superbugs which kill or maim healthy patients, as cleaning budgets and food budgets and training are slashed on the altar of efficiency and 'service' standards. Non-priority compliance is a simplistic control regime and forces a 'one size fits all' solution which takes away human discretion and compassion where it is most needed. Failed core standards often scupper the compliance of the rest and flexibility with training is needed to suit local conditions.

Planning approval is charged to vet applications on a compliance basis which is thought to give equal chance to applicants on a 'level playing field'. The target system to get applications decided within a given time period has led to distortions as many good applications are being rejected rather than amended in order to meet tight target times for processing applications. This has been wasteful and has led to mainly simple applications getting through first time because of the complexity of the requirements. Limited extra resources have been allocated to pre-application advice.

Recently, on the quality front, ethical objectives to require more sustainable design and carbon-free operations have been made the responsibility of the development control process. The guidelines for this have led to an improvement regime which has become competitive between the authorities and produced a 'postcode lottery'. Further guidelines are introduced to control the rate of introduction to deal with community complaints that development is deserting their area for less sustainable outcomes elsewhere. Both of the above question the micro management of efficiency and quality regimes centrally and the need to encourage voluntary ethical change on a competitive business basis, with a strong public lead. The leverage of public client exemplary behaviour has always been a powerful one and in the UK has worked much better where evidenced. However, there are also examples of compliance with heavy boots where certain health and safety imperatives have overruled other ethical considerations out of all proportion. This is evidenced in the condemning of playground equipment so that parks have to be closed and the knee jerk reactions to require expensive kitchen equipment which is unproven in its operation and use. The requirement for sealed buildings which exclude healthy fresh air flows. All of these have good foundations, but have been enforced in heavy-handed ways, reducing the quality of life and eliminating minor or unlikely risks, while introducing quite different ones such as youth boredom, lack of community facilities to build community spirit and unhealthy buildings which are energy efficient.

Marginal quality and health and safety requirements are adversely affecting the quality of life where priorities have distorted the outcomes and there is a moral obligation to bring back a utilitarian ethic that provides a wider happiness ratio to those disappointed. The Kantian approach to health and safety is essential, but when matched with the quality requirements in a more integrated approach, there are side effects which outweigh the harm done by some possible hazards.

## Changing culture

Culture is defined in the *Oxford English Dictionary* as 'the trained and refined state of the understanding . . . prevalent at a time or place'. It is also referred to as the 'way that we do it round here'. It emerges from traditions and customs, but can be changed. It is part of the TQM approach to health and safety.

A well-known phenomenon called risk homeostasis theory (RHT) is that workers may adjust their behaviour if the risk is reduced because of the machismo associated with risk and because of the level of risk habitually accepted. This means that risk reduction must be managed and training given to indicate dangers caused to others. RHT suggests that there is a natural level of risk that an individual or group feels comfortable with and that to go above *or* below this level of comfort will induce compensatory behaviour. An example of this is the reduction of speed because of road narrowing due to road works. However, it will also work the other way round that if the road is widened and straightened, then the average traffic speed will also increase. RHT is often used to exercise traffic control by painting lines on the road to make it look narrower to give a sense of restriction and thus reduce speed. It can be associated with the use of many warning and speed restriction signs and cameras to try and adjust driver behaviour subconsciously.

Some traffic authorities, however, feel that control of this sort is divisive and takes away the responsibility of drivers for their own safe speed. This has led to experiments to do away with road signs and traffic lights and to create enabling structures such as roundabouts to enable drivers to join busy traffic and to encourage courteous driving by letting people in. This approach could well be accompanied by training.

This approach has been trialled for several years in the Dutch town of Drachten[17] with a population of 50,000 where there are now only two sets of traffic lights out of fifteen originally. Instead, roundabouts have been installed at junctions and road signs and most road markings have been removed. Motorists have noted that tailbacks are now almost unheard of and although you drive around more slowly and cautiously sharing the road with many cyclists, they believe they get around faster. Hans Monderman, the architect of the scheme, is backed by the EU and he claims that it works like a skating rink where people learn to avoid each other with no lines and in crowded conditions and very few rules. There are one or two small collisions, but he believes that this has reinforced the learning process as people then try even harder to avoid them. Cyclists and pedestrians have both said they feel they have priority, although pedestrians say they more often wait to cross the road with others, before they venture out, cars have given way.

This represents a different way to attempt a culture change for poor, dangerous or discourteous behaviour to induce better health and safety records and pleasant conditions. In the control method, the criticism is that the creeping level of control has led to a confusing level of instructions that has led to motorists taking their eye off the road to read sometime conflicting or unclear signs. Speed cameras are respected, but many see them as a revenue raiser and lead to the wrong attitudes for compliance and technical evasion of fines and a 'smart' driving to ignore restrictions.

On construction sites there have been similar moves to reduce health and safety by control – to enforce the use of CSCS cards, to give workers and visitors lengthy induction and to patrol sites on a regular basis, zero tolerance and draconian fines and instant dismissals to deter others on a 'example by punishment' basis. This has worked to a degree, but is in danger of setting up an atmosphere of fear rather than of co-operation in getting compliance at work.

The equivalent to removing traffic lights and installing roundabouts for construction sites is individual respect, team working and counter-training to remove bad habits. These methods have majored on the cultural aspect and have insisted on common knowledgeable starting positions for key supplier supervisors and managers. This has been possible with the use of strategic partnering in the supply chain so that projects use familiar core teams and contractors can insist on certain basic standards for which there are intrinsic as well as extrinsic standards taught and drummed into each worker.

The use of improvement techniques from the bottom up, such as quality circles and toolbox talks ensures the maintenance of these across larger sites where there are multiple subcontractors working together. Ideally it should be an integrated culture where the whole management team and design team are involved in implementation. Tours round the site for managers and designers can provide extra eyes as well as feedback to designers and estimators in seeing where there are problems so that they too can improve in terms of design choice and resource provision. Clients have responsibilities at the strategic level and the impacts of dangerous occurrences as well as accidents need to get back to them through reporting these as well as accident statistics. The reputation of a new culture depends upon safe outcomes and better results at the construction stage, without poor publicity for the project. The new culture is one of health and safety inclusive of quality. A site that runs itself with the involvement of all in the health and safety process is one that is knowledgeable and interested in the jobs of others on site and respectful to them and responsible to assessing its own risks in the light of that knowledge, and proceeding with caution as they would do in unknown traffic conditions. In this sense, others and not health and safety are the main focus. Risk assessment in a vacuum for each contract lacks the visibility of quality and so workers need to communicate their risk mitigation to each other where they interface and raise the visibility and democracy of the process. This takes away management dependence on dedicated health and safety managers, giving responsibility back to supervisors and managers who must look at the bigger picture of site-wide co-ordination. In addition, the managers need to manage the process of training and dedicating resources to health and safety management.

### Protection of the public

Under these conditions inclusion rather than exclusion is an important concept for encouraging public support for new construction which, after all, will be a nuisance for many. Considerate contracting encourages the distribution of information to neighbours about the site and makes an allowance for cleaning up, reducing noise and dust, restricting working hours on urban sites. Order and site tidiness, traffic control of deliveries and parking and good interaction between all workers and public are basic to safety culture, but often neglected. On larger urban sites there is a need to think about the use of pictures on hoardings and scaffold netting to raise expectations for improvements. The visibility of viewing panels allows the public to see the progress, together with notices that explain dangers and future benefit. In severe disruption, as in the case of constant road works, an explanation of how nuisance has been avoided is good psychology. Allowing visits to the site with outreach to schools helps gain public interest at key points in the development and gets people on side in the development. On really small sites resources are scarcer, but the common rules of respect are a guide for reducing and explaining nuisance caused.

Work carried out at night on the roads follows the same reduced disruption logic as work to occupied airport terminals. Health and safety measures that extend misery on the roads for longer periods can be shared by using night work in rural areas. Reduction of noise and well-planned diversions can help reduce day time work. Good planning can promote more efficiency and use more resources to do it more quickly. New technologies can induce less intrusive road works and control traffic effectively and safely. Health and safety measures which are applied indiscriminately create frustrated and less safe drivers, who are less likely to be responsible drivers.

## Ethics

The ethics of culture are about the importance of a means to an end in health and safety and the term welfare and quality of life can be extended to workers, managers and the public who need to maintain their enjoyment of work and life throughout the construction process. The domination of construction nuisance is no longer acceptable as it is such an integrated part of our lives. Nuisance is not the fault of the worker but it might be aggravated by the planner and designer.

Health and safety wise, it is more important than business profit that users are happy and have a safe building and it creates a climate of security and well-being. The perception of danger in the office or school is not as important as the perception of health and quality of life, because absence of an efficient safe and interesting environment can create its own stress. More hours are lost to employers through stress than they are to other sickness except back pain.

## The psychological contract

The psychological contracts 'define the informal beliefs of each of the parties as to their obligations within the employment relationship' (Herriot 1998).[18] This means that employer and employee are able to redefine their relationship as the job evolves and both sides are probably looking for something more as the employee grows into the job. This will have an effect on the level of health and safety that is expected by an employee from their employer and the project culture. As technology develops, expectations grow in maturity. When a staff member becomes a member of the project team, the culture found will affect the level of expectation and more may be expected from the employer. If the job is more risky, they will look for danger money and if the project culture expects a certain level of behaviour and commitment, they will have no choice but to comply. This may be affected by conditions in general such as longer hours of work or co-location in the same office with other organisations, e.g. contractors and designer. This sense of secondment to the project where safety standards are high should pull the standards up where there is strong leadership of a high standard or down if there is a 'laissez-faire' attitude to health and safety.

## Teamwork

A high ethical standard for the project manager is even more critical. Team working can, at its best, draw out a synergy for creative and productive thinking in
overcoming joint problems for drawing a high health and safety standard. Loosemore et al. (2003)[19] identify the issues for maintaining motivation. These are:

- an equilibrium between workload and energy;
- control in establishing priorities;
- sufficient rewards for the perceived effort;
- a social community;
- fairness and resolution of conflicting ethical values.

Where these are substantially out of balance, then the motivation for contributing to a happy and a safer team is at risk even if only one or two members of the
core management are affected.

## Conclusion

Health and safety in construction has not been up to standard[20] and more
people are killed and injured than should be the case. This is a moral problem
and the number of accidents needs to be reduced in construction particularly
because there three times as many deaths and serious accidents than in manufacturing. This can only be done by changing the culture, which has remained
stubbornly traditional with an addiction for a high level of risk, passed from
old to new workers which some construction companies have tried to break.

It has been argued that the range of health and safety responsibilities enshrined in law are more easily achieved by integrating them with a quality
approach to health and safety. It is also argued that health and safety or quality
is a moral as well as a technical requirement. The client has a key ethical
responsibility to create a motivating environment for safety, health and quality
and this is the key to success. Quality is a promise to supply a building or
structure fit for purpose. Ethically it needs to suit the users. This is integrated
with the need to maintain a building that feels healthy, safe and secure in use.

It must also be a design that does not endanger the life or safety of construction workers and again can be built and co-ordinated by the principal contractor
in a way that reduces the conception of danger to the construction worker and
motivates them positively through the quality of work they do. From the contractor's point of view, they have a complex task to co-ordinate a better health
and safety environment and need to mobilise the whole management team
into the risk planning and management and include the workers in the risk
assessment.

Clients have greater responsibility under the new regulations to make sure
resources and competence are invested in the construction process to ensure

health and safety is integral to everything that is done, both to stop poor health and safety management and to get more management input. This applies even more so to small contracts where awareness is less and resources for health and safety are proportionately going to be a bigger part of the budget. They also have to ensure that a health and safety plan is in place before work starts. More clients than any other category have been convicted of CDM contraventions and have argued their ignorance of construction health and safety.

## Notes

1 *Pocket Oxford Dictionary* (1934). Oxford: Oxford University Press.
2 Usability.gov (2007) *What is Usability?* Online. Available http://www.usability.gov/basics/whatusa.html (accessed 28 December 2007).
3 ISO 9241–11 (1998) *Guidance on Usability*. London: British Standards Institution.
4 HSE (2006) *Health and Safety Statistics 2005/06*. Online. Available www.hse.gov.uk/statistics/overall/hssh0506.pdf
5 HSE (2007) *Fatal Injuries*. Online. Available http://www.hse.gov.uk/statistics/fatals.htm (accessed 28 December 2007).
6 HSE (2003) 'To Obtain a Robust and Validated Influence Network Tool for Construction', *Health and Safety in Construction, Phase 2, Depth and Breadth, Contract Research Report 387/2001 RR231–236*, Bomel Ltd.
7 Reporting of Injuries, Diseases and Dangerous Occurrences Regulations SI 1995/3163 (1995).
8 Department for Business Enterprise and Regulatory Reform BERR (2007) *Construction Statistics Annual 2007*. Norwich: The Stationery Office.
9 Campbell, H.F. and Brown, R.P.C. (2003) *Cost Benefit Analysis*. Cambridge: Cambridge University Press, pp. 284–6.
10 The Construction (Design and Management) Regulations 2007, Statutory Instrument 2007: 320.
11 Gotch, S. (2005) 'Consultation discussions', Aug. 11. Online, Available. http://consultations.hse.gov.uk/inovem/consult.ti/conregs/messageShowThread?threadId=398&includeId=3819 (accessed 26 October 2007).
12 CIOB (2003) 'Improving Site Conditions', proposals from two workshops held with SME and major contractors. March and July.
13 European Union (1992). Available at: http://www.usha.europa.eu/en/good_practice/sector/construction
14 Pratley, P. (1995) 'The Essence of Business Ethics', in Harrison M. (2005) *An Introduction to Business and Management Ethics*. Basingstoke: Palgrave Macmillan.
15 Hanson, D. (2007) 'Total Quality Management Tutorial'. Online. Available at: http://home.att.net/~iso9k1/tqm/tqm.html#Introduction
16 Smith, N. J. (2002) *Engineering Project Management*. Oxford: Blackwell Science.
17 Millward, D. (2006) 'Is This the End of the Road for Traffic Lights?' *Daily Telegraph* 4 Nov. http://www.telegraph.co.uk news/main.jhtml?xml=/news/2006/11/04/ntraffic04.xml
18 Herriot, P. (1998) 'The Role of the HR Function in Building a New Proposition for Staff', in P. Sparrow and M. Marchington (eds) *Human Resource Management: The New Agenda*. London: Pitman.
19 Loosemore, M., Dainty, A., and Lingaard, H. (2003) *Human Resource Management in Construction Projects*. London: Spon Press.
20 HSE (2002) *Revitalising Construction: A Discussion Document*. London: Health and Safety Commission.

# Chapter 7

# The planning ethics

The central purpose of the planning system is to identify and plan land use for a local community and a national infrastructure and to provide a fair development control system for planning applications and arbitration in case of dispute. It is considered in the UK and many other countries, that both the land use allocation and the development control system should be decided democratically or at least with significant consultation.

The democratic planning system is run by the local authorities balanced between appointed planning professionals advising and elected councillors who make the final decision. Planners are guided by central government policy statements who also deal with the planning appeals system. Planning is one of the local authority services that generates the most vocal feedback, as decisions made to allow certain types of development affect house and business values and individual development control decisions are directly connected to the quality of life of those neighbouring new developments, be they infrastructure or buildings.

The planner is in a unique position of public trust and should exercise a commensurate high level of understanding and impartiality in the public interest. Their advice role means they have unique access to and influence on the elected councillors through their authorship of the reports and have a further responsibility to present a balanced argument and contextual interpretation which must remain clearly objective in each application. The outcomes have a big impact on the community.

The American Planning Association[1] states that 'ethical principles derive both from the general values of society and from the planner's special responsibility to serve the public interest'. They further point out that societal values often compete with each other. Socially they need to 'strive to expand choice and opportunity for all people' with a special responsibility to plan needs for disadvantaged persons and groups. This challenges the value free, facilitator/mediator approach often propounded, with a social concern agenda (Brown et al. 2002).[2]

In land use, planners will deal with supporters and objectors and will need skills in stakeholder management. In addition, they need to present material transparently for public consumption. They will also need to deal with the interrelatedness of community and social matters and prioritise applications.

The challenge is to maintain public respect even though there may be aggressive campaigns against their policy-making. To avoid being accused of conflict of interest, they will need to keep the highest standards of integrity and respect in the face of opposition.

## Questions

Fundamental questions about the efficacy and legitimacy of the planning process arise. These relate to the effectiveness of land use planning to produce a fairer allocation and to judge the real issues of importance to development control. Who should make the decision in order to ensure basic human rights for members of the community? Thomas and Healey (1991)[3] claim that ethics, legitimacy and validation of the skills and knowledge of the planning system and those who operate it, are insufficiently debated. What skills and ethical standards are considered valid in determining acceptable development and ensuring fair decisions? Who has the right to make these decisions? The present system of most market-based economies is based on local democracy to avoid party political loyalties, but would development be better controlled by central government to ensure parity between regions and remove subjectivity? Should it be a simpler, more transparent, organic process justified in terms of whether the proposal generates economic wealth, whether it represents minimum loss of private amenity for the region, or in terms of whether the proposal is a sustainable solution? Are there sufficient checks and balances in place to avoid bribery and corruption to the decision-making process?

## Development planning in the UK

Land use planning and development control is established widely round the globe, but has certain local characteristics. Initially a regional structure plan is produced, identifying major policies and major infrastructural development which must be taken into account in local plans. Prior to planning application decisions in the UK, all authorities – counties, cities and unitary authorities – are required to set out a core strategy for a development framework, based on consultative processes and indicating the planned long-term development of the areas under their planning control. With a county authority a structure plan is produced which is further developed at a secondary local plan level. The plans for the unitary and city authorities are at one level called the Unitary or City Development Framework. The plans provide a strategic overview and provide constraints for development decisions by identifying the 'site specific allocation' for various development types. They have to go out to consultation before the plan is adopted. A plan cannot be adopted until objectors' concerns have been addressed or a public local enquiry has resolved ongoing issues.

More recently local community involvement has been restricted to truly local matters with action plans where development is higher impact. Other 'low impact' zones have been given special status such as business development zones and regional development agencies for the particular purposes of fast tracking development for regeneration and economic improvement of rundown areas. Special grants are available and in the case of regional development agencies separate development control powers are given for regional infrastructure. Pre-defined planning criteria are used to speed up planning permission and deal with broad areas of development. Limited permitted development for domestic property, such as domestic wind turbines, is to be allowed at the time of writing without an application, to speed up bottlenecks.

In order to give some equality of basis for both plans and the decisions based upon them, Planning Policy Statements (PPS, replacing planning policy guidance) are produced by central government which are enshrined in the plan to ensure minimum standards. In more recent times factors such as sustainability and transport planning have become more closely related to development planning and local transport plans and sustainability strategies have also been built in. Transport for London (TfL) is responsible for a combined plan for the whole of London. Regional Development Authorities (RDAs) at the time of writing are also responsible for longer (20-year) plans covering a region. Greater London has a strategic approach to the whole area and there is centralised strategy with local development control by boroughs.

PPS documents deal with issues such as housing provision, sustainability, out of town supermarkets, archaeological arrangements and developments on ground which may be flooded. Other agencies such as the Environment Agency, Highway Agency and English Heritage may make recommendations on these issues in more detail, but do not have the power of veto.

## Development control

The key personnel in the development control process (Figure 7.1) are the development control officers (DCOs) who devise and moderate the development plan documents and evaluate and recommend an outcome. The elected councillors who sit on the district-based planning committees actually make the final decision.

A recommendation is made by the DCO after inspecting the local and any regional plans that have set out land use policies and seeing if there is compliance and that the central PPG policies are served adequately. This is further viewed by looking at any constraints or covenants that affect building development in the location and taking a view on how the building will fit into its surroundings – height, massing, access, colour, etc. This can be helped by looking at precedents for similar development in the area and considering overall cumulative affects such as road congestion. In practice, there may be very little guidance as to previous development and strong points made by the developer

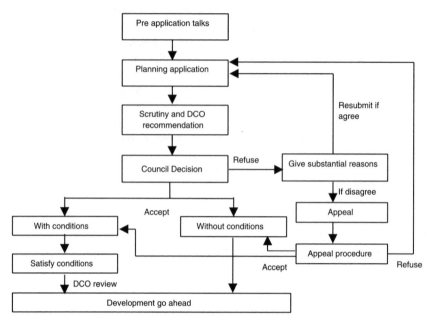

*Figure 7.1* The development control process.

for a design that contrasts to the existing are difficult to assess. Buildings also exist in the context of what they provide and in the way in which they sit in the urban or rural space.

A case may be made for variation from the local plan if additional benefit is found, e.g. employment, for a larger than normal footprint on the plot to accommodate the size of company into the plots available. Officers may advise on how to make the building compliant or to advise where recommendations might be made to overrule a constraint or not. Planning authorities will also need to see overall intentions for the operation of the building such as travel plans and measures to make the building sustainable. The latter issues come as a result of negotiation between the DCO and the developer, but the DCO, although knowledgeable, can only make suggestions in good faith and cannot guarantee any decision by the LPA. Again there is a good sense that professional and ethical guidelines would be useful so that each party is clear and properly and equally advised. If a decision is made to refuse the application because of non-compliance or relevant objections, then each party will have recourse to appeal. The appeal is dealt with by the planning inspectorate who will appoint a suitable neutral inspector who will make a report which is binding unless a further amended application is made.

Where the planning regime is complex as above, there are particular challenges to the consultative and democratic process where many would be referred to different parts of the system because it is unclear at what level they should lodge their complaints. Other major issues exist such as the level of housing provision nationally and the release of sufficient development and the reuse of 'brownfield' land at an appropriate rate to ensure recycling of obsolete past developments.

## Ethics of development control decisions

Ethically, apart from ensuring an appropriate public voice, there is a need to ensure equality of opportunity where it is possible to 'play' the system by the more experienced or streetwise developer at the expense of those who are less in the know. A balance is needed in an iterative system where a spiral of consultation is allowed, ensuring progress to the policy and decision-making rather than bring it back to square one with a rejection. There needs to be a cut-off point for each level of decision and 'checks and balances' for a minority who want to impose their views by using blocking procedures. It is also important to allow for visionary action, ensuring that development does not land up as a stifling and bland common denominator of compromise. An example of this would be the delivery of prescriptive standardisation of provision in the central business areas of towns and cities which seems to have happened in many of the urban shopping districts.

Market demand needs to be balanced by interventionist polices to ensure sufficient affordable housing, accessible shopping for those on public transport and long-term provision for climate change and flooding. In addition, land use should not be blind to the economic factors that drive development and learn to provide incentives for desired development in otherwise uneconomic sites. There are a lot of responsibilities in co-ordinating an overall desirable community that has not advantaged some at the expense of others. Is this level of control possible or indeed desirable in the long term? Will freer market forces find a balance in the end? The development control framework is an attempt at some agreed level of control. Hopefully some level of flexibility can be built in to bend to the market force which is able to sponsor such development without losing sight of the plan.

DCOs and elected councillors have complementary roles which reflect the analytical and social aspects of the decision. Clearly they need to work together in understanding the issues, but need to be separate in the sense that councillors should not have a role in the forming of the application and maintain their distance so that they do not have a conflict of interest. Both have a role for the best good of the community and arriving at just and equitable solutions. Physical and social information presented to the councillors should be technically competent and should provide a fair picture which is reasoned, complete and unbiased. There is a key ethical issue here in transparency and honesty and in developing trust.

The DCO responsibility for scrutiny and recommendation to councillors for application decisions is an important and influential one and may also be guided in certain parts by other 'expert' advice such as conservation officers and other specialist consultants and organisations. They interpret their obligation as 'value neutral' in terms of local plan compliance, but may also see themselves as an advocate of a just distribution of the social impact of individual applications. In practice, there are dominant applications and lobbies which provide benefits or point out harm which may be long term or short term. Short-term expediency may erode long-term community good, but short term is very difficult to resist if it is in technical compliance with land use planning. The DCO juggles PPG, client pressure and local plans in making their recommendation.

Commercial benefit, party political voting power and community social good may often be in tension and councillors may wish to override land use planning for a duty to understand local public opinion short-term political expediency. Other factors include the power of lobbying pressure groups, the possibility of corruption and the personal values and precedents that exist in the department.

### Planning gain

In addition, authorities have power to require a Section 106 agreement to put something back into the community directly in return for the impact of the building or indirectly due to perceived extraordinary profits received because of market demand – a sort of 'development tax'. These measures are often controversial as there is a concern that there is a double charge on the buyer with council/business rates which should supply these services and a premium price for the property to cover costs of the agreement. Others believe that a development needs to pay for the development's initial impact and that the council tax/ business rate is for the maintenance cost of services.

Who decides tangible impact value? The size of the 'development tax' is an ethical issue and there is an argument that it should be imposed proportionately to impact. Also it should be applied equally to competing developers with a rationale for its use by the Council. Rationally monies should be used for the benefit of the development, but this is not compulsory. Developers may be asked to provide the amenity themselves, such as children's play areas and traffic lights, or a contribution for the service is paid to the council and used as it sees fit. The latter is particularly applicable to an ongoing social service such as pre-school provision. Where a clean-up service is being provided, such as decontamination of the site, then this may be chargeable to previous owners under the Environment Act 1992.

If a benefit to the community specifically arising out of the development is tangible, such as employment in a rundown area, then grants might be available to make development viable, but this is generally outside the planning process and ethically should be kept so. However, the same planning department might have a close liaison with the economic development department in order to co-ordinate this aspect. Planners themselves need to be aware of the terms of these grants and have a role to advise potential developers equally. Some planning departments see this as a conflict of interest. This is determined by the different management approaches to the planning role below. Again there is room for bribery and corruption for whatever officer is responsible for the advice where big money is at risk to the developer. A separation of role between marketing and granting planning advice may or may not help.

In the UK, there is a requirement for each authority (under the Freedom of Information Act) to provide basic public information about the development and if asked, to discuss the development in detail with third parties but some confidentiality will be important to the applicant where specific business plans are discussed in the application. Advertisement of the application is initiated by planning notices posted adjacent to site and in the local paper. The process prior to the planning application allows negotiations with local development control officers which provides essential information to the applicants and encourages applications compliant with the latest requirements. However, it is rather less public and therefore open to some abuse in terms of the equality of the advice and access to DCOs. Many feel that influential parties may be given privileges over and above others in terms of access to information and the interpretation of the land use and form of development which would influence the DCO recommendation to the Council. Councillors and DCOs take account of the number and seriousness of community objections. Objectors have to refer to local framework issues to protect their interests. Applicants can appeal against the council decision.

### Delays to applications

There are additional funds for an authority able to deal with planning applications within a government time scale. This was imposed by the government to deal with valid complaints about unfair delays in the time taken for councils to make decisions. This has had the effect of councils refusing permission for applications on the basis of minor technical flaws so that average decision times can be met and incentive monies claimed. To the applicant, this can appear unfair and prejudiced as they can be put into a position of serious delay and extra expense disproportionate to the shortfall in the application. Applicants have a responsibility to discuss their applications before submission with DCOs to ensure technical compliance, but their design and business preferences should also be genuinely unaffected by any personal preferences of the DCO. It is obvious that there can be a subjective side to planning and some applicants may receive more help to get their applications through first time.

One of the key issues to overcome this is the setting out of substantial reasons for planning refusal and appeal against non-substantive refusal, which could have been presented as condition or reserve matters. Appeals are clearly expensive and time-consuming. It is also wasteful for all concerned to go through the application process twice. It is also hard for the DCO, where they clearly have their recommendation overruled, to set out 'substantive reasons' for the refusal – which they are required to do. A case study is set out in Chapter 17.

## Basis of planning decisions

### Community

There are many different angles on the fundamental ground of planning decisions. One of these has been put forward by Blanco (in Hendler 1995)[4] describing the 'four jewels' as reason, democratic process, community and nature.

These are described as more than rationality and democracy and are bound together by the context of the decision in the community which is affected. In this sense the democratic process is the inclusion of all the stakeholders in that community where a decision on development is being made. Community is defined as having a common past, a memory, that leads to a common present where the decision-making has some base and common language for communication and a commitment for a common future which is infused by values and ideals. Nature provides an objective ground for the culture of the place where the community is based and not just the generic requirement to be in touch with the environmental imperatives which sustain us and need to be preserved.

This is an interesting perspective on the usual call for a just and fair solution to the consideration of the planning application and provides a more community orientated framework decentralising the planning decision firmly in the hands of the community where most are affected by it. It recognises the need to bring in social as well as rational aspects and recognises the distinctiveness of a case for a local environment and location. It demotes the role of a council vote.

### Equity

Another approach would be Lucy's five concepts equity approach which Lang espouses in Hendler (1995).[5] This provides the opportunity to distribute benefits fairly among society. These concepts are:

*   *equality*: everyone should receive the same service within reason;
*   *need*: those requiring a service should get it;
*   *demand*: those showing an active interest in a service should get it;
*   *preference*: to include others not using that service, but whom have an interest;
*   *willingness to pay*: only people who use a service should pay for it.

## Case studies

In practice, some of these concepts conflict and there will be competing claims between equity and efficiency which tries to maximise benefits to the community as a whole.

### Case study 7.1  Flood defence wall

An example of this would be the building of a flood defence wall where there will be a decision which is resource limited. The usual way of assessing the need for a wall is on a cost benefit basis that provides some weighting for various benefits that are obtained in terms of protection of property of different uses and land they are differently weighted. The cost benefit analysis therefore brings up an optimisation efficiency model and takes no account of who wants it. Here there will be those who have different views. Some

- will clearly suffer from flooding without the wall;
- will see the wall as not protecting their property, but would like it to;
- will not wish to pay additional taxes to cover the cost of the wall because they will not gain any benefit from it.

### Exercise

Evaluate the case under each of the five principles in the equity model given that there are scarce resources for flood defence.

Case study 7.1 illustrates the need to make choices where there are scarce resources and conflicting benefit between different stakeholders. Choices of this type require experience and possibly a model. The model which says you save houses and not fields from flooding is too rough an instrument and weighted benefit is often preferable. Planning decisions are often made on the basis of right or wrong and this is misleading and there is no process for softening the blow of an outright refusal except to wait and come back with major modifications.

### Case study 7.2  Hidden development

A farmer who hid his bungalow,[6] which was built knowingly without planning permission, in a barn behind straw bales, was not caught until four years later when a guilty conscience and lack of light made him admit to his work in the hope that he would get retrospective and compassionate planning permission. However, it cost him dear as he had to knock it down and lost his wife in the four-year process of deceit. This retribution seemed very harsh at the time as it took four years for anyone to notice.

### Questions

Would a community have condemned this even though it was on green-belt land? Should the rules be more important than the impact of development to encourage order and discipline in society? Was honesty properly awarded?

Interestingly, there were sixty-one responses to this article when it appeared on the web and there was more of a preponderance of council 'red tape' and bureaucracy comments 'than keep to the rules' comments.

### Case study 7.3 Shopping development

A developer was granted permission to build a large store adjacent to a few houses, because this is one of the few sites in the area for such a large store and there are many new houses that need a shopping amenity. Is this a fair decision?

Either the amenity of existing residents should not be affected or they should be compensated for any loss. The additional amenity of the shopping facility may offset an increase in traffic congestion. In practice, people in a community will be affected unequally and will value the amenity differently.

Using the equity principle, it would be approved with conditions that balanced the liberty – to give people shopping facilities, but to adjust traffic control so as not to disadvantage neighbours and control noise so as not to deafen them. A certain low level of nuisance would not have to count in gaining an amenity. Rawls' principle of justice is twofold – the equal liberty principle and the difference principle. The difference principle is to make improvements to one area without disadvantaging another area in doing so. Ideally the difference should not be extended either.

Using the community principle, it would provide a larger argument for the good of the community which needs a shopping facility and there are no other sites which are close to the new housing. It would also be a utilitarian principle that the happiness of the few would override the misery of a few. The community would try and come together to try and make a proper proposal that would not isolate a few.

## Case study 7.4 Transport planning

This concerns the permission for a traffic dampening scheme through a residential area where a road is being used by through traffic. The scheme for installing chicanes and bumps to stop speeding and danger to children and pedestrians is inconvenient to residents (as well as to 'rat runners'), making it difficult for them to park and making access to their homes in a car less comfortable. It is also inherently safer and means that on balance children can play more safely and pedestrians are less likely to be in RTAs.

- On balance, has there been an equitable benefit for residents who only drive in cars?
- Has the community gained as a whole when balancing benefit and loss of benefit?

The benefits would need to be weighted with an inordinate weighting towards injury and loss of life reduction. The inconvenience of not parking would be weighted lightly. This rationalised approach has not made reference to the community who may have strong feelings about the danger perceived from 'rat runners' or stronger feelings about bumps and chicanes in their effectiveness in reducing accidents. A factor for the democratic process and community can be factored into the weighting or full choice can be given to those most affected – the community. As it is at present, the third party, i.e. the pressure groups, are likely to get much more clout.

This case study is interesting in that it is a clash between community democracy and the desire to cut down accidents which is not an issue covered in the planning policy statements. It is, however, an issue that might have split the community and the case for vulnerable people and children is unlikely to be made strongly by others. The different planning approaches below might throw further light on the subject.

## Different planner roles

The above methods for making decisions are affected by the role that planners are given. Thomas and Healey (1991)[7] describe dilemmas with reference to the role of the DCO.

- *Public bureaucrat role*: This may be a stifling straitjacket or it may be a reassuring framework to the planning officer. It provides a rule-based approach with a strong precedent culture and a fairly inflexible approach to each application with minimum pre-application contact and negotiation. It can provide a just, but limited creativity and frustrations for a client. Lang (1993)[8] also says to expect the unexpected. In this case we could make a plan incorporating our predictions and the solutions, but if we get the unexpected, then the plan may even be harmful unless the plan becomes dynamic in response to change and gain it is possible to gain some flexibility to deal with unexpected outcomes.

- *Policy analyst role*: This is based on technical neutrality with negotiation based on the application of policy and with a limited consideration of the human and community impact. It allows for an arm's length application, but is not particularly ethical as it has not worked through the politics or social contexts which are the reason for placing development control with officers who have local knowledge. It is a good position for an appeal officer who may want to keep distance from the politics.

- *The political policy analyst*: This is a similar approach with an additional political awareness dimension to distribute social benefit more equally. It is a more proactive stance allowing officers to negotiate with different parties prior to land purchase and get the best community deal. It may also involve an astute judgement of the political climate of the type of development which councillors are minded to grant permission for.

- *Social reformer*: This is similar to the above but with a much more socialist approach to distributing benefit. It could appraise itself in a much more directive way and provide an acceleration to social betterment of urban areas. In its worst form it appeared in the 1960s and 1970s as a quite arrogant imposition of developments which in hindsight were considered inappropriate and socially obsolete in quite a short time. An interventionist policy such as this needs confidence and a long-term rationale.

- *Developer and marketeer of economic assets*: This is a much freer market approach to regeneration. A council merges the economic development role with the planner role to work with the developer to make the development attractive where formally it would not. The development corporations were given specific planning powers to fast track urban renewal in areas of industrial wasteland such as the extinct shipbuilding areas along the River Tyne in Newcastle and Gateshead. Often significant progress is made in limited time and vast amounts of government money are available for infrastructure restructuring and decontamination. It has sometimes been criticised as a quick fix which has no roots in the future which takes resources from other areas which have an equal need. This may be unfair where long-term solutions and reinvigoration are patently successful and can be a catalyst for other adjacent development.

## Case study 7.5  Newcastle Development Corporation

Newcastle Development Corporation (NDC) lasted for a 10-year period from 1987 to 1997 and imaginative planning has engineered creative public buildings and infrastructure as well as attracting much needed private capital and business into the area to counter the decline in shipbuilding. NDC was able to commission the Millennium bridge and to ensure that many older landmark structures such as the Co-op building were retained and recycled as office buildings. It also led to reciprocal development across the river in Gateshead where the Baltic Wharf flour mill was regenerated as a landmark public gallery and innovative buildings like the Sage symphony concert hall were commissioned for what was now an attractive waterfront environment. Public access to the river front was ensured by drawing contributions towards a continuous riverside walk. Once the public was attracted, leisure and eating facilities were built which could serve both the daytime office and night time leisure activities and provide a new viability and life to the area.

The ethics of such developments are often criticised for being artificial and short term. The narrower focus of control has often meant that the NDC and its clients can often see 'eye to eye' much more easily and developers know the conditions to expect. Conflicts of interest should be easier to sort out with less complicated arrangements. A narrower clientele in the frame will need an audit trail as a balance against any favour given.

Each of the approaches may be adopted by an authority or a mixture of power. Ethically different issues will typically come to the fore. As personal values are important, there may be considerable unrest amongst officers who feel that they are unable to support objective approaches based on a pragmatic, utilitarian, moral or justice accountability. Clearly the methods are mainly based on rational decision-making methods, but the involvement of councillors injects a political dimension as they may also have taken a poll on opinion in the more controversial cases. They are likely to look to the social equity and the on balance popularity of the policy as well as the utilitarian approach where there a vote that benefits the most for good.

### Sustainability agenda

Although this post-dates Thomas and Healey there is now a significant target-based approach which has been born of more recent concern of the macro impact of buildings on the environment and climate globally. Through regulations at both planning and building level there is a necessity to demonstrate more sustainable and long-term buildings which use less energy and have a low/ zero carbon profile. Planning permission depends upon suitable whole life cycle measures which demonstrate this. This approach implies a subsequent check of the management of buildings and opens up a negotiation approach due to the way that compliance legislation has been written. It is essential that applicants and their agents discuss the constraints. Designers are anxious that the approach allows them leeway to comply with high standards of 'green' without being straitjacketed to standard solutions. Scoring schemes like BREEAM and Eco-Homes are a new area for planners to understand and assess.

Sustainability has been forced into the community consciousness in usually quite rigid applications. Permissions are heavily dependent on the demonstration of low carbon or other factors which are on the current government agenda. These factors should be given broad flexibility to foster the uniqueness of the community. For example, a rural estate with lots of thatched roofs in upgrading the stock should not have to demonstrate the same degree of air tightness as it already has a high score on roof insulation and aesthetics. Tearing down existing solid walls should not be a feature as the windows could be double glazed and are likely to be smaller in area.

On the other hand, conservation is an important part of community, but planning permission for changes or extensions to graded stock should not be made impossible or too expensive to achieve. If rural community was given more say, they might like to protect stock from price rises by making some stock inaccessible to rich incomers. This could be ethically incorporated by having restricted deeds of sale and incorporating planning restrictions on alterations by non-residents.

## Facing moral problems

There are a number of ways of dealing with the moral problems that a planning officer faces. These are disruptive, but to some extent may be necessary in a complex decision-making process where no two decisions are comparable. However, similar treatment has to be given to similar applications.

Faced with uncertainty about the right decision or course of action, it is important to bring the team into the discussion so that a conflict of interest may be declared or a concern shared. This method depends upon a willingness to see issues from a view which may be different from one's own and to be perceptive to the barriers. The solution may need more experience in applying existing principles or a reflective change to existing practice. It also depends upon who is ultimately accountable for the decision and whether it is serious enough for an officer to disagree with the rest of the team. This could apply to moral or political differences (Thomas and Healey 1994).[9]

An officer may decide to leave in the case of continuing disagreement. Alternatively they may change the direction of their work to take them out of that type of decision-making in which they are at odds with their colleagues. This is an unsatisfactory but ultimate solution when all other avenues have been explored.

Compartmentalism is a way of separating different aspects – professional, personal views – and in an imperfect world it may be necessary to compromise for the sake of unity in a team. This would be inappropriate in the case where morally dubious action is confirmed and again should be tested with others in order to ensure a position is not compromised. Another form of this is to work with clients in relationships based on trust where a planner compartmentalises a different moral stance according to context and justifies it in a serious of separate relationships. Again, although pragmatic, this may be difficult to live with and to justify in terms of equality and justice.

Reference to an over-riding principle such as sustainability may help to arrive at a decision in the case of conflicting principles. A code of ethics in itself has principles which in certain cases conflict in a given decision. These ethics cover the essential relationship ethics of openness, honesty, integrity and fairness which allow for a respect for the person(s) you are dealing with and give a relevant people perspective. However, conflicts of interest vary between the good of society and client demands quite often. Stating a loyalty in a conflict of interest exercises openness and retains respect for the client.

## Case study 7.6 Kitchen's[10] principles

These principles are on the basis of developing close emphatic relationships with the client (applicants). The first principle is the acceptance of *representative democracy* in planning decisions. The second principle is that *all clients deserve respect* and this should be represented through honesty, openness and fairness in dealing with them.

These two principles acknowledge the special relationship of the DCO with the client recognising them as having a unique need and dealing not only with applicants, but also with third party agents who may inform the planning decision such as conservation bodies and rural affairs. Also building up skills to reassure third parties' neighbours and interest groups who are concerned by the new development, and negotiating concessions to mitigate the effects. Providing specific briefs for the councillors to give the facts and the context of the application allows them to ask specific questions to play a particular role. Pressure groups may obtain an inordinate amount of officer time and this might have to be re-balanced in deciding materiality to give other parties sufficient weight. The business community is concerned with some stability in committing to future plans that need to make assumptions about future demand and availability. Housing developers are specific applicant clients and will be looking to buy land that can be developed in the future, but will be in competition with others. The wider public are often consulted for a general view of an impending development. They are concerned about traffic congestion with new business and housing and want to be reassured as to the strengthening of the infrastructure.

## Housing supply and sustainability imperatives in the UK

Scarce housing supply is increasingly coming under criticism in the UK for raising house prices and the industry in turn have blamed the planning system and government policy for reducing land supply. In 2007 the UK government was exercised to increase supply substantially from under 200,000 to the predicted demand level of 223,000 houses per year. This had shown an escalation during the five years up to 2007 of 10 per cent. This, if achieved, would reduce prices and help first-time buyers. Funding is also needed.

Supply is also perceived as dependent on the supply of land. This supply has been squeezed by government sustainability requirements to rebalance the use of greenbelt and brownfield land (26 per cent reduction of greenbelt in the past seven years compared with 9 per cent brownfield increase in same period). To support this, the government has also increased the density levels for housing in order to make supply less scarce. As a result there is a stand off between government who have made land supply more scarce and want cheaper, denser solutions, and developers who are being blamed for profiteering and hanging back.

The Office of Fair Trading[11] also believe that major developers have slowed the supply of 'permissioned' land onto the market in order to falsely keep prices high and is investigating housing developers. In justifying its involvement, it identified a weak supply of housing land, which had been responsible for bolstering price increases ahead of the market and stagnating the first-time buyers market.

### Sustainability

The government has targeted new housing as a significant contributor to the production of carbon and greenhouse gases and codes to force house builders to meet tough new reductions in carbon have led to a target of carbon neutral by 2016.[12] This has followed a plethora of regulation such as the Sustainable Housing Code,[13] the stiffer requirements for energy saving in the Building Regulations[14] and the introduction of PPS3[15] to inject a degree of sustainability into the planning requirements for housing. There is a desire to show progress towards a zero carbon level on a cumulative basis which is not opposed by the house building fraternity. However, this has had a further supply effect because the the HBF[16] predict it will drive out some of the smaller suppliers. In any case there is a possibility that complex regulation may drive down supply because of the greater effort in achieving it.

The HBF also point out that although there is a difference between the general public interest which pushes for higher sustainability for the good of long-term objectives and the individual buyer primarily value for money interest. They argue that it is important to have regulation which provides a level competitive playing field for suppliers' sustainability and that suppliers should be left to do things voluntarily in a cost-effective and operationally effective way to match their business model.

### Ethical issues

Ethically there is a desire to see that market competition is working properly so that there is not supplier control of the economic factors including land and funds, and technical innovation which could falsely reduce the supply in order to increase demand and raise prices. However, the government desire to bring about a more sustainable product through the planning approval system, including brownfield land use, may in itself bring a restricted supply, or at least partly move the blame for the constrained supply to the planning authority who require to apply difficult conditions, which cost more to build. There is a particular problem with the different standards being applied in a postcode lottery so that applications which are the same type are treated differently from authority to authority. There is then a conflict between objectives and an equitable balance has to be reached between the two.

## Conclusion

The planning process has been well developed over the past 60 years, but has been faced with significant changes in the past 10–20 years which has bought about a lot of policy changes. This has meant both land use planning and development control have needed to take a long hard look at their ethical stance. The role of the officer has been under scrutiny for returning to a more co-ordinating role which has produced less holistic solutions.

The sustainability agenda has meant that the DCO has to take quite an expert view of these factors, which should include a deep knowledge of economic, environmental and social issues in order to compare similar schemes and not to reject plans on the basis of applicant ignorance. It may also be producing a bottleneck and a greater element of government intervention in the development process impacting directly on past working practices.

The pressure to get a throughput has distorted the way that approvals are made and planners are being strongly criticised for delaying the harder decisions but this also means applicants need to see the process as more interactive.

The issue of morality and fairness is critical and automatic appeals are not helping the system to remain local.

# Notes

1 American Planning Association (1992) *Ethical Principles in Planning*. Online, http://www.planning.org/ethics/ethics.html
2 Brown, C., Claydon, J. and Nadin, V. (2002) *Background Paper 2, Planning, Planners and Professionalism*. Centre for Environment & Planning, University of the West of England, UK.
3 Thomas, H. and Healey, P. (1991) *Dilemmas of Planning Practice*. Aldershot: Avebury Technical.
4 Blanco, H. (1995) 'Planning Ethics: Community and the Four Jewels of Planning', in S. Hendler (ed.) *A Reader in Planning: Philosophy, Practice and Education*. Center for Urban Policy Research. New Brunswick, NJ: The State University of New Jersey.
5 Lang, R. E. (1995) 'An Equity Based Approach to Waste Management Facility Siting', in Hendler, op. cit.
6 Martin, A. 'Farmer Tries to Dodge Planners by Hiding Bungalow in Barn', *Daily Mail*, 7 March 2007.
7 Thomas, H. and Healey, op. cit.
8 Lang, R. (2003) 'Expect the Unexpected', *Ontario Planning Review*, 18(5).
9 Thomas and Healey, op. cit.
10 Kitchen, T. (1991) 'A Client-Faced View of the Planning Service', in Thomas and Healey, op. cit.
11 Office of Fair Trading (2007) *House Building: Reasons for a Market Study*. June. London: TSO.
12 Department of the Communities and Local Government (DCLG) (2006) *Building a Green Future: Towards Zero Carbon Development*. London: DCLG.
13 DCLG (2006) 'Code for Sustainable Housing: A Step Change in Sustainable Building Homes Practice', Dec. This replaces PPG3 (2000) and is supported by Good Places to Live by Design guide, Regulatory Impact Assessment and Equality Impact Assessment.
14 The Building Regulations. Available: http://www.ukbuildingStandards.org.uk
15 PPS 3 Housing (2006) *Communities and Local Government*, November. This replaces the PPG document.
16 House Building Federation (2007) 'Submission for the Calcutt Report'. April.

# Chapter 8

# Ethics of sustainability
## A UK example

This chapter is concerned with the interaction of the built, intentionally organised and engineered environment with the natural spontaneously organised environment. In this interaction there is an increasing ethical concern about our ability to conserve and maintain the beauty of the natural world around us because of our lack of control of the impact of the built environment on our natural world. It considers the particular issues that constitute a sustainable built environment.

A new, more prescriptive ethical paradigm is emerging in planning the built environment which has gathered pace in the sustainable development movement. The underlying ethic is justice for a secured future for our children summed up in Prince Charles's plea that,

> I don't want to see the day when we are rounded upon by our grandchildren and asked accusingly why we didn't listen more carefully to the wisdom of our hearts as well as to the rational analysis of our heads; why we didn't pay more attention to the preservation of biodiversity and traditional communities, or think more clearly about our role as stewards of creation.[1]

This is an eclectic, but powerful view of the type of justice ethics which merges the common good and the personal factors which motivate vocal public opinion.

## Sustainable development

In the past three decades the concept of sustainable development has come to the fore and has a critical relevance ensure the development and maintenance of the physical and social environment around us is sustainable. This is tackled at a micro project level and at a community, urban and global level in this chapter. Sustainable development is at the root of Girardet's question (2000: 20)[2] 'How can people still lead lives of continuity and certainty whilst still achieving compatibility with the living systems of the biosphere?' This is an ethical question which taxes our intellectual and instinctive facilities for survival in the future as well as in the present. It deals with the fundamental area of people, their inter-relationships and inter-dependency. From the ethical and built environment point of view, interventions need to create living environments that are pleasant and impact favourably on the natural environment and ecosystems. Lee (2000)[3] distinguishes between our attitude to the artefactual – that which we have created – and the natural, and insists that there are different ethics to our intervention. In the manmade, such as a building, he argues, we have a right to adapt for our purposes and preserve it at the height of its beauty as a heritage in the making. In the natural, although we may wish to preserve a certain beauty, there is a dynamic to nature which, it is argued, will allow nature to regenerate in spite of imposition by the built environment.

The ethical issue is whether we are destroying the sustainability of the natural environment and *not* whether we need to maintain it exactly as it is. The ethical argument for the artefactual environment is how do we need to develop or maintain it and fulfil our responsibility to remain sustainable in adapting the environment to suit our living needs? How do we ensure that we do not destroy the enjoyment of others in making it sustainable for our own ends?

The UN defines sustainable development as 'Development that meets the needs of the present, without compromising the ability of future generations to meet their own needs' This definition emerged from the Brundtland Commission (1987).[4] The UN attempted to bring together the various agencies for development and challenge both developed and developing nations to tackle the problem of limited and disappearing resources. Climate change, pollution and over-stretched resources were threatening the future of existing and future development. The report urged institutions and nations to work together and recognise their interdependence and ethical obligations to counter the current destructive trend that led to ecological, economic and social systems' break-down. The Chairman's foreword recognised it as an ethical and political imperative to muster the resources for change of direction: 'We became convinced that major changes were needed, both in attitudes and in the way our societies are organised.'

It was also recognised as making 'painful choices' because it sought to rebalance a system which was not being replenished for survival in the future. Socially distribution was skewed in favour of the rich nations.

Sustainable development is an ethical choice based upon Aristotelian principles. The most difficult thing about it is that major nations who most upset the balance have to work together. One nation working on its own (e.g. the UK releases about 2.5 per cent of the world's emissions) would be ineffective and unilateral action makes that nation less competitive in interdependent world markets, because the initial capital cost of cutting back on emissions would not be recouped unless all nations were to follow suit to gain a level playing field. Brundtland's catchphrase is 'The earth is one, but the world is not'.

The UN as an organisation has the potential to bring nations together – how successful can they be, given the egocentric utilitarian approach that pervades much of the Western political system? Will nations whose progress is relatively low on the developmental scale accept caps on their development for the common good or calculate that they should gain at the expense of those who have more? Will developed nations accept their collective responsibility to reduce their dominant share of scarce inputs and harmful outputs? Will they be smart enough to do so without affecting standards of living or will a change of lifestyle be inevitable or even desirable? How will emerging economies feel if they are asked to cut back? How painful is the process going to be, how should it be done and what is the priority?

The World Commission on Environment and Development (WCED), which commissioned the Brundtland Report, brought together twenty-one nations which agreed priorities along which the change should be drawn. Twenty years later, the UN Sustainable Development Commission[5] further breaks down the issues for sustainable development into energy, industrial development, air and pollution issues, climate change and a consideration between issues that overlap. They consider a long-term view in each of these areas and meet annually to establish common approaches and best practice for the mutual good of all, but not at the expense of others.

As well as defining sustainable development which is at the heart of the built environment we shall look at the associated concepts of urban sustainability, sustainable construction, sustainable living and working places, community participation and social sustainability, efficient and effective uses of resources for property and infrastructure development and the sourcing, manufacture, disposal and recycling of materials and waste in the process of construction and in building and infrastructure use. We are also concerned with climate change and the impact of greenhouse gases and other toxic emissions caused by buildings and property development. Socially we will consider how all these also have an effect on the quality of life and our interaction with each other in communities and globally.

There is now an admitted 90 per cent probability of climate change due to temperature rise because of carbon emissions and ozone depletion being a result of the manmade activities (IPCC 2007).[6] Governments are setting out agendas for the reduction of greenhouse gases and resource efficiency in construction and energy use in buildings where 40–50 per cent of emissions are shed (Godfrey 2006).[7] In the UK there is 1 per cent of the world population and 2.5 per cent of the world's emissions. Worldwide the construction industry is responsible for £31.6 trillion worth of output worldwide and employs 111m people. It is estimated that 20 per cent of the output is wasted. This represents 40 per cent of all waste generated, while the industry represents at most 8 per cent of GDP. The 4th IPCC report indicates that, in addition, buildings in use generate about 33 per cent of carbon-based greenhouse gas emissions.

The UK has committed to all new homes being carbon neutral by 2016 CIOB (2007)[8] and the government has set out zero carbon construction for 100 per cent of new build by 2020 and all existing homes by 2030. They also set targets for 100 per cent recycling of construction materials by 2020; 40 per cent reductions are considered a minimum.

### The three areas of sustainability in built environment projects

Conventionally, governments and organisations see sustainability as a triple bottom line of accounting for the social and environmental costs of our actions as well as the economic outputs which are essential to the continuance of modern society and business (Figure 8.1). Sustainable business is tackled in chapter three on business ethics.

The environmental area of sustainability is well rehearsed and covers our concern with issues such as climate change, energy efficiency, pollution, ecological balance and excessive waste and the need for recycling scarce resources caused by physical development. The economic area of sustainability is the long-term financial balance for development, so that whole life costs are considered with capital costs to give excellent overall value to the client and give efficiency in the context of maximising benefit and saving waste. Certain upfront payments may be a good investment for the future and need to be recognised beyond temporary ownership while still making a profit. The social area of sustainable development is a an issue facing people and communities concerned with the impact of development on the quality of life and its integration with the community needs and the wider social issues. This is the least tackled issue in development because of the overarching desire to satisfy a business plan with tangible objectives. For example, the insistence on higher density dwellings in the 1970s later created effective slum areas in inner cities in the UK which led to the demolition of tower blocks in the 1990s as social factors such as safe play areas and access for family living had not been thought through and communities became isolated and crime-ridden. Government regulations and taxation have subsequently been used to incentivise action in the social and environmental areas, which is vital to quality of life for the wider community.

Ethically there is a need for an integrated stakeholder approach to development presenting the three sustainability areas so that transparency and trust can

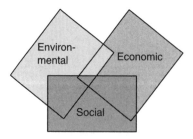

*Figure 8.1* Three areas of sustainability.

be built up. One of the key complaints is lack of consultation and communication. If there is consultation, it is ignored and rationalised by unprecedented economic and environmental factors. Of course, some social issues coming out of consultation are short term so a mechanism to recognise those which contribute to sustainability and quell short-term fears is a good social contract approach to consultation. Further conflict may be negotiated through the planning appeal system. Is this ethical or is it a mechanism for a dominant developer to bury consultation issues that were difficult to comply with? How much control does the developer have? Does the planning system properly deal with sustainability in all three areas and manage any monies elicited from developers for social and environmental purposes appropriately? How are those most affected by change consulted in ways that mean their concerns are incorporated?

We have frequently seen sustainability as being an 'either–or' with efficient production, but Day (2000)[9] argues that we do not need to choose between 'humanity or nature, utility or beauty, ecology or beauty', suggesting that we can have both by looking at buildings and design in a holistic, socially cohesive way. In other words, there are material synergies with ecology such as natural ventilation, the use of minimally processed materials and the reduction of energy use and other wastes in the more proactive design of a building to 'sustain us'. It is a nice thought that sustainable design makes it more sensitive to the well-being of users and may be in itself associated with greater productivity. However, achieving this ethical balance may not be easy without the right incentives to and commitment of developers.

## Environmental accounting and motivation

Many developers are designing longer-term strategies to recognise their corporate social responsibilities and to manage them as a benefit rather than a drag to their prosperity. This can be done by building up a reputation through justifying their environmental credentials that gets them more work in the pervading climate once a critical mass is achieved. Public clients *must* be leaders in sustainable design.

Dalton (1998)[10] identifies some of the drivers which would persuade companies to improve their environmental record. These are

- investment allowances for clean technology;
- changes to advertising regulations to stop misleading claims;
- compulsory environmental liability insurance;
- compulsory environmental reporting;
- full public access to information about pollutants produced by individual companies.

Dalton's drivers indicate an expectation for government financial incentives as well as penalties in the case of non-environmental compliance. It also opens up the potential for public accountability through a key performance indicator (KPI) based reporting system. However, if reporting is compulsory, it may produce grudged and massaged figures that confuse those who are trying to make real progress.

Regulation to increase compulsory minimum insulation and low carbon emissions seems to be making some developers commit themselves to a competitive higher attainment of what will in the future be a standardised expectation. The other driver for energy saving is the increasing energy-saving requirements of building regulations and for developers to present a strategy which meets increasing standards of sustainability. Existing buildings not included represent 90 per cent of stock and need to be retrofitted to the same standards to make overall environmental improvements. Standard housing currently costs about £500/year more in energy costs than the zero-rated BedZED development in London and this could offset one authority's estimation of a 2.5 per cent extra cost[11] for green design.[12]

The ethics of this exercise are interesting. The driver for governments is credibility to meet previous commitments made in the international and European arena (currently 80 per cent carbon reduction by 2050) and remain a trend setter. Ethically they are driven by such things as an equitable planning system which meets sustainability targets without disadvantaging the poor. The developer's ethics are encouraged by their perception that sustainability legislation will be expensive to meet over a short period and to build their experience early to cut the extra costs of sudden requirements. Contractor drivers are to remain on selection lists where sustainability requirements are specified by public and reputable clients. They have a strong role to play in making the front line of development implementation acceptable. The drivers for sustainability are mainly business and not ethics orientated. Public opinion beyond the pressure group has been mobilised by recent compelling scientific agreement (IPCC 2007)[13] for future disaster if the trend of 'business as usual' continues. This evidence has identified energy use in buildings and living habits as a focus for change. Public opinion has rising expectations for sustainable improvements. Some countries have seen rising green votes and some are voting with their feet to improve their quality of life by downsizing. Energy-guzzling development is no longer acceptable. Green taxes have support where the revenue is ploughed back into incentives for sustainable development.

### Design

Guy and Farmer (2000)[14] identify some interesting conflicting views of the logic for producing green buildings which they call 'competing logics'. Six logics are presented in pairs as three polarised dimensions below:

- The building is a negative impact. This is tackled from either an eco-centric or a smart futuristic ethic. The first is to minimise the impact of building on the eco environment and to use sympathetic materials and design. The second is to use technology as a scientific solution to the environmental crisis. It recognises a need for an 'incremental, techno-economic change' for which there is a solution. These buildings last because they were designed to.

•   Green building as an appropriate form. Form has either an aesthetic or symbolic ethical rationale. In the former, both function and image celebrates the environmental message. In the latter, the symbolic is a mechanism in the design process to touch the cultural awareness of the local community in reaching environmental goals. Form is further differentiated as the ethic of the building form and the ethic of identity. These buildings last because they are yours.

•   Green building as a social concern. Social justice and quality of life are represented in a building emphasising either comfort or community logic. In terms of comfort, they would reduce pollutants and promote the healthiness of buildings in their physical, psychological and chemical impacts becoming 'living environments' in their own right. The community ethic is a way of using buildings to bring communities together so that buildings are 'participatory and flexible' and serve the needs of a range of users. These buildings last because they are good for you and the community.

These different logics for looking at environmental sustainability also point to different ethical concerns which relate to naturalism, efficiency, visual impact, justice and social contract. It also suggests a wider approach which is closer to the different issues for sustainability. It should be noted that there is an environmental and social aspect to these approaches and the economic is not discussed.

In each of these logic categories, there may be objections. In the first, the technique used, the buildings may not be respected or seem out of context. In the second, there is a danger that the symbolism or aesthetic is personal and not universally appreciated. In the third dimension, there is sometimes a reaction to others choosing what is good for you – there is a desire to challenge the senses rather than affirm them. The economic aspect needs to be applied in the context of the whole life costs as normal.

## Implementation

Our ethic for sustainability may also differ. It may primarily be a justice ethic to ensure survival of the human race in equitable environments, or it may be for qualitative and social benefits as Day (2007)[15] suggests, or driven by innovative economic improvements that we see we can contribute to ensure future availability of resources. This pragmatic utilitarian ethic is often propounded by the government although the utility is long term. Different peoples and cultures could be argued to have different values and judgements.

Third World development has many dramatic examples of seemingly 'good for' developments that provide a Western technological solution to real problems such as water supply or overcrowded substandard housing. This may be short-lived if it fails to deal with the cultural issues, or ends up destroying the environment. Ethical sustainability is substantially a virtue principle, but aims and needs should be clear and not at odds with one another.

In the building process of client briefing, design, construction and use there will be different ethical perspectives held by the designer, the client, the contractor and the user. Their values will be associated with their professional education, business needs and work ethos. In one sense there is a need for an ethical compromise. It is common practice to work towards the values of the client as they are the prime movers and hold the budget, but there is often an ethical conflict for professionals who have committed to a more explicit sustainability virtue ethic in their code of practice. The end user will be most affected and, where they are united, they may be able to apply a market force for sustainability. As a whole, ethics is an underlying principle for making right decisions and these may not be popular decisions. These may be corporate, personal/professional or community/government led.

The Kyoto and Rio Protocols for the reduction of greenhouse gases have been a critical focus, but have struggled over 15 years and still do not have some of the biggest players on board. Critically large blocks such as India and China and the South American continent see themselves as catching up with the living standards of the West and are using carbon-rich fuels to do so. This logic suggests bigger reductions in the West to allow an equitable distribution of carbon-based emissions, which in themselves should be producing energy which does not depend on fossil fuels. Of course, the new technology is more expensive and requires binding agreements to ensure a level playing field. Even more generous aid is not going to solve this problem overnight!

### Economics of sustainability

The Stern Report (2006),[16] commissioned by the UK Treasury, is an economic view of sustainability looking at the global situation. It is based on pricing the cost of reducing carbon emissions and the cost of delaying that reduction. Stern concludes that unless an early attempt is made to reduce carbon emissions by all countries, it will be almost impossible to get stabilisation of emissions and the later it is left, the more expensive it will be to achieve without irreversible damage to the environment. The argument about equitable reduction between rich and poor nations is not fully discussed.

## Case study 8.1  Stern economics of climate change

The Stern Report (2006) gives a snapshot view of projected measures and costs of taking action to contain the emissions of greenhouses gases which are agreed today to have an effect in raising the Earth's temperature.

### Temperature change

To give some scale, it is estimated that we are 5°C higher than the last ice age. Since the Industrial Revolution we have increased the concentration level of greenhouse gases by 50 per cent (280–430 parts per million (ppm)) which has raised the temperature by ½°C. He suggests that if we acted now with some controls, there could be some stabilisation at a slightly higher temperature in 2050 (see Figure 8.2). At current rates of increase (business as usual (BAU)) scientists estimate that we will reach a trebled level of emissions by the end of this century that will have the effect of raising temperatures by a further 5°C which is disastrous for life on this planet. This would manifest itself in increased sea levels (currently forecast at 4.5mm/annum[17]) with flooding risks and changes in weather patterns which would affect fauna and flora in all parts of the world. Other natural phenomena such as the ocean Gulf Stream, the Niño effect, hurricanes and monsoons may also vary and have a deleterious effect on climate. In short, we are in unknown territory.

Working back from BAS, the Report looks at the scenario of containing the current levels of greenhouse gases which means that we will need to reduce the 2005 emissions levels in the UK by 25 per cent by 2050. In addition, other countries will need to follow suit and if some are unable or unwilling to cut their levels, then complying countries will need to take up the slack by reducing theirs by the absolute amount to compensate.[18] If we were only able to contain

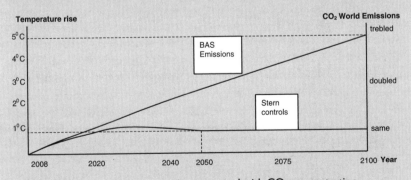

Figure 8.2  Projected temperature rise compared with $CO_2$ concentration.

emissions at current levels, then there are serious risks of the temperature raising by 2°C so that there will be irreversible changes to climate as concentrations of gases still rise. These relate to irreversible melting of ice caps, destruction of coral reef systems, water supply problems where small mountain glaciers disappear, and continuing spread of desertification especially in the Sahel region of Africa. The only offset to this will be the rising crop yields in high latitude countries. These are not net gains as there are decreasing crop yields in already hot countries.[19]

## Reducing emissions

Stern connects sustainability or more specifically climate change to the cost of reducing today's emissions by 25 per cent by 2050. Because of an expected 3–4 times growth in world GDP by 2050, then this is equivalent to 70 per cent reduction on today's rate of emissions, i.e. per unit of GDP. In a realistic scenario of control, the report recognises that world emissions will continue on up in the next 10–20 years with a cost for slowing down the rate of increase to nothing. After the 20-year mark, there must be a real drop of 1–3 per cent emissions per year giving the required real drop by 2050. This drop is where temperature can be stabilised (i.e. the level at which the Earth has natural capacity to remove green gases from the atmosphere). This model represents over 40 years of effort to stabilise at between 500–550ppm which is considered to be an attainable and sustainable level. The other message is that the longer we leave it, the more expensive it will be to stem accelerating emissions and to take the larger financial 'hit' of climate change impacts which are both direct in terms of environmental damage and indirect in terms of health deterioration of large numbers of people, loss of life and loss of quality of life.

## Economic

Stern's economic evaluation is that in the BAU scenario the cost of climate change will be an annual charge of between 5 and 14 per cent of per capita consumption (pcc). The higher figure of 11–14 per cent pcc is more realistic as it takes into account the non-market costs of human misery as indicated above. BAU is also going to have a disproportionate cost upon those who can least afford it, which is going to multiply the human misery effect pushing the pcc up to 20 per cent when the global effect has raised the temperature above 5°C.

His proposal for immediate action and expenditure using the integrated scientific predictor models has an estimated cost of 1 per cent of GDP which is costly and will be contended by some, but if it is to be believed, the cost is affordable for developed countries and can be shared with developing ones. Emissions can be reduced by new low carbon technologies (especially energy sources such as wind, tide and solar power), increased efficiency, reduced demand for emissions intensive goods and carbon absorbing by re-forestation or underground storage (Stern 2006: xii). These represent additional cost but offset carbon emission levels. Most of these are at an early level of development, but other technology may also be developed. The report has considered the aggregate effect on all the economy sectors. Non-energy emissions make up one-third of the total of greenhouse gases with two-thirds for energy use for buildings and transport. Therefore, for radical emission cuts of 25 per cent large-scale take-ups are required of 'clean' energy for transport and reduced carbon construction and outputs for new buildings and other structures. Government incentives to bring down costs and to penalise carbon are vital in reaching these targets.

Radical proposals to cut carbon use are argued by many to connect to change of life style with less use of energy for industry, cars, air travel and 'gadgets'. Without these, they say the targets are unachievable. Others argue that efficiency gains, greater use of non-carbon technology and trading carbon use with carbon absorption will be sufficient. Stern is keen to demonstrate that it *is* possible with changes that do not affect quality of life factors on balance, but recognises it will need a huge attitude change to energy use by consumers and suppliers. Stern terms a successful reduction policy 'establishing a carbon price' (Stern 2006: xviii).

Depending on the success of reducing new technology cost, an economic model will predict a greater or lesser uptake. Governments need to show their commitment to penalising and incentivising new technologies and there are mixed views about this with discounts still being given by energy suppliers for greater use and not less use of energy. Government tax incentives are not seen as decisive to change people's minds. Cars are more efficient, but not at levels which they could be, because manufacturers fear depressing the market. Recycling or more durable goods are not the main focus of any modern marketing because of the prevailing throwaway society. New buildings are being targeted for zero carbon in eight years time, but house builders are moaning about the cost, and existing stock (90 per cent) is already less efficient and will dominate the housing and other stock for many years beyond 2016.

For this reason and for the uncertainty of long range forecasting, Stern has looked at the macro economic climate scenarios and has suggested that the cost could vary from as much as −1 per cent to +5 per cent GDP in the periods up to 2050 with an average rate of 1 per cent. This would give stabilisation at between 500–550ppm based upon 'the scale of mitigation required, the pace of technological innovation and the degree of policy flexibility'. World economic conditions would also seem to be important.

The BREEAM scheme has provided a points scoring system for grading good, very good or excellent for a range of different building types. These schemes have been adopted by developers and public bodies as a way to sell their properties.

House buyers have not shown so much enthusiasm to pay more for an environmentally sound house which currently has a 5–10 per cent premium, but the BREEAM and standard assessment procedure rating (SAP) schemes have created a sort of eco labelling scheme called Eco-Homes, which is available for interested buyers. The new code for sustainable housing[20] (PPS3)[21] is based on the Eco-Homes guide in the UK and is compulsory for public homes and private homes. The key design categories included in the code are: energy, water, materials, surface water run-off, waste, pollution, health and well-being, management and ecology. House builders are unlikely to eat into their profits if they do not have to reach a standard or it does not give them a competitive edge in selling. As a result of Friends of the Earth (FOE)[22] lobbying MPs, the code has been applied universally and the requirement for an Eco-Home has been upped from very good to excellent as a basic requirement. The current situation in the UK is that *all* new housing has a step change to have carbon neutral rating by 2016. These schemes have rapidly moved from a voluntary to a compulsory requirement and the government has ensured that the target standards are continuing to be challenging.

### Social sustainability

The social aspect of sustainability is not well represented in the Stern model which has focuses on the economic and environmental aspects. Vancouver City Council (2005)[23] describes social sustainability as the relationships, networks and norms that facilitate collective action taken to improve quality of life and to ensure such improvements are sustainable. It talks about the four guiding principles of 'equity, social inclusion and interaction, security and adaptability'. This is a good basis for testing the social aspects of sustainable development.

In the original definition of ethical, legal compliance is duty and action taken beyond compliance by developers and clients is an important part of their approach and ethical reputation. There is also a way in which developments are connected to the whole urban and community picture. For instance, increased traffic due to two or more developments is not an issue on its own and needs the input of urban planners.

## Urban planning and sustainability

Whitelegg (2000)[24] gives some excellent case studies of how communities have become disenfranchised by authorities that have been set up to control development and drive through policies that improve certain aspects of city life, but end up helping one community – the affluent – at the expense of the other. Poorer communities, as in the case of failing to stem car usage and privatising public transport, are made poorer when the fares are driven up by the investment costs. Higher fares lead to the decline of the transport system and less use by those who used it before. Ecologically the resulting greater use of cars is a disaster. Ethically this is an equity problem which should be considered in interventionist policies – this may mean taxes and incentives.

Development which is moved to the edges of the city results in the greater use of cars and fossil fuels. This accords well with Guy and Farmer's different logics. The technological solution would be to introduce hydrogen fuels that are carbon-free in public transport. The social logic would be to build a 'Bourneville' type village (see Chapter 5) where facilities are decentralised and work is integrated. The aesthetic/symbolic design solution would be to forge the identity of new communities. The comfort design logic would be to ensure pollution control for the extra traffic and the community logic would be to consult communities and meld the differing needs of stakeholders in reaching a suitable compromise and a feeling of ownership. Whitelegg also makes a case for eco-diversity which recognises different ways of solving the same problems without forcing different communities into the same mould. He also makes the point powerfully that built environment development along 'consumerist and industrial' lines is unethical and is rapidly becoming dominant in the developing countries. He claims it is creating 'future inequalities, inequities, social injustice, loss of community, loss of place identity and loss of a spiritual dimension to life'. His final point is that professions have a particular ethical responsibility for development in the built environment.

### Ecological footprint

The concept of an ecological footprint refers to the amount of resources used per person translated into the land needed to produce those resources. The developed world's rate of using scarce resources equalised across the whole planet would not be sustainable for very much longer. This projection is based, not on world population growth, but on the bigger use of resources by city dwellers and the rapid urbanisation of the world population, currently taking place. Cities create more unrecycled waste per person. China alone is projected to increase its cities from 600 to 1200 by 2010. This trend is seen the world over as a way of improving living standards, but it is unsustainable because city living uses an unproportional ecological footprint. In Case study 8.2 Giradet emphasises we are interdependent and 'one planet' living is the aim if we wish to be sustainable. It is a Kantian ethic of a community able to work towards a fair ecological footprint. A justice ethic might go further in requiring an equality of footprint.

---

**Case study 8.2 Ecological footprint**

Giradet (2000)[25] defines a sustainable city as 'a city that works so that all of its citizens are able to meet their own needs without endangering the wellbeing of the natural world or the living conditions of other people, now or in the future'. Over 50 per cent of the world's population lives in cities compared with 15 per cent, 100 years ago. In 1800, there was only one city with over 1m people (London). Today, with 8m inhabitants, it is much smaller than the largest world cities, but it produces 60 million tonnes of carbon dioxide per year and 20 million tonnes of sewage sludge and waste. It is unsustainable, because with 12 per cent of the population of the UK, it uses up the equivalent land area of the whole of the British Isles in supplying itself with resources such as fuel, food, building materials, etc. that it uses annually. Giradet has worked out that the London ecological footprint is 125 times the size of the area within the city boundaries. Other cities are the same, but some are much more resource-efficient than others and have proportionately smaller e-footprints. He believes that ethical city planning needs an ecological footprint proportional to its population, given the increased squeeze on resources globally. He believes that this can happen by using the knowledge capital of cities. Together with the relatively efficient use of land in the dense development of cities, which cover 2 per cent of productive land with 50 per cent of the world population, it should be possible to provide creative solutions that reduce their e-footprint to sustainable levels. In London's case this is a big task as it needs to reduce its footprint to one-twelfth.

---

The health of a city also depends upon public health issues such as contamination of water supplies and on reducing pollution and harmful emissions which are immediate (killing off fish in a river) or cumulative for a later date (asbestos and heavy metal use). Many of these causes are due to buildings, such as heavy metals which leach out of many traditional metallic building materials over time, e.g. lead, and create a build-up in the soil or watercourse pollution resulting from careless construction work. Airborne pollution often ends up as acid rain which attacks stone and metals, shortening the life of buildings. Pollution is controlled by regulations using environment agencies to monitor and there are fines for companies or contractors who contaminate land and water in this way. The sustainability of manufacturing processes and transport of materials is measured in terms of energy used – this is called embodied energy. The use of local and minimally processed materials gains in this respect.

Energy consumption is connected with source, but most is from carbon-based fuel and contributes to carbon dioxide emissions. It is estimated that 50 per cent of greenhouse gases and pollutants come from the actions of those constructing or using buildings. Some building materials can also be toxic and cause later health problems and building itself can contribute to greater run-off and flooding. The location of buildings in flood plains is a short-term measure in the light of climate change sea-level rise and sensitive ethical decisions have to be made whether to design in flood proofing or defence systems or to limit building to certain locations. The Environment Agency responsible for flood defence deals with this as a risk assessment process which uses a system of scoring against critical factors. Other location issues are associated with the land use planning system, but more attention is being taken by planning authorities to assess sustainable improvements in planning applications (see Chapter 7).

The ethical challenge then is to reduce harm to the environment and to work for the equalisation of consumptions levels particularly by reducing the disproportionate use of resources by Western cities. There is a duty to future generations to make sure that resources do not run out as the population grows and as developing countries demand a fairer distribution. This will impact on built development in a big way, because of the way that it is responsible for a large proportion of resources and carbon emissions. The harm and cost of climate change have attracted popular attention, but it is only one of the future hazards to plan for. Ethically sustainable controls raise the problem of equality of living conditions, but in the case of climate change there is an imperative for universal agreement for impact to be realised.

## Sustainable procurement

Sustainable procurement is the starting point for sustainable construction and has an exponential rise in effectiveness in the ability to run and maintain the building sustainably. Ethically there is a wide range of sustainability issues that can be encouraged by the client which relate to all three areas of sustainability. Sourcing of products is a typical area of concern and this means strategic conditions to control the supply chain and a best value selection to include a significant weighting for a sustainable bid. Economically, large sums are involved in construction projects, but whole life costs must consider the intangible benefits of sustainability. There is a temptation to choose the lowest priced supplier, but the capital building cost is only about a 20th part of the ongoing maintenance and running costs[26] and this is in turn only a fraction of the business benefit of a commercial or public building. Ethically, money spent on quality in reducing user and maintenance costs at the procurement stage is therefore well spent and produces greater returns.

Environmentally, there is a design concern with reduced water consumption, carbon emissions, energy efficiency and increased biodiversity and waste/recycling. There is a thin line between supplier discrimination on a prescriptive basis which disadvantages certain suppliers and allowing suppliers to make their case for ethical and sustainable trading. Partnering with approved suppliers is another way of properly vetting suppliers, but can exclude smaller suppliers who are not geared up to complex selection procedures. Diversity in making some procurement outside the framework may open the door for new and smaller businesses to get onto the learning curve and 'grow' their worthiness. These people may also prove more pliable and economic to meet client objectives.

The Constructing Excellence initiative with Sweden[27] covered use of design quality indicators, planning issues covered public space in high density development and the introduction of affordable housing. In the construction process, it suggested respect for people indicators, a considerate constructors scheme, labour and training opportunities, a healthy environment and encouraged community use of buildings and space. Buyer and project staff need to understand sustainable procurement. The client has some influence over a sustainable specification.

Public procurement clients have to be sustainable in the UK.[28] New public buildings should be zero carbon by 2012 and there is a programme for the running of existing buildings to be made more sustainable and to reduce emissions by 180,000 tonnes of carbon dioxide by 2020. The office estate must achieve carbon neutral state by 2012 by offsetting carbon-based emissions by up to 550,000 tonnes per annum. Government suppliers need to reduce their carbon footprint and will need to demonstrate how they are achieving this. A BREEAM rating of very good or excellent is needed on all new school buildings with training for school inspectors to reward sustainable schools in their OfSTED inspections. In the universities and colleges sectors, funding incentives for sustainability are given to increase funding up to 10 per cent and to cover the cost of the Building Regulations Part L requirements in addition. In the NHS, a rating of excellent on all new build hospitals and very good on all refurbishments is required. In addition, there are some target goals for energy usage per 100 cubic metre of building volume.

In the recent National Audit Office Report (2007),[29] few public building projects considered energy regeneration from renewable sources, monitoring of energy and water use with a view to improvement, social issues such as childcare, local employment and community consultation. In terms of take-up from the 2005 Report recommending 'quick win' solutions such as combined heat and power (CHP), air conditioning and central heating efficiencies and energy control, only two out of 11 suggestions had more than a 50 per cent uptake.

## Sustainable construction

Sustainable construction is a recognised term focusing on the environmental performance of new and refurbished buildings. Construction uses 90 per cent of the non-energy minerals quarried, constitutes 25 per cent of the waste delivered to landfill and 49 per cent of the emissions come from buildings in use. The discussion for an EU Thematic Strategy on the Urban Environment (CEC2004)[30] uses the term sustainable construction to mean, 'integrating the functional, economic, environmental and quality considerations to produce and renovate buildings'. The aim behind the strategy is to enable responsibility for the implementation of numerous EU environmental directives by an integrated approach to sustainable construction and its connection to sustainable transport and urban design. In the UK, construction wastes 70m tonnes of material annually and throws away 13m tonnes of the materials delivered for use to sites. A CIOB survey[31] shows that construction professionals believe that green building is the future of the industry, but two-thirds feel that regulations are not firm enough to get the best results and half feel that there is insufficient financial incentives and client demand.

The NAO (2007) identifies six environmental impacts for sustainable buildings which are mainly implemented by the BRE, BREEAM and Eco Homes schemes:

- reduce energy consumption and associated $CO_2$ emissions;
- minimise the use of resources;
- reduce the release of pollutants;
- maximise recycling and sustainably sourced materials;
- promote sustainable materials;
- conserve or enhance bio-diversity.

Their audit on public buildings is critical of the lack of impact that sustainability codes have had on building stock. The NAO discovered in their 2005/6 survey that only a few projects were being checked (35 per cent of 106 new build and 18 per cent of 335 refurbishment projects) and when they did their own tests on 45 projects, 80 per cent did not reach the test for excellence and some even failed. However, the UK government has yet to successfully implement policy which is to require an excellent rating in the BREEAM assessment for new build and at least a very good rating for refurbishment projects, but some projects have aspired to this level to gain tax breaks and kudos. Interestingly NAO also criticised the BREEAM credit collecting approach as allowing even excellent and very good ratings to concentrate on limited aspects of the spectrum of assessment to obtain these scores. However, they sympathised with rural projects as having a harder job in attaining higher ratings because of the high drag effects of little public transport compared with quality of environment. NAO recommend that one agency be responsible for implementation and that a single person be made accountable such as the Head of the Civil Service.

**Case study 8.3  A social approach to sustainable construction**

The UK Government *Building a Better Quality of Life* consultation (2000)[32] on sustainable construction indicates a social and environmental approach, but also covers respect for others, and working with communities and producing more efficient and effective buildings. This underlines the social and economic aspects of sustainability. It recommends respect for employees and external stakeholders and better relationships between the project team, which tunes in well with the message of the Egan Report (1998).[33] It also uses a best practice benchmarking system and recommends the use of environmental management schemes (EMS) such as ISO 14001.[34] The report summarises the UK government strategy in such areas as tax levies and capital allowances to incentivise energy efficiency and conservation of resources and to bring in voluntary action through getting commitments from corporate and industry bodies to design and police codes of best practice in sustainability and to subsidise training. The three major tax levies introduced are:

- *Landfill tax* (1996) encourages the recycling of material waste an annually increasing charge/tonne.
- *Climate change levy* (2001) which is a charge against the use of fossil fuels energy use in businesses. It also reduces the levy for those who have above average energy efficiency and use alternative energy sources.
- *Aggregate levy* (2002) which is a charge against the use of newly mined aggregates.

These three tax levies are designed to give negative or positive incentives to reduce waste and mineral extraction. The Quality of Life indicators have been developed separately from the report, but work well in giving a more holistic view of sustainable construction. They are best represented in such schemes as the Considerate Constructors Scheme which is described in Chapter 3.

In the UK, a further plethora of documents appear for sustainable building including public procurement from the Office of Government Commerce (OGC) including the Achieving Excellence guides. DCLG has the responsibility for planning and building regulations. DCLG supports an approach to use the planning authority to manage sustainable construction, by requiring suitable proposals in the locally policed planning application for approval. The Building Regulations set out definite 'deemed to comply' standards for minimum energy efficiency and other aspects which are regularly updated reflecting the current state of the technology. They are also responsible for the housing standards. BERR holds the responsibility for the sustainable construction strategy.

In a new house, it is a value decision. Does paying more now affect the payback and increase resale value when I want to sell? Am I changing house to make a life change statement? Regulation for new house standards is also easier, though it may affect the supply and demand equilibrium. Many housing developers believe it is competitive to beat the 2016 zero carbon target and the government is offering challenges to beat the target by offering easier planning terms to designated eco villages. In Germany and France, there are reportedly more incentives. China has designated pilot city developments[35] with zero carbon emissions ready for 2010.

## Cost of retrofitting

The cost of making existing housing stock sustainable has been analysed by the CIOB.[36] They give three figures and also distinguish between the urban and rural circumstances, considering a turbine to be unlikely in an urban house. Travel without a car would also be unlikely in a rural situation. In Table 8.1, low costs represent minimum effort, the medium costs represent some significant

Table 8.1  Retrofitted measures and cost comparison

| Option | Measures | Cost (£) | Saving on | |
|---|---|---|---|---|
| | | | Cost/year (£) | Carbon (ton) |
| **Low cost** | Turn heating down, lag tanks + pipes, energy lights, lights + standbys off | 350 | 255 | 590 |
| **Medium cost 1** | Above + Replace boiler (condensing) and change to AA* appliances, double glaze | 9,000 | 255 | 1,900 |
| **Medium cost 2** | Above + solar (PV) water heating and tank | 11,000 | 255 + 265 + 150 | 2,300 |
| **High cost** | Above + PV supply for lights and power, rain water harvesting, heat pumps (GSHP) and underfloor heating, heat capture ventilation | 47,000 | 255 +265 + 150 + 44 +75** = | 5,800 |

Notes: *Best energy rating.
**An attempt has been made to not double count savings, however, the cost may also be reduced because of, say, either UF heating or GSHP.

capital outlay and the high cost is the optimisation of measures in terms of reducing energy, i.e. equivalent to a low energy new house.

The costs above are related to a three-bed semi-detached house, but capital costs and savings will change depending upon the style of house. It is obvious from Table 8.1 that the higher cost of a full low energy house is a much longer payback period than the lower and medium cost measures and makes less economic sense when trying to convince users. The solar heating is only effective on south-facing roofs at a 40° slope. However, individual measures such as double glazing in the medium cost can take as long as 55 years payback, but are also popular in temperate climes because of additional benefits such as sound reduction and condensation control. The equation then between doing it because it pays and doing it because it is ethical and sustainable is extremely complex. There is a conscience premium that will be paid and this is dependent on funds and on the resale value of the house. Not many people will only do it to reduce climate change so a sales pitch or government incentives will need to be developed at least until there is a critical mass. Sustainability may also create its own social pressure.

The Blue Angel Scheme in Germany is a simple long-standing eco-labelling scheme that has worked well promoting information about products. Compliance is voluntary, but it also gives kudos to the manufacturer. A culture of sustainability needs to build up so that those who benefit in the short term by not complying become pariahs.

### Impact of construction on the neighbourhood

The impact of construction work on the neighbourhood is obvious and is an inconvenience to be minimised but must be tolerated for the sake of change, but environmentally most would agree there is a deeper issue as to the enduring impact which comes out of the process as well as the use of buildings. These relate to the introduction of hazardous building materials, pollution of the air, soil or water courses, the generation of dust or other unhealthy emissions which inflame or create new respiratory diseases and the increased danger to the public of the building works themselves such as being hit by falling objects or the risk of explosion or collapsing temporary structures. Is the standard absolute or are there certain guiding conditions depending upon the situation? Clearly there are different levels of tolerance by neighbours, some are resigned and some protest. They blame different causes. What is the ethical position of a contractor – to spend money, time or effort to reduce nuisance and inconvenience and to what level is it considered considerate?

In construction, noise is a temporary inconvenience but can have health issues if it is prolonged for sensitive members of the community. If very loud or of a certain frequency it may even set up vibrations in a building which cause cracking. What is the ethical level of noise pollution that is acceptable over short periods of time for the neighbourhood? Should contractors bring in acoustic screens around all sites that are within certain distances of occupied premises during the ground works or other noisy activities? Is compensation for inconvenience more ethical?

The CIOB and the Strategic Forum for Construction have both run campaigns for the improvement of site conditions (CIOB 2003)[37] and there is a belief that the old way of a 'pig in muck' approach is no longer acceptable for the image of workers and the need to achieve better targets as a considerate contractor in the community. In order to change, they believe that a programme of education needs to be sustained and for sites to be shamed and named by reporting them to the CIOB. A change of attitude is best obtained by the inclusive ownership of the target to 'clean up' the site by the workforce. A considerate culture can be brought about by the reciprocal provision of better welfare and a clean kit to show employer commitment.

Do the community blame nuisance on the contractor or the commissioner of the works (the client) for putting the building there in the first place? An extension is blamed on the neighbour or the planning authority for allowing it. An accident or a large number of lorries down the street or working out of normal working hours is blamed on the contractor. The building of new warehouses overlooking your house is blamed on the greed of the developer.

## The environmental management ethic

ISO 14001 or EMAS (Environmental Management and Assessment System) is an international environmental management standard that has been developed to give a set of tools for businesses to manage their environmental impact and to implement specific procedures that can be audited by a third party. It is becoming important to have these systems if a company wishes to compete globally. Many clients who already have EMS to guide and monitor their own business are also keen that their designers and contractors should implement environmental management in the projects that they commission. It is possible to do this on an 'ad hoc' project basis or to adapt formal company-implemented systems that have committed the organisation to an holistic auditing system. It is much more likely to get top management commitment if the company as a whole is committed. In construction there is a particular problem of waste generation and measures are being introduced to ensure that all large projects have a site waste management system in place to show commitment to recycling materials and cutting down the amount of waste that is transported to landfill. Because of this, more construction waste recycling collections are available for segregated waste.

## Measuring sustainability

One of the key ethical aspects for sustainability is the credibility of what people say they are able to achieve. This requires some trust as well as having a measurable system that can be checked.

## Environmental

An environment management system has a duty to measure outputs so that improvements can be validated and it is this aspect which gives the EMS potential to prove an ethical and transparent approach. The BREEAM assessment also gives a tool to clients to set a tangible level of sustainability though it has been criticised in that the predicted level and the actual level of performance are quite different because the way the building is actually used is not factored into the equation (NAO 2007).[38] This calls for a more whole life approach to management as well as to predicted costs. This failure to reach expected standards of sustainability could be disastrous for the credibility of schemes if there is no user education. It also begs the question as to who monitors this and whether the monitors have the required neutrality. The voluntary aspect of environmental improvement done in the spirit of improvement and not just marketing and compliance is another aspect of an ethical approach so that trust and reliability can be built up in the figures.

There are also KPIs produced by Constructing Excellence as a basis and motivation to match industry-wide average performance and move beyond to get market leadership. These emerge from strategies such as the DETR (1996)[39] guidance to map the level of environmental practice against a grid of criteria. BRE promote knowledge exchange in the environmental best practice programme by encouraging sharing between organisations at workshop level through the CIEC network. Chen and Li (2006)[40] would argue strongly for a quantitative integrative approach to measure pollution and environmental hazards so as to incentivise waste reduction.

## Social measurement

Designers might also consider the importance of a more organised assessment of the quality of their own design through a rating system such as CABE Creating Excellence design awards[41] and predicted BREEAM assessments for the design. The CABE system allows self-assessment of the effect of different design criteria and can be used on a 'what if?' basis.

The findings of the SusCon team working on a London study (Cooper 2006)[42] argue that there is a sensitive balance between the technical compliance with sustainable construction and the social and business issues which make the development viable. There is a feeling that sustainable construction should be a broader concern to take into account the social issues of sustainability and to integrate better with urban design and the planning system. The fragmented emphasis by the DTI for economic efficiency and the DCLG for environmental requirements seem to squeeze out some of the social issues connected with quality of life. Cooper (2006) also reports that extra sustainability, as measured by a formal ecological footprint approach (REAP),[43] has been maximised at a 38 per cent reduction using the London Mayor's code.[44] Reductions were mainly achieved from recycling household waste and the introduction of CHP. Neither of these were the responsibility of sustainable construction and they questioned the heavy dependence on building codes for achieving government targets when it was more to do with changing household attitudes. There is a case that the impact of sustainable construction, e.g. a good BREEAM rating should be interpreted by reference to subsequent lifestyle changes. For example, the supply of coloured recycling bins is not a guarantee that owners and tenants will use them.

### Socio-economic measurement

Pearce's (2003)[45] approach to a socio-economic rationale is equally convincing and backs up the arguments with statistical data which focus the main problem areas for increasing a sustainable construction effort. Pearce argues that measures of achievement are confusing and need to be defined properly.[46] The firm size distribution indicates that 0.5 per cent of companies employ 1,200 workers or more and they manage 12 per cent of the output. These are the easiest to reach. There is a big skew towards small companies (90 per cent under 7 employees) and these 90 per cent carry out 30 per cent of the work and are the hardest to reach. There is a roughly equal balance between the value of new build and refurbishment type work, but smaller companies are more involved with refurbishment and domestic work. Value added (GDP) in the UK compared with other European countries and the USA is high except for the Dutch per capita GDP value.

The implication for a sustainable transformation according to Pearce (2003: 23)[47] is that the fragmented nature of the industry is not conducive to efficiency and the introduction of sustainable construction with lots of small companies is initially going to be unaffordable without subsidy. There are benefits for small firms who have better customer relationships and the industry has a favourable comparative level of productivity with other construction industries. As buildings last a long time, the sustainability factor of running costs is more critical. The annual rebuild rate in the UK is 0.83 per cent which means that new standards trickle in comparatively slowly, this is up to a third less than most European countries and the USA, so upgrades of existing stock are relatively more important overall. Longevity of stock in itself is not as sustainable as new technology. This analysis identifies the slow process of making buildings sustainable.

### Ethics

Ethically, the implementation of a more environmentally sensitive regime is a proactive, duty-based approach to the better management of an area that has not been considered acute. Others would argue that an environmental ethic has simply been ignored on a mass scale before because the rationale has not been so pressing as today when climate change has created a rallying point for the wider rather than specialist focus groups. They would argue that there is a natural environmental ethic about living in harmony with the world which has not always been recognised in the development of the built environment.

There is an absolute duty to maintain the health and safety of other people and it is not connected to economics in quite the same way due to the stigma associated with negligent concern for human well-being. This ethic is again based on duty and risk management, but is a moral commitment. The two ethics are differentiated on the basis of the degree of immediacy of harm, but the ultimate outcomes of both are still mainly argued because of their effect on the future quality of life. I would argue though that the difference is still fundamental as it is the difference between a proactive and reactive approach and a different way of connecting time scales and priorities. Health and safety have an urgent timescale. This is the difference between the health and safety and the environmental ethic.

The ethical challenge for sustainable construction is to use the planning system or other mechanism as a democratic and equitable way of ensuring a degree of improving sustainability for new and formally renovated buildings. The above provisions by themselves do not do the following:

- ensure a change in lifestyle to achieve sustainable objectives;
- tackle the 90 per cent of existing stock that is not improved;
- ensure a take-up which, because of the associated extra capital cost, may actually slow down the process of development temporarily;
- get commitment from landlords and businesses owning existing property. They need to lift their performance and invest in making their properties sustainable whilst at the same time not over-pricing the market.

The ethical challenge is not just complying with codes, but changing culture so that conscientious developers don't suffer at the hands of less conscientious ones and existing stock does not drag down achievements.

## Conclusion

This chapter has studied the various impacts of sustainability and has shown that there are a plethora of initiatives ranging from the UN-led summits in Rio and Kyoto down through the EU and to national guidelines, to reduce emissions and to deal with climate change. Buildings are important to the sustainability because they are responsible for between 33–40 per cent of carbon dioxide emissions depending upon which sources you look at. Since 2000, the construction industry in the UK has had to absorb a range of design constraints to reduce the emissions, which have moved from a voluntary to a legal compliance status because of the perceived urgency. This has not pleased the industry which has now got a lot of catching up to do. Some clients, however, have realised early the competitive edge that can be gained from a whole-hearted approach to beat the targets and build reputation. This is sold to their shareholders as a business rather than an ethical proposition. The recognition of a moral obligation is becoming so strong that clients, constructors and designers need to show their sustainability credentials prior to the output which is expected to comply with strict standards. This virtue ethic for the good of the community is attractive because it enhances the reputation of the client and the designers who are able to gain kudos for further commissions. Some of the targets for excellence, e.g. BREEAM, do not match the implementation even for public buildings because they depend on the way the building is used. For the contractor there is a Kantian duty ethic to work with the client and to maximise the implementation.

Other social aspects of sustainability are important and are mostly controlled through the planning approval system and the use of Section 106 monies by the local authorities. These social aspects are being measured through the use of 'quality of life' indicators and other KPIs are being used to measure environmental aspects such as energy reduction, waste, etc. There is a question as to whether planners and councillors have the right qualifications to make complex decisions on sustainability in order to approve and advise alternative planning applications. This also raises an ethical issue as to the equality of the approval system in that many developers automatically appeal, giving them more clout through their knowledge and size than those with smaller projects.

Environment and waste management systems help, but are no guarantee of the stated compliance as design solutions lack the innovative technology, and measurement systems fail to measure key aspects of sustainability such as building use. This has put the spotlight on new technologies and on the claims of the measurement systems. On a global scale the government targets to reduce emissions have not been met as they only target new build which is 10 per cent of stock. Also 30 per cent of construction work is carried out by small contractors employing less than seven people and there is a real danger of the message not reaching a fragmented industry with a stubborn cap on achievement. This is leading to a serious crisis of trust.

In order to achieve new targets, manufacturers need to innovate with new technology which has a premium cost and designers have a steep learning curve to understand performance and advise and convince clients. This technology needs to establish itself so that prices can be reduced to create better value. Contractors need to be serious in waste reduction in order to maintain value for their clients. The ethical dimension to respond to ensure future survival is very much led by compliance to regulation and more innovation needs to be led by the private sector and sold to consumers at an affordable price.

## Notes

1 HRH Prince Charles (2000) 'A Royal View', Highgrove Roundtable, Reith Lecture. BBC, 17 May, Highgrove, Gloucestershire.
2 Girardet, H. (2000) 'Greening Urban Society', in W. Fox (ed.) *Ethics and the Built Environment*. London: Routledge.
3 Lee, K. (2000) 'The Taj Mahal and the Spider's Web', in Fox, op. cit.
4 Brundtland Report (1987) 'Chairman's Introduction', in *Our Common Future*, World Commission on Environment and Development. UN.
5 UN Department for Economic and Social Affairs, Division of Sustainable Development. http://www.un.org/esa/sustdev/ (accessed 9 July 2007).
6 Inter-governmental Panel on Climate Change (2007) 4th Report, February. Online. Available http://www.ipcc.ch/
7 Godfrey, J. (2006) 'Systems Thinking Approach to Sustainable Innovation', in *Proceedings of CCIM Conference, Maastricht University*. Maastricht.
8 Dale, J. (2007) *The Green Perspective: A UK Construction Industry Survey on Sustainability*. Ascot: Chartered Institute of Building.
9 Day, C. (2000) 'Ethical Building in the Everyday Environment: A Multi Layer Approach to Building and Place Design', in Fox, op. cit.
10 Dalton, A. (1998) *Safety, Health and Environmental Hazards in the Workplace*. London: Cassell.
11 'Achieving Sustainable Communities – the ZED Challenge'. Online, http://www.zedstandards.com/
12 Hewitt, A. (2005) 'Presentation to the Welsh Assembly', 7 December 2005.
13 Inter-governmental Panel on Climate Change (2007). 4th Report AR Summary for Policy Makers, February. Online http://www.ipcc.ch/pdf/assessment-report/ar4/syr/ar4_syr_spm.pdf
14 Guy, S. and Farmer, G. (2000) 'Contested Constructions', in Fox, op. cit.
15 Day, op. cit.
16 Stern Report (2006) *The Economics of Climate Change*. HM Treasury. Online: http://www.hm-treasury.gov.uk/independent_reviews/stern_review_economics_climate_change/stern_review_report.cfm
17 DEFRA (1999) 'FCDPAG3 Flood and Coastal Defence Project Appraisal Guidance: Economic Appraisal'. London: DEFRA, p. 43.
18 If China has an emissions output that is 10 times bigger than the UK's and they want to reduce by 20 per cent instead of 25 per cent, then the UK will have to reduce its emissions by $10 \times 5$ per cent = 50 per cent more than 25 per cent of its own emissions. Clearly this is a collaborative process with all major players doing their bit and more to take up some slack for developing countries.
19 Stern Report (2006), op. cit. Figure 2, Stabilisation levels and probability ranges for temperature increases (p. v).
20 DCLG (2006) 'Code for Sustainable Homes: A Step Change in Sustainable Home Building Practice', 13 Dec. Available: http://www.communities.gov.uk/publications/planningandbuilding/codesustainable
21 Department of Communities and Local Government (2006) *Planning Policy Statement 3: Housing*. London: DCLG.
22 Friends of the Earth (2006) *Delivering Sustainable Housing, MPs' Briefing*. January. Online. http://www.foe.co.uk/resource/briefings/sustainable_housing_mps.pdf They estimated that building 2.8m houses by 2016 would put 142.9m tonnes of carbon into the atmosphere with an extra 12m tonnes per year in use (96m tonnes).
23 City of Vancouver (2005) 'Definition of Social Sustainability', in *Policy Report Social Development*. Policy Report Meeting, 24 May.

24  Whitelegg, J. (2000) 'Building Ethics into the Built Environment', in Fox, op. cit.
25  Girardet, op. cit., p. 22.
26  Adamson, D. (2007) 'Getting the Best Value out of Construction and Building Maintenance', paper presented to Conference on Best Value Construction, UWE, 20 August.
27  Constructing Excellence (2007) *Sustainability in Constructing Excellence*, Joint UK-Sweden Initiative on Sustainable Construction.
28  DEFRA (2007) *UK Government Sustainable Procurement Action Plan*. On behalf of UK Government.
29  National Audit Office (2007) *Building for the Future: Sustainable Construction and Refurbishment of the Government Estate*. London: The Stationery Office.
30  EC (2004) *Towards an EU Thematic Strategy on Urban Environment*, European Commission. COM(2004)60. Adopted on 1 Jan 2006, it sets out aims in four areas of sustainable construction, sustainable design, sustainable transport and urban management.
31  Dale, op. cit. This was an online survey of 850 construction professionals, 80 per cent CIOB members.
32  DTI (2000) *Building a Better Quality of Life: A Strategy for More Sustainable Construction*. London: DTI.
33  Egan, J. (1998) *Rethinking Construction*. London: Department of the Environment and the Regions (DETR).
34  ISO BS EN 14001, *Environmental Management*. International Standards Organisation.
35  Arup (2005) 'Dongtan Eco City'. Arup Online. Available: http://www.arup.com/east-asia/project.cfm?pageid=7047
36  Lemon, A. and Keech, T. (2007) 'What Price a Sustainable Home?' *Construction Manager*, 16 October. CIOB. Online http://www.construction-manager.co.uk/story.asp?storyType=143&sectioncode=12&storyCode=3096557
37  Herridge, J. (2003) 'Improving Site Conditions: The Construction Manager's Perspective', Findings of a Workshop 5 March, CIOB, Ascot.
38  National Audit Office, op. cit.
39  DETR (1996) *A Strategic Approach to Energy and Environmental Management: Energy and Efficiency Best Practice Programme*. London: DETR.
40  Chen, Z. and Li, H. (2006) *Environmental Management in Construction: A Quantitative Approach*. Abingdon: Taylor & Francis.
41  CABE (2007) 'Public Building Leads Way on Sustainable Design'. Online http://www.cabe.org.uk/default.aspx?contentitemid=1958&field=filter&term=Education&type=8
42  Cooper, I. (2006) *Sustainable Construction and Planning: The Policy Agenda*. London: The LSE Centre for Environmental Policy and Guidance.
43  Stockholm Environment Institute Resource, *Energy and Analysis Programme*, cited in Cooper, op. cit.
44  *London Mayor's Draft Code for London*. London: GLA.
45  Pearce, D. (2003) *The Social and Economic Value of Construction. The Construction Industry's Contribution to Sustainable Development*. London: nCRISP.
46  The traditional narrow GDP value (aggregate of net outputs) in the UK construction industry is 5.4 per cent. A broader definition would include the construction activities of supply, and design, and the DIY and informal construction sector add up to as much as 10 per cent of GDP and give a truer impact of the poorer performance of this sector.
47  Pearce, op. cit.

# Chapter 9

# Trust and relationships

Trust is tied up with ethics in a fundamental way. The operation of ethics concerns behaviour that is beyond compliance. Trust means that there is a move away from action guided by fear of penalty or punishment to action which is guided by belief and understanding, so there is an implicit connection with ethical behaviour. Ethical behaviour depends upon a reciprocal behaviour which recognises the integrity of 'keeping your word' (one area of trust), reliability and loyalty. In our context of the built environment, there are many situations not tied by contract and many situations which, when they are, are not successful in obtaining the outcomes which both sides had envisaged. Trust remains critical even where a contract exists, although a contract may help to initiate actions which are understood to both sides. Situations such as submitting a planning application or signing a contract also necessitate a relationship to be developed which requires the building up of trust.

This chapter will primarily relate to project work to give a context as this is very common in the built environment development cycle. Trust also relates to companies, for example, dealing with shareholders and other stakeholders, and accountancy methods are one area 'under the spotlight'. Trust broken is reputation lost.

## Contracts and trust

A contract formally exists when: (1) there is a mutual objective, e.g. to provide a building; (2) there is a reward in return for providing the service; and (3) there is an agreement between two or more parties to meet the objective. It is normal for a contract to be written down on paper, with a signature or seal, which can be presented in a court of law in the event that the terms of the contract are breached. A set of conditions may accompany the contract. However, a contract does not have to be written down for that contract to be valid in a court of law, or in ensuring that a normal obligation is carried out. Conversely, no amount of careful drafting of terms is likely to be sufficient to stop a party intent on violating their intended obligations. There is, of course, a heightened risk in this opportunistic behaviour. The point is that some degree of minimum trust is needed to be able to establish a basis of a relationship which will carry out useful work. Both contract and trust play their part. The role of trust is to build up a relationship where a level of expectation is more certain. It is clear that the level of trust can be on a sliding scale according to the cultural expectations of the two parties. It is also clear that there is plenty of opportunity for those expectations to differ or clash so that the more trusting party is more vulnerable to the other party, but there is still a case to examine for the role of trust.

## Definition

The *Oxford English Dictionary*[1] describes trust as 'Firm belief that a person or a thing may be relied upon; feel sure of the loyalty of; treat accordingly; accept as true without testing'. It can also refer to confidence, conviction and dependence.

In the light of the dictionary definition, we can see that some of the key concepts of trust have been identified.

* *Firm belief* gives a reasonable level of confidence that will guide a party to take an action in support of the belief and acceptance of something done or said as true. It may be qualified of course! The level of belief will establish the perceived degree of risk that the truster has. Giving trust without belief would be an irresponsible action.
* *Reliance* is the quality that a party will carry out what they said they would do. It can be an absolute personality or company trait, but it can also be a reciprocal reaction to another's trust and reputation. Relying upon a party is the basis of the 'gentleman's agreement'. It is a choice of who you work with as well as how much you monitor their performance. For example, reliance on a subcontractor is often based on past experience and is a mitigation of risk in that work not performed to standard is less likely. Again, being able to rely upon someone is an essential element in developing business relationships and in choosing a partner to work with in, say, the procurement process (for both sides).
* *Loyalty* is the attachment of supportive behaviour to a person, organisation or a product and is helpful because it helps to predict future behaviour. Loyalty may of course be misplaced or given only for a while. It is recognised in business as giving competitive advantage, because it is a quality that can bring back future custom, but can also help to retain employees who may otherwise take their company-learned skills and experience elsewhere. At the heart of loyalty is a long-term relationship.
* *Treat accordingly* is a way of reciprocating action at a similar level of trust. Mutual understanding of the other party's intentions is essential to trust. The perception of the other's actions may also be linked to leading indicators of behaviour that will trigger off actions to close down vulnerability in a relationship. This type of language appears in the collaborative and

partnering contracts where the will to build relationships 'i.e. good faith' is a point which may also be examined in a court of law (but see also Chapter 11).

- *Accept as true without testing* means that the relationship has reached the point where a statement or a promise is accepted at its face value. There is an increased vulnerability of the truster in trusting someone else and the trustee in accepting the responsibility, which, if it works, will give opportunities for better productivity as less monitoring of work is required.

### Development of trust

Lewicki and Bunker (1996)[2] classify risk at three levels of development and imply that there is a movement from the base level to the top level as trust develops. They term them calculus, knowledge and identification based trust (CBT, KBT and IBT).

CBT is trust based on calculation and exists in a relationship where the two parties try to ensure consistency of behaviour. They will trust the other because there are deterrents which are sufficiently clear, possible and likely to occur. A contract with recourse to remedial measures, liquidated damages, and retention monies are examples of deterrents. They also acknowledge that trust of this type can be based on reputation and other mutual benefits which emerge from trusting each other, such as incentives.

KBT is based on better predictability of the other party and a deeper knowledge that has been built up, generally in a special effort to collaborate or in a special relationship such as partnering. In this sense, KBT relies on knowledge rather than deterrence. Parties may have a history of working together or a general expectancy is built up that the other party is predictable. KBT is primarily based on information about the other parties or about the situation. It could be familiarity and past good experiences or a more formal charter committing each other to act in good faith. In the definition above, the last four attributes are more in play in this type of trust. This predictability does not mean the other party is to be perfectly trusted, but that their weaknesses and strengths are known so that partners can work together and the level of trust can be judged. A party that is known to always keep to their promises is likely to become a key partner. One that is good on quality, but is one or two days late can be treated with a sort of qualified trust with less control.

Partnering is a special case for organisational trust as it is likely to be connected with longer-term organisational relationships. Some of the new collaborative contracts use a project risk profile earlier on in the contract to build up trust. Alternatively or simultaneously, parties develop relationships with the use of value management workshops which provide opportunities for getting to know each other's values. It is easier to set up teams where KBT exists or is engendered, because teams work best where people trust each other and need to depend on outputs. Team building exercises in their turn increase knowledge-based trust among individuals and help to feed back to develop the top management commitment of organisations to deeper collaborative activity. Bennett and Jayes (1998)[3] allude to this in the move from first to second generation partnering they describe.

The third level of trust, IBT, is based on people sharing common values. It requires a meeting of minds and a common concept of moral obligation (Child and Faulkner 1998).[4] In most business relationships, there are some differences of interest. To develop IBT, there would need to be a merging of the various organisations' interests and objectives. In doing this, there needs to be a common statement of values. Some values are cultural, some are archetypal, i.e. they are either locally or universally accepted. Archetypal values relate to the sanctity of life, the need for survival and the need to protect one's own kin. They refer much less to a universal business approach. Some cultures will emerge from common codes of company ethics or from a project. Company codes are not always explicit or written down, but implicitly there are expectations and a sense of the 'way we do things round here'. Public codes emerge from expectations in a society in general, but they may still be cultural. Increasingly, there has been a need to work on a more universal code to deal with international relationships and projects, for example, the UN Convention Against Corruption (UNCAC).[5]

To bring a merging of identity in a project requires common values and objectives. In construction projects, stakeholders are likely to have some mutual objectives, but the business needs of a client and their external supply chain are different. There is room on the part of the project team for an understanding of the client's business better and an attempt to read their requirements from their point of view. This is becoming more common as long-term relationships develop through partnering and framework arrangements, but it is still well below the level of a merged identity (Fewings and Hodgson 2005).[6] At the development stage, the interactions between the developer and the community are much more remote and it would be hard to imagine that consultation with the broader and sometimes conflicting aims of the community will be developed beyond the second level of trust.

At the design stage, there is a certain attempt by the designer to visualise the client's business in the building, but there are many cases where the client simply wants to specify functionality and would prefer to leave the business of asset development in the hands of professionals. This rules out IBT because mutual objectives and culture are not shared.

At the construction stage, we have a history of adversarialism and the 'them and us' mentality strikes and it is difficult to speak the same language, let alone merge the identities of the designer, contractor and client, because of a broad difference in the educational background between built environment professions. There is also a divide between commercial and professional objectives, but common ground can be found where the client's interests become a binding

force. It may be better to distinguish this level of trust in the built environment by working *towards* a common identity. Framework suppliers may co-locate within the client's premises so that they integrate with client systems, or in FM contracts seconded teams may work in the client's premises as part of the team and come into contact with the users.

The areas of IBT which might be merged are open communications and meaningful consultation, a project ethical code or integrity pact which helps to link values, the broader use of value management meetings throughout the development process, and the formation of an integrated team composed of all parties.

## The case for trust in construction

The Engineering and Physical Sciences Research Council (EPSRC)[7] Trust in Construction project also summarised six areas which consistently emerge as relevant to construction participants in all areas of the project:

- *Honest communications* – can they be trusted?
- *Reliance* – what do you do when you trust someone?
- *Outcomes* – what happens when you trust?
- *Building trust* – trust in relationships.
- *Levels of trust* – different understanding of trust.
- *Reputation* – trusting people and organisations.

Interestingly, these six points have strong overlap with the issues dealt with in the definition above. The importance of reliance – doing what you say you would, depending on the outcomes and able to develop good communications that could be depended on as being clear and full – is particularly corroborated. The other three points are in terms of building up trust and reputation.

The authors report that trust is seen as vital to working in construction projects and they come up with some benefits when working in trusting relationships (Swan *et al.* 2002: 1).[8] The benefits are the ability to be more flexible, the overall reputation for trusting site staff, the motivation of financial gain through building trusting relationships, the ability to respond flexibly to change, and how trust can reduce risk and uncertainty. They believed that money was saved in trusting by:

- reducing conflict;
- reducing risk and uncertainty;
- removing poor communications.

The areas of trust that were highlighted varied depending upon the position of the respondent within the organisation.

## Team and trust building

There are four pillars of team work described by Lipman-Bluman (1996) and Owen (1996) as described in Gardiner (2005).[9] These create the right environment, create effective communication channels, create strong personal and interpersonal values and develop flexible leadership. These are quite closely related to the concepts of trust discussed above in terms of believing in others, reliance, loyalty, accepting as true and acting accordingly.

In a team situation there is particularly a need for trust and support, according to Lipman-Bluman (1996).[10] He believes getting to trust and support others in the team develops as we learn to trust ourselves and build up confidence in our own role. He also believes that total trust and support lead to commitment and open communications. Open communications refer not only to the sharing of knowledge, but also to constructive confrontation and is all part of creating a conducive environment (Pillar One). As part of this, the leader needs to set a prime example in trust and support which others may follow.

The EPSRC (2002) project identified the importance of building trust with people in teams, and at an operational level this had to be done quickly. This was often done by team problem-solving exercises that resulted in mutual satisfaction. At a strategic level, it was about solving problems without resorting to contract. Team building was also seen as a way of creating shared goals and mutual understanding which led to building trust. The conscious building up of relationships made trust in the teams a reality. Trust building occurred when reciprocal trust was displayed and tolerance (i.e. forgiving of minor non-trust actions) could be built up as parties allowed themselves to get to know each other better over a period of time.

Another relationship issue was reasonable behaviour which was not necessarily seen as friendly or non-confrontational, but meant 'pulling one's weight', being 'easy going' and 'professional'. The reasons given for breakdown of trust were mainly to do with closed communication when things went wrong, not allowing fair representation and creating an atmosphere of blame. All underline how easily and quickly trust can be betrayed and broken to disastrous effect, because broken relationships can take even longer to build up again. In projects, such broken trust can easily spread to other projects even where the personnel are different.

## Trust in delegated relationships

Handy describes trust in the context of leadership as 'consistency and integrity, the feeling that a person and an organisation can be relied upon to do what they say, come what may' (Handy 1993).[11] He précises this as 'courageous patience' because it is also a risk to trust someone else to take on something that you are accountable for and it really matters if it does not go well. In the context of delegation, Handy comes up with the notion that trust is cheap. This refers to the trust control relationship, where:

trust + control = constant.

Any wish by the manager to give their subordinates more responsibility then means a loss of control which most managers find hard to do at first. Control is expensive because it means that something is done twice – it is done by the subordinate and then checked by their superior. Where trust is exercised, there is less checking and repetitive work. Of course it is comfortable for a supervisor to check the work of their subordinates and it may be necessary for them to do so for a period while the work is new to them, but it is not efficient or motivating to be constantly checked. The supervisor has built waste into the system and the subordinate has learnt to be *less* responsible by relying on the final check they know will take place. They are likely, when demotivated, to push deadlines to the limit, reduce standards and do as little as possible.

### Case study 9.1 Delegating and trust

A manager delegates the task of taking and writing the minutes of a meeting to a junior member of staff. Scenario one is where a manager requires to see the minutes, reads them and then makes extensive revisions to then be resubmitted for checking before circulation. This will barely save any time and will induce demotivation by the third meeting. The manager is using the junior as a secretary only and a trainee needs room to progress. A good manager could jot down action points during the meeting and instruct the minute taker to put action points in a list in front of the minutes. The first time he will read through the whole minutes to check the style and accuracy. After that he will then be able to glance at the action list to check nothing is missed and consequently either praise the subordinate for their work, or discipline them.

Handy's theory recognises the two-sided nature of trust and the transfer of the responsibility for learning the job as well as doing a task (Figure 9.1). If ability and trust are to built up, the manager will able to distribute the minutes believing there is no need for checking and passing the responsibility to the subordinate to put things right if there is a mistake. This will induce care and maturity in the subordinate and develop their confidence and role in the eyes of others. Trust has increased and control has reduced. The final accountability for quality and effectiveness remains with the manager, but they have freed up time for more proactive activity when they reduce the control elements of their role. Delegation is likely to be good for building trust in the team and develops when members of the team are given specific and demanding responsibilities and

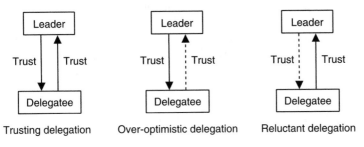

*Figure 9.1* Individual delegation and trust.

accountability. Trust needs to be accepted as well as given. There is also an ethical aspect to delegation in giving respect to a person's abilities and providing opportunities for career development.

Long acquaintance is often the basis for giving trust, but this is not always possible in project teams, which are formed for relatively short periods with new members who have not worked with each other. Each team is unique even if some have worked together, as the task and mix and context can be different. Trust is a greater challenge and depends more on the leader's ability to bring diverse groups and objectives together. Handy mentions four characteristics of trust in effective delegation:

- The superior has to have confidence in the person and this is more difficult with unknown people, especially if they are not in the same organisation.
- Trust is initiated from the leader, so must be explicitly given and withdrawn if unsuccessful.
- Trust can more easily broken by the receiving party and yet trust should not be withdrawn spuriously and it is better to define circumstances for withdrawing trust so that the position does not turn unmanageable or cause expensive misunderstandings.
- Trust must be reciprocal so that the trustee is also committed to trust the leader.

These are basic rules and they are made easier by the two parties having some choice in their working together. A forced position, where either side would have a vote of no confidence in the other, makes trust building difficult and the result is reduced productivity. He also suggests that it works best if control is connected to results and not to means and Handy connects it to management by objectives (MBO). This is a challenge to the manager who has recently been promoted and has strong views on how the job should be done. The culture of innovation can thrive in this situation.

MBO requires a programme for action by the subordinate to the targets given by the manager and agreed by the subordinate. Gardiner (2005)[12] calls this type of style the entrusting leadership style where the leader has complete faith that their team will not let them down.

All of these qualities have a role in building up relationships and building up trust. Subtly, trust can be perceived to be lowered by MBO-type delegation if it is used as a means of increasing control, enforcing targets or in a system for applying performance payments. All these indicate an underlying hidden agenda to collect evidence that the subordinate has not been made party to and represents a lack of transparency. Confidentiality is a key aspect of the personal appraisal meeting also. Agreed targets are the main public outcomes, together with agreed training programmes for development.

## Project trust

The EPSRC project (2002) has shown that larger and more complex projects are likely to need trust more, because a breakdown in communications is much more likely and greater sums of money are at risk. The Wembley Stadium project indicates the way in which trust relationships can break down. Wood and McDermott (1998)[13] found that there were five emerging behaviours in their pilot study where they wanted to test for effectiveness in building up trust. These were:

- sacrificing behaviour;
- problem solving;
- the establishment of a reputation;
- the development of social relationships to encourage open behaviour;
- the implied desire for longer-term relationships.

The need to develop behaviours to build up trust is a step of faith and these acts of good faith are shown through sacrificing behaviour by giving something away on the basis that something else will be gained later.

## Inter-organisational trust

In construction projects, members of the team usually come from a range of organisations that are trusted to commit resources to the project. Individuals do not necessarily have senior management support for the promises they make.

Individuals differ in their values and so the building up of inter-*personal* trust is the prime area of contact, but for decisions to be made that affect use of an individual's organisational resources, they need to have appropriate authority and senior management commitment in order for others to trust their ability to deliver. Individual respect and trust are therefore dependent on inter-organisational trust and multiple links between organisations for different levels of decision-making. The organisational structure, communications and hierarchical accountability need to be clear. In construction, there are many different organisations and there is a complex web of supply chain expectations. It only needs one organisation to go out of line for trust to be compromised. The first line supplier's reputation is dependent on the supporting chain of

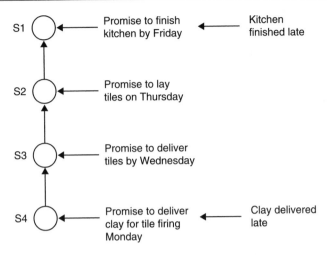

*Figure 9.2* Supply chain trust.

promises below. In Figure 9.2, you will see that one promise broken affects the whole chain.

If an architect asks for a change under the JCT contract, they will need to know that the client will pay for it. If an architect then changes their mind and this has, for instance, an adverse financial or disruptive impact, they will have a professional liability. If the client changes their mind, the client will hold an ethical liability for disruption, cost or both. Inter-organisational trust needs a framework of values and responsibilities that engenders open communications. A project charter is an example of a strategic framework document that gives scope for the project in the light of the requirements of each of the main stakeholders. In addition, it could be usefully used to assign responsibilities and authority.

The EPSRC (2002) report[14] only identified two company factors that were important for trust. These were the cultural reputation and experience of working with the company in the past and the financial position of that company, which sometimes forced it into under-bidding and being under-resourced.

### Partnering

Partnering is a particular situation where organisations come together with a specific intent to collaborate trustfully with each other. A simple open-ended charter is proffered and signed by participating parties in the primary supply chain, providing a framework for the trust. In some cases, a partnering contract is used which makes a commitment to working together and deals with issues such as value management, assessing risk early, and managing risk and ordering provisions for change management. Gardiner (2005)[15] sees this as 'fostering a mutual independence' and trust that tries to create more open communications and institute a 'no blame' culture, where people feel safer to make suggestions and mistakes are learnt from rather than demonised. In this context, the higher level commitment of the organisation can give authority for a non-traditional view in construction projects and should ensure the training and resources to do so.

Many organisations cite 'use of an open culture' as a marketing tool to get on to partnering tender lists and secure future work with trusted clients. In order for an open culture to work, and avoid conflict, operational staff need to have training, as they may be unfamiliar with the expectations of a no blame and open communication culture. The responsibility for creating a 'no blame' culture is also firmly with the project manager who will also provide a robust example to follow. Resources should be made available up front to ensure savings down the line.

The EPSRC project (2002) noted that partnering contracts were helpful to trust but not exclusively so and said that in situations where ongoing client time and cost savings were becoming more difficult to achieve and positions where contractors had 'bought work', there was additional pressure on maintaining trust in the contract. Interestingly, the issue of open book accounting often used in partnering was considered by some of the respondents to help, but was not really about trust. They claimed that trusting each other should be about respecting a certain amount of privacy. Theoretically, this argument ties in with reducing control and increasing trust by the client towards the contractor, but being prepared to open books on the contractor's side is part of the open communications demonstrating trust.

Bennett and Jayes (1998)[16] come up with six areas where tests for partnering can be made (Table 9.1).

Table 9.1 Tests for partnering

|  | Probably partnering | Probably not |
| --- | --- | --- |
| How and what you tell your people | Deliver what best capable of – enabling by organisation | Train personnel on a need to know basis by organisation |
| Tendering selection criteria | On value so that client meets with project team | On lowest cost |
| Client briefing | Clarified objectives and know client strategic goals and share decisions | Comply with specification and BOQ and make as much as you can |
| Site meetings | Discussion and openness | Risk avoidance |
| Standards | Continuous improvement | Minimum required |
| When things go wrong | Clear procedure to resolve quickly involving CE if necessary | Try and sort out then claim |

Source: Adapted from Bennett and Jayes (1998: 2).

It is important to note that these tests are really a test for a collaborative approach rather than a formal partnering agreement. They do, however, indicate the involvement of the client in the process and the ability of parties, especially contractors, to be involved in the earlier project stage, so certain types of procurement ease this process.

The terms of column two in Table 9.1 show an enhancement of the level of trust between the client and contractor and the consultants and the contractor and client and consultants. In Lewicki and Bunker terms, these represent a move from CBT towards KBT.

### Joint venture (JV)

Joint venture (JV) involves the coming together of separate organisations as a single entity in order to build capacity and specialism in the delivery of large projects. Examples of this would be:

- The Channel Tunnel where French and British private contractors joined, sharing a common profit level and bank account.
- Public private partnerships such as build-operate-transfer (BOT) and private finance initiative (PFI), where ownership passes to the JV for a concessionary operating period.
- Consortia come together without sharing common profits in order to manage a large capital investment, such as the Channel link rail project, for the construction period.

All forms of joint venture are the result of a lengthy period of negotiation to build up trust and ensure that each party is allocated risk they are best able to manage. Rowan (2004)[17] differentiates between integrated and non-profit-sharing JVs. Motivation for JVs could be to build size and capacity, to share different expertise, to partner with a local contractor to get a foot into the market and use their networks and 'local' knowledge. These motivations are important to the way that trust is built up. If you are going into a long-term relationship, Rowan suggests that you might start small to test the compatibility of the organisations and to see if trust can be built up.

### Case study 9.2 Prime contract

Prime contract is an example of a trust-based culture consisting of a prime contractor working on behalf of the client and gathering a team of consultants and suppliers who are able to work as a project team inclusive of client participation. The method was developed by the MOD to put in place better control over the supply chain.

The MOD Andover North site[18] was an example of this, where there was an agreement to work as a virtual joint venture with a single project account and project insurance cover. Neither of these was a contractual requirement, but they were proposed because, in a trusting atmosphere, they encouraged interdependence between the key players. The fortunes of all depended upon the trust that everyone would pull their weight in the contract and would make a good level of profit. The client also encouraged an early and equal involvement of the key players in the supply chain and this encouraged the innovative approach.

## Trustworthiness, values and ethics

It is clear at this point that the exercise of trust is an essential part of any contract between parties and the basis of co-operation, meaningful collaborative action and partnership deals. It is also the basis of the reputation that a company is able to build for itself. In construction, the old non-trusting behaviour of conflict and adversity has led to much unpleasantness and inefficiency. Clients and contractors have ended up in expensive litigation which has either benefited one side disproportionately, endangering future relationships, or has neutralised both in a principled and expensive legal battle adding value only to the legal sector.

However, the exercise of trust may not be a rational step to take for its own sake because the other partner may not agree with the essential morality underlying trust. Hollis (1998),[19] in his book *Trust without Reason*, grapples with the concept of whether trust and reason actually are in accord with each other. Trust can be read as being a code of pre-modern social conduct which is based on morals and therefore is dependent on the exercise of certain social habits which are patently being challenged by modern practice, such as the gentleman's agreement. Reason can be seen as the new enlightenment which supports the scientific view of 'how?' rather than the moralistic view of 'why?'

Hollis considers whether there is a way in which reason can underpin trust so that it can lead to mutual self-interest that can rationalise trust in modern terms. This could be similar to the egoism discussed in Chapter 1. He proposes and explores game theory as the basis of reciprocity and the common good to see how this can be harnessed to support a win–win situation which will meet ethical ends.

## Ethics or trust in relationships?

If ethical behaviour can be distinguished by the golden rule 'treat others as you would like to be treated yourself', then there are certain business behaviours like this that may put managers in a dilemma as to the right thing to do. The question of whether trust clashes with trust is most likely to be connected to relationship issues. Alex Todd (2007)[20] relates a dilemma:

### Case study 9.3  Ethics or trust?

As a front line manager, you learn from higher up that there are to be redundancies that you are sworn to secrecy about, but you also learned that one of the employees in your care is about to invest in a new house and he is also on the list for redundancy. In this case should you keep the trust and not tell the employee that they may not be in a position to afford an increased financial commitment, or would you feel that ethically you should speak to the employee in confidence knowing that the redundancies will probably get out as a rumour?

### Questions

- Should the manager do the ethical thing and tell the employee and probably put their own job in danger or at least put the company shares at risk? If the roles were reversed, they would most certainly like their manager to do the same thing for them. They would, however, have broken the trust of senior management.
- Should the manager exercise loyalty to the company and toe the company line and so attract the further trust of the company?
- Is the betrayal of trust an ethical action in the first option and, if so, is trust always a good thing?

This dilemma highlights a conflict of trust as well as an ethical dilemma. The personal situation of an employee may be justified by the company as not being the concern of the company so the manager as an employee may be considered responsible to do the best for the company.

It is probable that there would be varying actions depending upon the personality and the attitude of the manager and the historic actions of the company towards its employees or to them. So, if a manager decides to do the easy thing and just keep quiet, have they also been ethical or, as Todd would suggest, have they just been loyal to the company and asked to do something that is unethical, i.e. keep quiet about something that ethically should have been brought out into the open in the spirit of trust? Do you tell the employer you are *not* prepared to keep the news quiet and it is your duty to your fellow workers and they can sack you, but you will argue it at an industrial tribunal as unfair dismissal?

The issues here for the company are, can they reasonably trust this manager in the future or have they been unreasonable in their lack of transparency to those most affected? Are their employees more dispensable than their shareholders?

### Efficiency and trust

Efficiency does have an effect on trust given, through its link with competence. There are limitations to the application of lean principles in building up trust. Staff like to be trusted to drive efficiencies, but also a wider than a purely commercial efficiency is achieved. Benefits in building satisfaction and personal gain in the job are motivators. Trust is lost where the benefits of efficiency are not perceived to be shared fairly and the responsibility for risk (losing your job) is transferred to a weaker party without the prospect of a reward. Personal gain for shareholders and directors/partners is often seen as disproportionate to the effort that is believed to have fallen more heavily on the lower levels of employment.

### Economic theory and other possibilities

Economic theory assumes that each company will act in the best interests of their own business. For example, in the theory of supply and demand, if the demand increases and the supply stays the same, it is assumed that the supplier will increase the cost, as is the usual case in the supply of private housing. A moral action in the case of housing might be considered by some to supply at a standard profit which means that no one party gains at the expense of the other, so the price will only be increased if the material or labour price increases, and the other risks are reasonably managed. This is normally termed an altruistic action, which is defined as an action which is within the control of the giver and does not maximise benefit to them on account of the better good of others. Economic theory would not predict such action in the light of other influences such as competitive benefit or planning constraint to provide affordable housing. Economic theory is harder to interpret when such action might be for the longer-term rather than the shorter-term good of the organisation. Reputation and image are marketing terms which often use the rhetoric of trust to bring sales benefit for the organisation. They also may represent the key messages about company values that are believed to represent consistency and reliability. The visibility of connecting trust to these values is important to influence customer choice and continued custom. This is the language of mutual self-interest and represents a way of combining economic rationality and the importance of trust in maintaining relationships.

There is a certain efficiency that exists in acting morally in an attractive way, such as innovating more sustainable construction, which ensures future business. However, other organisations may benefit, in the short term at least from the more bullish actions of maximising price and minimising cost. Certainly, the concept of 'lean construction' (minimising waste) is a popular one and perhaps is another way forward in being able to combine trust and reason by offering better value by improving the product for the same price and increasing profits. Alternatively, by giving a lower price for the same product, more profits can be attractive.

### Public authority and trust

Many rule-based and benchmarking targets are seen to be missing the overall point, with too much emphasis on efficiency. Efficiency is connected to rationality but organisations which have been set up to ensure more ethical action, such as public authorities, may be seen to take a political line which is not trusted, because it is not seen to be fair or it tips the balance of power. Decisions are seen to be competent because they solve the problem rather than keep to the targets. Targets are of secondary importance if local authorities have promised more than they can deliver.

For example, a local authority may decide to insist on a certain proportion of affordable houses being built into every application for residential development as a condition of planning. This may be inefficient in cost to the developer because of reduced returns or, as is more likely, passing on a higher price for those needing to buy a new full price house. This may be seen as unfairly loading the social cost on a few. Parity may be a strong factor in acceptance (trust) here and it is clear that market forces may lead developers to build in areas where local authorities require less. These authorities could be seen to be less ethical, depending on who you are concerned for and whether you are a buyer or not. Trust could be an issue of economic necessity where the social problem of housing is more acute in some areas, and consequently affordable housing should be prescribed more in some areas than others. The community expects a rational action which is not just about how quickly or cheaply you can build houses, but about how good the houses are for the area. They are also likely to hold a selfish view that biases the strength of their judgement towards how much they are affected.

In essence, community trust is built up in public authorities with a mixture of

- transparency – knowing what the rules are;
- fairness – being able to bend the rules so that local issues are taken into account;
- efficiency – so that action taken is value for money, given that clear objectives are being met.

Community objectives may be organic and grow with awareness. Trust is based on being able to manage change efficiently.

On a business basis, there is a much greater drive for efficiency to satisfy the shareholders and to remain competitive so that they can stay in business. Trust is built up when there is the ability to build a case for a development and to negotiate a compromise to benefit the authority and the commercial objectives of the developer. A predictable and efficient decision engenders trust and helps with corporate planning. In short, they need a rational decision without ongoing interference.

Clients will be looking to optimise value for money and will buy or rent on the basis of their belief that the promises are credible and that they will be able to run the building or facility efficiently. Developers need the trust of their customers so that they can sell their developments. If it is believed that they cannot be trusted to meet the targets they have promised, then clients may lose money and miss their own targets.

Contractors will be in the same position with their clients and can expect to build up trust in meeting programme dates and providing value for money for their clients in the competitive marketplace.

## Trust in professionals

### Relationships, trust and professional standards

Relationships are affected by personal and company values. However, there is a further consideration in the development of trust through the professional membership of consultants and the individuals they employ. These are based on the commitment of the individual to common high standards of behaviour called professional codes of conduct. These are also available for the membership of organisations who improve their attractiveness by committing to similar minimum best practice. As discussed in Chapter 4, a professional who is selective about where and to whom they apply these codes of behaviour and in the way they select work, is open to the accusation that they are merely a club protecting their position and monopoly to practise. This is a serious issue of public trust, and complacency in following up complaints and censoring non-ethical behaviour will damage the trust not only of the individual, but of the whole profession. The value of these standards is that they have a hierarchy above company values. It is a good reason for organisations, especially consultancies to encourage professional membership to boost their reputation. In other cases, the professional may be employed by organisations that have considerably lower ethical standards than the institutions they are professionally committed to. This will cause an ethical conflict of interest.

Hollis (1998)[21] refers to trust built on predictive and normative foundations. So a client chooses an architect because of their knowledge of their previous performance, and ability to keep promises, but the client also trusts them to the level of normative ethical behaviour in the industry that s/he will be truthful, trustworthy, maintain confidences and fiduciary duty. There are levels of trust depending on:

- the recommendation on the ability to deliver;
- the standard of professional conduct s/he observes in the industry;
- the personal faith s/he has that s/he would comply.

This 'honour' bond is only partially backed up by the contract and so there is therefore a risk on the part of both parties that professional behaviour will also be reciprocated by the client. The commitment is likely to be greater on the part of the client, but the risk of trust for payment and reciprocal treatment also needs to be assessed on behalf of the architect. So what are the values connected with trust which are considered important?

The RIBA summarises the personal ethics side of professional codes as honesty and integrity. Bowen *et al.* (2007)[22] refer to the fundamental level as being able to trust a professional to operate in the client's best interest. This, however, seems to have been disproved in practice in their study where they found that personal ethics took precedence over business ethics and that breaches occurred most commonly in conflict of interest, and divulging of confidential and proprietary information. Environmental damage is a key area of conflict between client and professional obligations. This suggests that conflict of interest is a key issue in professional behaviour which needs to be understood and declared in building up trusting relationships.

It has been observed (EPSRC 2002)[23] that relationships that show or build trust need to be professional and reasonable and are not about avoiding positive confrontation. However, some projects may need more trust than others.

## Trust in practice

It is useful to look at two different situations in the development life cycle in which trust plays an important role.

### Trust in planning decisions

When an application is submitted in accordance with the terms and requirements of the planning authority, it is expected that the permission will be given where there are no major objections by the public and it conforms to the area plan. There are a lot of guidelines and, although some of the provisions are complicated, it is generally possible to explain the complexities of a particular site. If this is turned down or there is a major delay, it is expected that you will have received good technical advice from the planning officers. An applicant will know that the elected members might vote it down due to lobbying and social factors, but for you to keep trust with the system you will need to know why and you will need to think that the reason is objective, impartial, substantial and logical.

In the system in the UK today, there is a lot of frustration about the way decisions are made because it is believed that they are not always logical, have been subject to an unforeseeable change of heart, or even that they have been subject to elected member interest clashes or even payments by third parties. In some cases, there may be competition between two different developers either for the same site or for adjacent or overlapping sites, with clashing designs and objectives. In this situation, there needs to be sufficient transparency of guidance with some less complicated decisions being technical and not subject to elected member scrutiny.

### Trust in customer–supplier relationships

In construction supply chains, there is a strong tradition of short-term project-length relationships with suppliers who are mostly small and competing with many others. In this situation, it is easy for the dominant contractor to impose draconian conditions of retention, long payment terms and penalties in the case of perceived non-performance, especially in adverse supplier markets. This has eroded the trust of the supplier. Government regulation has never been able to entirely eradicate this abuse of trust.

The poor relationship between a contractor and their supply chain was questioned by Michael Latham (1993)[24] in his interim report *Trust and Money*. Sahay (2003),[25] from a study of the literature, cites the benefits of trust built up in longer-term and stable relationships as lower transactional costs, desirable behaviour, reduction of the extent of formal contracts, facilitation of dispute resolution, continuous improvement and frequent and transparent exchange of information. Jones (2005)[26] agrees with many of these benefits of trust in his application to construction supply chain management, and adds that the forecast of future events is more reliable and realistic and fewer controls are needed to measure, monitor and control performance that tend to be more legalistic, such as accreditation and insurance. This idea of building up the trust relationship due to longer-term relationships is a common one and is similar to Lewicki and Bunker's (1996)[27] model in developing trust with less reliance on contractual control so there is room for continuous improvement, innovation and productivity savings. Jones calls this goodwill trust, in that it was not originally contracted for and has a greater focus on mutual interests and makes synergetic savings, with teams working together better. However, these supplier relationships take time to build and they may have to suffer opportunistic behaviour by one or more of the parties in the process, making the parties vulnerable at the start of a relationship.

Sahay warns that partnerships get built up in the pervading marketing conditions of supply and demand, which tip the balance of the supplier–customer relationship and may lead to an erosion of trust for one party as the conditions change and the supplier or customer feels pressured into commitments that are one-sided. In a boom situation, conditions suit the supply side, which can control the terms. In a less buoyant market, the customer can demand more of their suppliers who are short of work. However, Ellram (1995)[28] found that lack of trust was ranked high by both suppliers and buyers, in making inefficient partnerships. Further, several see trust and commitment as being the most important factor in the longevity of a relationship (e.g. Morgan and Hunt 1994).[29] Trust is therefore a vehicle for overcoming short-term deficit because of a belief in higher long-term benefits. Maintaining trust, then, is seen as a basis for enduring relationships, but some writers also see it as a vulnerability that may be exploited by other parties, e.g. goodwill is accepted, but not reciprocated at the expense of the trusting party (Baier 1986).[30]

## Case study 9.4  Trust and partnership

A supplier has done many profitable jobs with a main contractor and they have a good relationship. On this occasion they are not paid by the main contractor even though they have completed the work and it has been duly valued. They realise that the main contractor has not been paid by the client and now, as a small contractor, they are having cash flow problems and are unable to get any more credit from the bank to purchase expensive materials for the next stage of the work. They are being told by the contractor that the payment is not too far off, but they have a dispute with the client that has to be resolved. They ask the contractor to purchase the materials if they supply the labour. The contractor says they cannot afford to do this and will be forced to invoke the liquidated damages clause against them which they signed if they delay this work which is now critical to the finish date of the overall project.

### Questions

• Who is in the wrong ethically for this impasse? Should the contractor pay? Should the subcontractor pull out of the project until they are paid or should they keep to their programme?

• If an atmosphere of trust is to be maintained in this situation between the two contractors, what can be done to resolve the situation amicably, assuming that both parties would like to do so?

However, more recent changes in procurement methods have led to opportunities to build up longer-term relationships by having select supply chains, where subcontractors are pre-qualified and promised work on several projects over a period of time. This affords the opportunity to develop trust relationships, first on the basis of competence (pre-qualification) and, second, by building in goodwill by the mechanism of continuous improvement. However, it is still easy to abuse this goodwill by the excessive expectation of the dominant party for continuous financial benefits, using the effort of the other. This could be considered a worse ethical state than before as trust that has been built up has now been broken, as expressed above.

## Trust and risk

Risk is about dealing with uncertainty which may result in a 'down' or an 'up' side. This is recognised by the Project Management Institute (2000: 207)[31] and they define it as 'an uncertain event or condition that if it occurs, has a positive or negative effect on a project objective'. Risk is widely connected with adverse events and many of them are considered to be external and outside the control of the manager and are a part of life. However, trust is often connected with risk and depends on people having the right communications, the right mix of complementary strengths and weaknesses in working relationships and teams, and also with the degree of trust that risk will not be at their expense for the advantage of another. Mutual advantage, due to risk-taking, has to be negotiated and understood on an equal and consenting basis. External risk should be evaluated and communicated to those affected. Ethically, all risk should be managed by those best able to do so, which depends on experience and how much control they have over it. In short, risk-taking may affect colleagues, customers, suppliers and supply chains unequally, causing a lack of trust in each other. This risk is tangible and can be measured in lost future contracts, poor service and poor productivity, because synergy has been lost. In the public arena, a lack of trust in the planning system may lead to lost development opportunities or a tendency to bend the system (or try to) for selfish ends at the expense of others.

The amount of risk that is taken is also an ethical matter and in the built environment there is either trust or scepticism that the appropriate risks may be properly managed. Risk management is defined as the systematic process of identifying, analysing and responding to project risk. It includes maximising the probability and consequences of positive events and minimising the probability and consequences of adverse events (PMBOK 2000). It is this dichotomy of maximising or minimising an event and recognising the choice to decide on the relative probabilities which is an ethical decision. Most people have a risk behaviour profile which tolerates more personal risk and less risk to others. However, certain profiles such as an entrepreneur may put others more at greater risk than themselves. They may see opportunities to buy property and make money which is good for them, but then this has adverse effects on others. Their enthusiasm for making money blinds them to the risks of the project and they are seen to make claims for the benefits of the development that put other people's quality of life at greater risk and trust is lost in the benefit of development.

Chapman and Ward (2003)[32] argue for a balance of risk between efficiency and incompetence by looking at the opportunity area. Excessive grasping of opportunity represents a move into an area of incompetence because the risks are raised too high for reasonable control. Excessive caution leads to inefficiency and lost opportunity. This may have led to high risk management costs for a small risk or it may have meant lost opportunity for a small risk premium over and above nil or minimum.

## Case study 9.5 Property development

The decision to grant planning permission may bring new facilities such as homes and workplaces and airports. However, it may also bring about more congestion on the roads, more noise, lower house values and restrict views for those already living there. The resistance flows from a belief that the new development will bring large profits for the developer, but at the same time threaten the amenity of existing residents. NIMBYism,[33] on the other hand, could be considered to be a selfish reaction to inevitable change and the general societal good of more homes for those without. Ethically, there is a risk that deeply held anxieties may be long term or that house prices will come down due to the nature of the development taking place and it will affect the quality of life there.

Risk assessment is built into the level of return on the development. The higher the commercial risk of the project, the more 'profit' will be built into budget to offset the risks taken. Some uncertainties such as delays in planning, contamination, or difficulties in future market sales are always there, but some are more predictable than others. Poor planning consists of taking short cuts because risks were not properly assessed prior to commitment.

The budget allocated or time scale may actually restrict amenity or good design or community acceptability and these external, less powerful stakeholders may feel disenfranchised and sold out by the process and/or the product. Consultation and careful planning and design do take time and money which conflict with maximum development profit, especially if the developer involvement is short term. However, public appeal and actions taken to stop development by pressure groups may actually be quite successful in bringing about 'justice'. Some argue that nothing good comes out of minimal effort, and welcome the debate and wider management of stakeholders.

Risk-taking applies in the delivery of new buildings and infrastructure too – in design and construction – and these risks have to be balanced ethically against the overall good and right and the short-term risks which may jeopardise future good.

### Design risk

The issue of good design is a fraught one as aesthetic aspects of design differ because of taste. It is essentially a question of trust, because risks are generated by the design chosen. Architects can argue for a radical design as they can see the bigger picture because of their experience and that there is a need for progress to build a modern society, and people will see the point after a period of 'settling in'. Others may argue an essentially egocentric experimentation is being undertaken at the expense of the sensibility and comfort of others. However, the lack of an opportunity for well-considered, but radical design appropriate to future needs and tastes is also an important ethical consideration. On balance, it may not be a majority decision, but a courageous one to ensure a vision and progress for the future. Design is both functional and artistic.

Building design covers sustainability as well as taste and this is discussed in another chapter. Guy and Farmer (2000)[34] discuss several different schools of thought and point out these will be in ethical conflict because of the aims they have. For example, a *smart* design approach is one which seeks to provide a technical solution to building sustainability, such as photovoltaic cells to generate electricity. In this approach, will cells generate the degree of electricity claimed, especially if the climate or orientation is not good?

In contrast, an *ecological* approach will seek to use natural, locally available materials to reduce embodied energy and will reduce use of highly manufactured elements such as glass and steel which appear in most solar panels. The risk in these solutions is whether they work. Is there a sustainable supply and storage of biomass fuel? Can these fuels be produced on a large scale without affecting essential food production? How does natural or techno fit in with the surroundings? Should we worry?

Design decisions are fundamental because they require a clear insight as to how the future might work. They are risky because each set of site conditions is unique and each new design is an untested prototype, so time to allow an iteration of design is part of the risk management process.

### Construction risk

Risks generated in construction revolve around unknown site conditions, access issues, and the impact of construction upon others – particularly health and safety and the environment. Construction health and safety risk management is well documented, but can contractors be trusted to look after their workers' and the general public's health and safety? Typical risks in construction that are not managed well are:

- A fatality and major injury record matched by no other industrial sector is quite unethical even if it can be argued that processes are inherently more dangerous.

- An excessive amount of unrecycled waste is also an issue that is more often seen as unsustainable. The use of a site waste plan to cut down on wasted and reordered materials and to recycle demolished materials is the solution.
- A shortfall in client's expectations to get their buildings on time and to budget.
- The building industry is particularly sensitive to changes in investment conditions, being a leading indicator for investment.
- Labour supply is not always sufficient to meet the variable workload, with impacts on programme and quality.

To maintain trust, these risks need to be managed better particularly in allocating risk management to those who are best able, and making clients aware of the level of probability where things might go wrong. So often false optimism is generated without the planning and assessment to back it up and respond to the risks and distinguish between responding and discussing the external threats and the internal controls.

### Trust and corporate governance

The development of trust by the ethical management of a company's systems comes under the term 'corporate governance'. The building up of trust in the presentation of accounts is a specialist as well as a sensitive area for the stakeholders of construction organisations. One way that has emerged has been the development of a combined code (2006)[35] which represents certain minimum standards of behaviour. These codes are replicated across the developed world. This has emerged because of the gross betrayal of trust that was instituted in such acts as the Maxwell misuse of pension funding, the Bering Bank–Leeson affair, the Australian HIH Insurance, Enron and Worldcom collapses and other situations where directors betrayed their stakeholders' trust. This resulted where they hid information or made their boards and shareholders insufficiently aware of who would be at risk. There has also been a need for greater transparency in the setting of remuneration levels and conditions because of the outcry over bonuses promised during times when performance has gone down.

Corporate governance begins with a prescribed format for accounts and the requirement for all accounts to be audited by a third party. The auditor has a responsibility to satisfy her/himself that the accounts are a 'true and fair' account of the state of the company. The need for the audit to remain objective includes changing auditors on a regular basis so that cosy relationships are not set up. There is a fresh requirement for price-sensitive reports to be 'understandable' and also for risk assessment and management to be carried out on internal financial controls by the accounting staff.

## Conclusion

Those responsible for the built environment have a sacred trust to manage their working relationships. Prescriptive compliance falls far short in delivery of value for money and it is not a way of getting people 'on your side'. The industry has realised that a confrontational style is often self-defeating, with conflict often ending up in the court and both sides losing out and relationships breaking up. Although contracts are necessary, it is a waste of resources to use them as a first line of defence. Research has shown that there is a benefit from partnering only if a new culture is adopted to integrate objectives and to harvest the new synergies available in a no blame culture. Trust can play a fourfold role of reducing supervision by improving reliance, using predictability by getting to know each other, making money by reciprocating a competitive and innovative edge for each other, and improving the efficiency by the synergy of better team working. Two best practice projects in Part II are used to exemplify this.

Collaborative partnering charters and contracts are becoming more popular, which require the use of better risk allocation at the beginning of the contract and have an expectation that different organisations will be committed to improve their team working and work through their differences. These partnerships have reached a second generation of partnering and trust has an opportunity to grow from the calculus level to the knowledge-based level. One of these contracts, the JCT Be Collaborative Constructing Excellence form, has been tracked through its use in a repetitive framework partnership as a case study in Part II. This has shown how excellent planning not only has improved relationships, but also has saved money, improved quality, increased value and made sure that improvements have been ongoing.

The management of risk affects all stages of the development life cycle, but putting more effort into allocating risk and taking responsibility for it is an ethical approach, so that trust can be built up with the management of stakeholders including community aspirations when the business case is being made, and allowing feedback through the design process. A case study is used in Part II to look at how one large central development consulted its stakeholders early in the process and maintained their support throughout the project, allowing them to have improved flexibility because trust had been built up.

## Notes

1  *Pocket Oxford Dictionary* (1934). Oxford: Oxford University Press.
2  Lewicki, R. J. and Bunker, B. B. (1996) 'Developing and Maintaining Trust in Work Relationships', in R. M. Kramer and T. R. Tyler (eds) *Trust in Organisations: Frontiers of Theory and Research*. Thousand Oaks, CA: Sage Publications.
3  Bennett, J. and Jayes, S. (1998) *The Seven Pillars of Partnering: A Guide to Second Generation Partnering*. Bristol: Thomas Telford.
4  Child, J. and Faulkner, D. (1998) *Strategies of Co-operation: Managing Alliances, Networks and Joint Ventures*. Oxford: Oxford University Press.

5  UN (2003) *Convention Against Corruption*, ratified by 30 countries in 2005 and by many others since.
6  Fewings, P. and Hodgson, (2005) 'Collaboration, Trust and Learning: A Business Integration Model', proceedings of the 21st ARCOM Conference, London.
7  Swan, W., McDermott, P., Wood, G., Thomas, A. and Abott, A. (2002) *Trust in Construction: Achieving Cultural Change*, Interim Report, Centre for Construction Innovation in the North West, EPSRC Project.
8  Ibid.
9  Gardiner, P. D. (2005) *Project Management: A Strategic Planning Approach*. Basingstoke: Palgrave.
10  Lipman-Bluman, cited in ibid.
11  Handy, C. (1996) *Understanding Organisations*, 4th edition. Harmondsworth: Penguin Books.
12  Gardiner, op. cit.
13  Wood, G. and McDermott, P. (1998) *Searching for Trust in the UK Construction Industry: An Interim View*. Online. Available http://www.research.scpm.salford. ac.uk/ trust/Documents/Thailand.doc (accessed 3 October 2007), Research Centre for the Built and Human Environment, Salford University.
14  Swan *et al.*, op. cit.
15  Gardiner, op. cit.
16  Bennett and Jayes, op. cit.
17  Rowan, V. (2004) *How Joint Ventures Are Organised, Operated on International Projects*, Construction Association of Japan, Pinsent Mason. Online. Available http:// www.constructionweblinks.com/Resources/Industry_Reports_Newsletters/May_23_ 2005/howj.html
18  Brown, A. (2002) Andover North Site. Online. Available http://www.strategicforum. org.uk/sfctoolkit2/help/andover.doc
19  Hollis, M. (1998) *Trust Without Reason*. Cambridge: Cambridge University Press.
20  Todd, A. (2007), 'Trust Without Ethics – Ethics Without Trust', *Canada Globe and Mail*, 31 January 2007. Toronto. Online. Available http://www.gather.com/view-Article.jsp?articleId=281474976899647 (accessed 6 September 2007).
21  Hollis, op. cit.
22  Bowen, P., Pearl, R. and Akintola, A. (2007) 'Professional Ethics in the South African Construction Industry', *Building Research and Information* 35(2).
23  Swan et al., op. cit.
24  Latham, M. (1993) *Trust and Money*, Interim Report for the UK Construction Industry. London: HMSO.
25  Sahay, B. S. (2003) 'Understanding Trust in Supply Chain Relationships', *Industrial Management and Data Systems*, 103(8): 553–63.
26  Jones, P. (2005) 'Supply Chain Management in Construction', in P. Fewings, *Construction Project Management: An Integrated Approach*. Abingdon: Taylor & Francis, pp. 323–5.
27  Lewicki and Bunker, op. cit.
28  Ellram, L. M. (1995) 'Total Cost of Ownership: An Analysis Approach for Purchasing', *International Journal of Physical Distribution & Logistics Management*, 25(8): 4–23.
29  Morgan, R. M. and Hunt, S. (1994) 'The Commitment–Trust Theory of Relationship Marketing', *Journal of Marketing*, 58(3): 20–38.
30  Baier, A. (1986) 'Trust and Antitrust', *Ethics*, 6(2): 231–60.
31  Project Management Institute (2000) *A Guide to the Project Management Body of Knowledge* (PMBOK Guide), Newtown Square, PA: PMI.

32 Chapman, C. and Ward, S. (2003) *Project Risk Management: Processes, Techniques and Insights*, 2nd edn. Chichester: Wiley.
33 NIMBY means Not in My Back Yard.
34 Guy, S. and Farmer, G. (2000) 'Contested Constructions', in W. Fox (ed.) *Ethics and the Built Environment*. London: Routledge.
35 Financial Reporting Council (2006) *The Combined Code on Corporate Governance*, June. London: FRC.

# Chapter 10

# Bribery and corruption

At the heart of ethics is a concern for fair transactions. Corruption is an emotive word and an accusation of corruption is a serious statement to make which could be strongly resisted and denied. It is widely applicable, for example, it can be political, justice, police, mafia, financial, tax, benefits, electoral, professional, educational, competitive or business orientated.

It is often connected in business with procurement and sales and in professional issues with conflict of interest. Examples of corruption are fraud, bribery, concealment, dishonest practices, nepotism, avoiding tax, conspiracy to mislead and false accounting. The list is endless and in its mildest sense is carried on almost as a matter of course in business. When it contravenes the law, then it is recognised as unacceptable in a given society, though it is often difficult to prove successfully due to the cover of secrecy that exists in many business transactions. In the public realm there is a particular effort to establish exemplary levels of practice to meet the demands of public opinion and stewardship of public funds. This chapter investigates bribery and corruption which are commonly faced by professionals in the built environment in order to upset the normal course of events, including fraud, underhand payments, unfair practices in construction procurement, property deals, distortion of competitive bidding, trading and selling. The planning process has already been discussed in Chapter 7, but is also open to unfair practices or even bribery.

The objectives are:

- to identify bribery and corruption in business and professional activity in the built environment;
- to understand the causes and why it is so difficult to eliminate;
- to consider how corruption, particularly in competitive contexts, can be tackled.

## Business and professional environments

Bribery occurs when a conscious attempt is made to influence another person unfairly either with a specific payment or gift, or influence is bought to bear to bring unfair advantage. Bribery emerges when a payment moves from an acceptable payment for services or a goodwill gift, to an attempt to distort the outcome which would otherwise be described as fair. Gifts and hospitality out of context may also influence conflicts of interest. Conflict of interest arises from providing services to two parties who need to preserve confidentiality or from a personal interest which might influence a professional's neutrality. The main weapon in the defence of bribery and corruption is the use of transparent practices and financial accounts. The main reason for resisting transparency is commercial confidentiality, which often neutralises the confidence of core internal stakeholders, e.g. shareholders, as well as the public and external stakeholders. The main reason for a company introducing more transparent practices is to gain a reputation for fair trading. This often places a company in a position of semi-transparency due to the continuing need to maintain client confidentiality, which is also a professional responsibility.

In corporate governance terms, the Companies Act and a combined code of practice have been used to specify non-executive scrutiny of sensitive areas of company business such as public reporting, appointments, remuneration, audit, governance measures such as rights of shareholders to know and powers to dismiss and approve directors as a whole who act on their behalf. These are much less onerous for private organisations that have a closer knit, naturally informed stakeholder community. However, quite large private organisations can exist such as private equity companies and large limited partnerships who have special exemptions. Their level of transparency has been questioned due to the perceived betrayal of trust which has led to undesirable outcomes. Anderson's is an example with their auditing support of Enron which did not stop the risky financial practices which impacted with unacceptable expense on the livelihood of many others, in shareholder losses and employee pensions.

Resisting bribery and not turning a blind eye to corruption lie at the heart of the professional responsibility to remain objective and be fair to all parties with whom a professional is connected. The concept of transparency is a key weapon in removing corruption, but this also needs a commitment to eliminating unfair advantage. Professionals also debate the balance between transparency and commercial confidentiality. Professionals have a commitment to fairness and honesty in their codes of practice and to the public good. This helps to make a broad definition of corruption for them, which exceeds the minimum legal compliance.

### Definitions

Corruption is defined by De Graaf (2007)[1] as 'behaviour of public officials which deviates from accepted norms in order to serve private ends', but the European Council goes further and defines corruption as 'requesting, offering, giving or accepting, directly or indirectly, a bribe or any other undue advantage or prospect thereof which distorts the proper performance of any duty or behaviour required of the recipient of the bribe'.[2]

This identifies an active corruption – offering or asking for the bribe directly or indirectly, and a passive corruption – accepting or not objecting to bribes. It also identifies non-monetary influence as corrupt and affecting future prospects.

According to Transparency International (TI), corruption can be divided into four categories:

1   *Bribery*, which is the offering of an undue award (supply side) or the demanding or receiving of a reward (demand side) in order to get favours and may be initiated by an individual or a corporation (institutional bribery). It may also be shared and secret kickbacks can be offered to certain individuals on the supply side in order to help its payment. *Facilitation* refers to relatively small payments to junior officials who have control of certain essential services such as customs officials to pass the goods through quicker or cheaper.

2   *Extortion* is blackmail where a payment is demanded where there is little choice but to comply, such as passing through police blocks. This could be similar to facilitation or it could be more sinister, connected to gang or mafia activity controlling certain activities or threatening certain actions, which might normally be termed *coercion*.

3   *Bribery and fraud*. Bribery is likely to include fraud such as the wrongful approval of work which is substandard or other deception to add false value or payments through the accounts that now do not comply with the facts. Tax evasion is a fraudulent activity, but is not necessarily connected with bribery if other benefits to the fraudster can be gained, such as more profit and less tax or payment due to others.

4   *Concealment* is a practice also associated with corruption and often involves the excess payment of an agent whose services are vague, who proceeds to pass on facilitation or bribe payments. The payments are likely to be into offshore funds or through subsidiaries into other less scrutinised countries or banking systems. False or inflated charges may also be made involving funding commission charges that are inflated and paid back by the funder through another subsidiary in a joint venture or multinational organisation. This creates a 'slush' fund for onward bribe payments.

The TI definition is fairly clear though there is less certainty culturally about what constitutes unacceptable behaviour and this probably revolves around the belief as to whether corruption is viewed as 'grease' or 'grit' (Langseth and Michael 1998).[3] By this they distinguish between whether a bribe adds value to a transaction, e.g. by unlocking bureaucratic deadlock (grease) or whether it is economically harmful and further slows the process and takes value and affordability out of the transaction (grit). They conclude that their research in Tanzania indicates that corrupt payments take value out of the system and make services harder to obtain in the long run. Rose-Ackerman (1999)[4] agrees about the general economic loss of corruption. However, she concludes that cultural views on corruption are centred on how self-interest is assessed so that interest for home and family matters is considered good in many Eastern and African countries, while the same provision might be considered nepotism in the West.

Corruption has been categorised into state capture and administrative corruption. Administrative corruption is the bribing and corruption of officials within a system to gain access or favourable conditions within existing laws such as paying off customs officials or bribery to win public procurement contracts. State capture is the more wholesale attempt by private individuals to change the course of legislation for their own narrow concerns, by bribing officials or fraudulently lobbying parliament to change legislation or to avert the course of justice. State capture results in corruption of political figures so that they act in a biased way to favour the agency lobbying or bribing them. It therefore militates for reform in the wrong way. State capture might help dominant or corrupt parties to meet their ends.

Administrative corruption has a propensity to operate at different levels of government departments and official agencies. It may be through the corruption of the law enforcement agencies, the judiciary, education entry and progression and health facilities. It is relevant in international procurement also where it is often termed high level corruption which affects the ability of junior officials and agencies because of fear of consequences.

### Reasons for corruption

The difference is made between corruption which occurs because of the *low salaries* of public officials who are tempted because of a need to supplement their income, and corruption which is *greed*, which satisfies the desire for status, influence and comfort that is given by being rich or powerful. The latter is notable in the higher position of public officials, private individuals and politicians who do not depend for survival on the bribes collected. A middle case is the temptation or blackmailing of individuals by the promise of betterment or threat of retaliatory action. In all cases it is the furtherance of *private* interest over public or corporate interest that brings about the corruption of the system. Rampant corruption is where corruption is so common and expected that it has shifted the control of transactions outside the proper authorities. In short, an unlawful, chaotic and unpredictable regime prevails. In the case of a corrupt company, the proper desire for profit and wealth creation has been replaced by an unreasonable drive to maximise profits for the benefit of a select few with scant regard for legality. Alternatively, the cause might be desperation to retrieve losses or as a passport to enter lucrative markets resulting in illegal or 'dodgy' dealings.

For Rose-Ackerman (1999),[5] the key point economically is the degree to which self-interest is channelled towards productive purposes or not and this is a factor that government structures should consider when dealing with corruption. However, it is important to see how structures provide incentives to corruption. For example, a tax system may be oppressive and will tempt many to try and avoid it. There will be many other types of incentives to corruption.

Corrupt power is also exercised in the distortion of competition by an influential few and this takes the form of a cartel or unfair practice in business. This may involve collusion to limit competition and falsely raise prices or payments, or agreements to exclude or disadvantage some competitors in terms of their access to tender lists.

### Legislation against bribery and corruption

Anti-corruption legislation outlaws the giving and receiving of payments or other gratification in order to influence the outcome of a decision and is most concerned with public officials who are in the position of gatekeeping favours such as planning permission, awarding contracts or releasing confidential information. It can also apply to any individual who may otherwise be bought, such as a witness or a politician or a professional to compromise integrity. The Prevention of Corruption Acts 1889, 1906 and 1916 are quite antiquated, but still in use in the UK. The 1889 Act dealt with public officials initially and then was extended to the private sector in 1906. The 1916 Act defined public bodies more widely and put the burden of proof on the public authority to prove they had not acted corruptly in the case of a prosecution. A new bill (1998)[6] is proposed for a modern Corruption Act to take into account the wider concern for international corruption, and also various agencies have been set up for specific contexts, such as the Serious Fraud Office and the Office of Fair Trading with the associated Competition and Enterprise Acts to deal with distortion of competition and fraud cases. New powers have been given to the police under the Prevention of Terrorism Act which also deals with fraud perpetuated to support acts of terrorism. The new bill defines corruption as:

- conferring *an advantage* on another corruptly (clause 1);
- receiving *an advantage* (clause 2) or;
- the knowledgeable agent of such a conferee or receiver.

This also considers performing functions corruptly. Agents who can prove they have no knowledge of the illegality of the payment are exempted.

In other countries it is common to have a single all-encompassing Prevention of Corruption Act. For example, the Anti-Corruption Act of Namibia (2003)[7] defines the following acts as corruption:

- direct, indirect or attempted corruption;
- offering or receiving a bribe;
- using public resources for private gain;

- conspiring to commit a crime;
- the corrupt acceptance or giving of gratification to an agent (Section 35);
- the corrupt acquisition of private interest by a public officer (Section 36);
- corruption in relation to tenders (Section 37);
- bribery of a public officer (Section 38);
- corruption of witnesses in a trial (Section 39);
- bribery of foreign public officials (Section 40);
- bribery in relation to auctions (Section 41);
- bribery or giving assistance in relation to contracts (Section 42);
- the corrupt use of office or position for gratification (Section 43);
- corruption in relation to sporting events (Section 44).

There is a great deal of debate about what is an acceptable influence, as even good service would influence the client for future work. Many would define normal competitive behaviour as ways of competing and selling so as to gain competitive advantage and clearly a boundary needs to be drawn as to ethical and unethical competitive advantage.

### Cause and construction examples

Rose-Ackerman (1999)[8] uses an economic argument to identify four key conditions that may create a corrupting climate:

1   The government allocates a scarce benefit using legal criteria other than willingness to pay. 'Bribes clear the market' because those who can pay will purchase.
2   Public officials responsible for key permissions are paid low wages and have little incentive to work. 'Bribes act as an incentive bonus' so that they can increase their income.
3   Onerous tax systems can cause those working in legal pursuits or business to seek to reduce the tax and other costs imposed on them. 'Bribes lower the costs' because they are able to avoid paying or reduce the tax burden or duty.
4   Illegal businesses frequently purchase corrupt benefits by corrupting others. In extreme cases a 'Mafia' where the law and order system is dominated by intimidation and violence emerges. 'Bribes permit illegal activity'.

These will help us see where the warning signals are and provide us with a diagnostic tool to identify specific cases in the development process that create the potential for bribery and corruption, though there may be other non-economic conditions, connected with the political system. The above conditions do to some extent exist in any system and so the challenge is to create checks and balances to ensure they do not become corrupt and self-fuelling. The following case study gives examples in the built environment field.

## Case study 10.1   Corruption examples

The first could be a shortage of land approved for housing development which in a legal market will put the cost of land up and also create a demand for housing so that house prices go up. Corruptly, it might also encourage developers to put additional pressure on development control officers and council members to free public land in favour of certain individuals at below market prices, so that better profits are possible at the higher house prices prevailing. An official may also create a delay by inventing their own conditions for extra payment of duty in letting through imported goods.

The second could be public officials, who are paid less than a living wage, will be tempted to use a bureaucratic system to delay legal benefits to those who won't pay a bribe such as the granting of a building permit by delaying inspection and approval of sections of the building.

The third could be avoidance of VAT which is commonly offered by the less reputable builder to the customer who is prepared to pay 'cash in hand'. Other instances may be the payment of bribes to customs officials in order to release imported building materials without punitive customs duty. The latter will apply where there is a heavy dependence on imported goods and duties are high.

The fourth refers to companies who run their businesses dishonestly to gain competitive advantage and are able to consistently find and buy out officials. This could apply to lucrative contract procurement where payment is made to give the company specific advantages to help them win contracts and to keep others out of the picture. These companies are prepared to devote regular proceeds to keeping people quiet, or artificially share out work between a few in such activities as supplier cartels. Proceeds are large enough to pay for major concealment activity or even intimidation.

Examples have been deliberately chosen to indicate a universal application in the first three cases, but the outcomes may well be quite different and honest in an effective controlling environment or culture which is seen to expose corruption and fraud and act as a deterrent to the temptation. As stated before (see notes 4 and 3), there is a net loss economically with corruption even though there are some gains. This is because the market is distorted and concealment activities and manager time in setting up deals and losses to further investment cancel out overall economic gain.

Large international companies are noted to change their ethics to suit 'country conditions' and this may mean selective corrupt action, which may be justified by them in business terms in order to be competitive and to work on 'the level playing field'.

## Dealing with corruption

In a national situation there is a need to define the local level of behaviour which is acceptable to build trust in the government and in reputable trading organisations. Failure to control political excess may also contribute to the breakdown and stalling of the economy and the inability of the justice and law and order mechanisms to be effective. In the worst case, this leads civil war when opposition parties acutely feel that the national inheritance is being squandered irreversibly. The sad thing is that civil war destroys the existing structures as well as delaying the change perceived.

Internationally there is a need to provide a more level playing field and to adhere to a common definition of corruption and ensure that international businesses do not play dirty away from home by requiring governments to take action against bribing public officials, although it is also tempting for governments to turn a blind eye because exports increase the balance of payments. In the UK, a police unit called the Overseas Anti Corruption Unit (OACU)[9] has been set up to investigate acts of corruption by British companies exporting goods and services to other countries.

### International bribery and corruption

International bribery is an ethical issue in the built environment because of the large amount of aid dedicated to the provision of *built* assets and the large failure rate of such projects either in inappropriate targeting of the aid or because of a haemorrhage of funding into unauthorised private bank accounts. The concept of the 'best bribe' wins has been determined as a waste of resources and at its worst it affects the livelihood of the most deprived sectors of the economy as it diverts funds away from welfare and development, education and health. As a future prospect, corruption diverts funds away from a sustainable economy for future generations, propping up personal gain at the expense of non-renewable scarce resources, pollution and irreversible climate change.

There is a need to build an international code which is acceptable in order to create a true level playing field for procurement and property deals and development control on a global scale. This has been taken up by the United Nations Campaign against Corruption (UNCAC)[10] and other international organisations as a key justice issue to eliminate value going out of the system of international aid. The Organisation for Economic Cooperation and Development (OECD) Anti-Bribery Convention (1997)[11] is a specific attempt to hold *industrialised* exporting countries to account and to control and investigate and penalise appropriately, any corrupt activities of national or multinational companies given export credits to do business abroad, especially in the developing and emerging economies, where the temptation is greatest to overcome the stiff competition and enhance trade balances.

Thirty-seven countries are committed to the OECD Convention (1997), because widening trade is a lucrative market to raise the economic trade balance of the exporting country as a whole and it is politically attractive to turn a blind eye. The top 10 OECD countries jointly export £1.5 trillion of exports each year and thus have a huge influence on world trade if bribes to overseas officials are not outlawed. The committing countries have agreed to monitor home companies that are working abroad and to provide export credits on the understanding that bribery of public officials is a criminal offence which will be prosecuted.

The difference in perception between the developing and the home country is evidenced by the Bribe Payers Index (see Appendix) which evaluates the supply side of corruption which is the propensity of companies from industrialised countries to pay bribes for services offered abroad. It is measured as the perception of expert business leaders in 15 large emerging economies for 21 major exporting countries and surveyed yearly by TI, which are a non-governmental watchdog. In 2002, bribery was described as 'disconcertingly high' (TI 2006).[12] For instance, on a measure of 1–10 of OECD countries, where 10 is clean:

- the USA and Japan received a score of 5.3;
- France, Germany, Spain, Singapore and the UK receive higher scores (5.5–6.9 respectively);
- Australia, Sweden, Switzerland, Austria, Canada, the Netherlands and Belgium received the highest scores (8.5–7.8 respectively);
- Russia, PRC, Taiwan and South Korea received the lowest scores (3.2–3.9 respectively).

The worst business sectors are construction/public works and arms (1.3 and 1.9 respectively) which again sharpen the focus for the built environment. This does cast a shadow over the behaviour of both multinational companies and the governments in bringing them to book under the convention and indicates the difficulty in controlling extra-business payments. The OECD writes a country report rating the efficiency of home countries in achieving their aim to reduce corruption and to identify problem sectors. The IMF 2000[13] has also participated by requiring their Fiscal Transparency Tool to operate in the projects which they fund. These include an integrity pact which is signed by the government and the contractor as a commitment to no bribes within the specified terms of the individual contract.

### National corruption

Corruption is a national issue because there is a distortion of *competition* created by unfair influence created by suppliers where the client, public or private, is being defrauded of significant value in their project, or there is an unfair distribution of work to inefficient suppliers for private gain. The definition of legal or fair payments can vary from one culture to another, based on 'traditional practice' but there is generally a perceived acceptable limit. Practice and research have shown (Geddes and Neto 1992)[14] that there is very rarely public tolerance for widespread corruption in any country and that traditional gift giving does not develop into widespread corruption (Ayittey 1992).[15] In construction projects and property deals, there is plenty of room for inflated valuations so that money can change hands privately behind the scenes. There is also a similar capacity for influencing the bidding process so that certain suppliers make payments to receive lucrative contracts or to influence outcomes.

There is also a difference in perception between the public and the government, and public pressure plays a large part in bringing about specific measures to eradicate some practices that have got out of hand. The TI handles another survey called the Corruption Perceptions Index (CPI)[16] which measures corruption within a country based on the perception of NGOs, resident and non-resident country experts and resident business leaders. The results of the 2007 survey (TI 2007)[17] indicate scores from 1.4 (Somalia and Myanmar) to 9.4 at the top (Denmark, Finland and New Zealand). The UK comes 12th with a score of 8.4. There is a strong correlation between corruption and UN-defined low income countries with 40 per cent of these scoring below 3.0 – rampant corruption. Deeply troubled states such as Myanmar, Somalia, or Afghanistan remain at the bottom of the scale which is not surprising as war is a significant result of rampant corruption charges. Some African states have made significant jumps up the scale with a genuine will to reform, such as South Africa and Namibia, together with some Southern and Eastern European countries such as Italy and the Czech Republic. Others, including Malta, Austria and Jordan and Thailand, have slipped down the scale. See Appendix on p. 275.

Fighting corruption includes in-country reforms where there are determined efforts to outlaw corruption, convict public officials and strengthen judicial powers against political corruption. There are measures to strengthen anti-money laundering measures so that offshore funds can be confiscated and return to a country's exchequer. In this sense rich countries 'sit on the wealth stolen from the world's poorest people', but they also make it more difficult to track down and eradicate corruption when country agencies are are doing nothing to stop money-laundering efforts. Countries which have a tradition of family and goodwill payments will need to concentrate on situations that are on a country-wide scale and concentrate on criminal activity.

There are also other reasons for national corruption which are non-economic and associated with history and culture. These are not traditional, but have been brought about by the forces of intervention. Transitional economies have adopted corruption in the period following central command economies and this has particularly affected former communist bloc countries in Eastern Europe and the former Soviet Union. These emerging free market economies have produced unequal distribution of commodities that has forced a whole culture of bribery for survival. These economies are being monitored by the World Bank (1992)[18] and the European Reconstruction and Development Bank (EBRD).[19]

## Corruption in the construction business and competition

An average loss of 3.7 per cent of turnover, which represents 17 per cent of profits, has been attributed to bribery payments in the construction industry world-wide. This is a serious economic loss leading to less investment and/or damaging company performance. The international construction industry is huge, with an estimated £3200bn turnover, giving a loss of turnover is up to £100bn. Rose-Ackerman (1999)[20] argues that there is a critical threshold level when the profit margin is eroded and private companies pressure the government to make reforms. This strengthens the arm of government to do something which will work. Corruption is considered by the TI index to be worse in construction and armaments than any other industry. The TI annual 2005 corruption report (TI 2005)[21] was dedicated to construction and post-conflict reconstruction, recognising the additional pressure of expanded demand after the double catastrophes of the tsunami and the Iraq conflict. This also gives an insight into widespread corruption in the sector globally.

### Case study 10.2  Delays in supply

Fewings and Henjewele (2008)[22] looked at a case study in Tanzania[23] indicating a real-life situation where the purchase of generators to cover electricity power during extreme drought conditions was let out to a supplier who was dependent upon supplies through a subcontractor. Although the order was urgent, no checks were made as to the availability before placing the order and a serious delay resulted, until it became clear that the correct equipment was not forthcoming. The order was finally taken from the first supplier and a new supplier was identified who could supply the equipment. Time passed with tendering formalities and the new supplier was appointed and duly delivered the generators at the end of the drought period. The generators were now no longer required as normal hydro electric supply had resumed. This case study was not strictly a bribe situation, but the lack of knowledge that caused a redundant order to be placed and the concealment which existed in pretending that supply was possible induced a situation where new expensive equipment was purchased with no expected use. The same wastage effect has ensued through the refusal to use procedures and criteria to control major purchasing. The main adverse effects have been suffered by the people who endured long periods with intermittent supplies and, interestingly, the electricity company did not lose money due to the liquidated damages clause. The report also indicates that there was a high residual level of distrust in the electricity company by the electorate who thought that there was bribery, patronage and incompetence.

## Professional and commercial corruption

Three key studies have been carried out among contracting and professional organisations in construction. Fan et al. (2001)[24] among quantity surveyors in Hong Kong, Vee and Skitmore (2003)[25] and Pearl et al. (2004)[26] who have covered a range of professionals' experience using a similar technique in Australia and South Africa respectively. These studies, together with Stansbury's (2003) report for the UK construction industry,[27] have identified widespread corruption including fraud, unfair conduct, conflict of interest, and collusion. They all found that bribery and collusion are the most commonly experienced forms of corruption among contractors and negligence is a discrete form common amongst consultants. Fan et al. (2001) also found that there is variation in behaviour between young and old. There are many forms of collusion between companies to share out work between them and restrict competitive pricing. They may be able to influence other parties to make it more likely that they will be chosen in practice, such as writing specifications or weighting criteria which favour the core products of certain companies so that they will have unique advantages.

Gillam (2006)[28] puts it this way, 'There is no right way to do the wrong thing.' There are lots of grey areas for ethical behaviour which applies unfair influence and many are reluctant to admit actions close to themselves. The Canadian Construction Management Association (2006) surveyed members that suggested that 85 per cent had personally seen acts that they considered unethical and the CIOB (2006)[29] survey of members in the UK records that 41 per cent of the respondents had been offered a bribe in the last year. The CIOB requires its members to consider questions of ethics in its professional interview

In the TI report on construction in the UK (Stansbury 2003)[30] states there is clear evidence for twenty-three different acts of corruption which, if investigated, could add up to criminal activity. Among them are:

- Submission of false claims and those which are reckless in their accuracy, which is a charge of false accounting.
- The refusal to pay all valid submissions, which is a charge of false accounting or attempted fraud.
- Concealment of documentation such as inflating figures of subcontractors, which is a charge of false accounting or attempted fraud.
- Cover pricing on a tender, which is a charge against the Competition Act.
- Specification of a design that can only be supplied by one contractor, which is a charge of conspiracy to fraud.
- Obtaining a price only for the purpose of comparison where the intention is to go back to the favoured supplier and negotiate the price down. This is a charge of fraud by misrepresentation.

These practices are often considered not to be corruption, but the TI have clearly identified practices that simply lie dormant as largely acceptable practices throughout the industry.

---

## Case study 10.3 The cost of bribery

Stansbury (2003) has demonstrated that a power station budgeted at £100m could end up costing £158m + an additional finance charge of £3.6 million per year. This is because of:

|  | £m |
|---|---|
| deceptive practices in a willingness to allow for over-specification of | |
| 20 per cent | 120 |
| bribes of 10 per cent | 12 |
| lack of trust to believe in the release of 10 per cent retention | 13.2 |
| (contractor overprices) | |
| Claim and counter-claim | 10 |
| **Total** | **155.2** |

There will be increased finance charges for the client which accrues because of the 55 per cent inflated figure. These show the inflationary nature of an environment without trust and commitment. At the end of the contract a late finish and poor quality may lead to a client's claim with likely inflated counter-claims to match the client's and later payment of legal costs in the courts.

The only people who gain out of the £55 million extra costs are the banks who take the interest, legal practices who gain fees and those who have taken the bribes. The contractor has also successfully protected their profit and the client has already been charged by the contractor for the unpaid retention as well as paying legal fees to defend their claim. This type of practice is badly inflationary and counters the claims that bribery is only 10 per cent.

---

Further bribes may be used to get clients to accept substandard work and product specifications which create wasteful failures later so that systems are ineffective or whole life costs become expensive or even prohibitive. These are often termed incompetence rather than corruption and expensive reworks or even total failure of the system are overlooked as a quality rather than a corruption issue. They are very hard to investigate because tracks are covered or obstacles are put in the way of low level enquiries. In many cases, tax has been avoided illegally, but it is hard to identify.

Clearly a transparent environment is one to be preferred where the specification is rationalised and there is a commitment to stand by payments by the client. This needs to be matched by true costs from the contractor with shared incentives to reduce costs and improve quality in a partnership where retention is paid. Partnering also has a better chance of finishing on time and preventing inflationary spirals and expensive delays. The trouble is that both sides have built deception and distrust into the process. Stansbury (2003)[31] also stresses the need for a benefits analysis so that assets are provided for need. Feasibility reports indicate the importance of value managing the process of building so that waste is squeezed out of the design.

The CIOB (2006)[32] survey indicated that 51 per cent of members who responded thought that corruption was fairly or extremely common in the UK construction industry and 81 per cent of these were senior or middle management. There was, however, some difference of emphasis on what was considered corrupt and Table 10.1 indicates this variation.

Table 10.1 shows some acknowledgement that issues are corrupt, but a significant difference between how blatantly respondents thought them to be – in particular cover pricing. Interestingly this is not the view held by the Office of Fair Trading in the UK which has focused on exposing the strategies which suppliers have used to restrict the market such as those marked * in Table 10.1. These have been given a lesser severity by the respondents in spite of high profile prosecutions by the OFT at the time of the survey and 34 per cent had come across cartel activity. In addition, the inclusion of false claims (** in Table 10.1) has also been given a lighter touch. Where fraud has been uncovered, it was mainly through regular checks or by chance. A tip off or reporting to the police was not a common occurrence. Some 41 per cent had been offered a bribe and 23 per cent on more than one occasion. Only 16 per cent felt that corruption had increased in the industry in the past five years, but they strongly felt that neither the government nor the industry was doing enough to combat corruption. It was felt that individual corruption at all levels was most likely with fewer possibilities at senior corporate level, local government and client levels.

These results are a bit worrying but may be associated with the level of regular current practice in the industry. The area of distorting the market in the tendering and pre-qualifying stages is an indication where the culture might be changing on a national scale. A section later in this chapter deals with this change in the UK.

### Gifts and entertainment

The giving of gifts and entertainment in small measure is a valid and traditional way in the UK and many other countries, e.g. Thailand, to show appreciation for services offered and for sealing a bond of friendship which in itself has enhanced understanding and the synergy between contractors and their sub-contractors or even clients and contractors. Modest gifts in the UK are a bottle of wine, a company diary or calendar at Christmas, or the taking out of a client

Table 10.1 CIOB survey on corruption perception of different activities

| Misdemeanor | Very corrupt (%) | Moderately corrupt (%) | Not very corrupt (%) | Not at all corrupt (%) | |
|---|---|---|---|---|---|
| Cover pricing* | 18 | 45 | 32 | 5 | NC |
| Bribery to obtain planning permission | 56 | 17 | 20 | 7 | NC |
| Employment of illegal workers | 52 | 32 | 13 | 2 | Common |
| Concealment of bribes | 57 | 22 | 17 | 4 | Common |
| Collusion between bidders for market sharing purposes* | 41 | 35 | 22 | 2 | NC |
| Bribery to obtain a contract | 57 | 18 | 21 | 4 | NC |
| Leaking information to a preferential bidder* | 39 | 40 | 19 | 2 | NC |
| Production of fraudulent timesheets | 53 | 31 | 13 | 2 | NC |
| Production of fraudulent invoices | 54 | 22 | 20 | 4 | NC |
| False/exaggerated claim to withhold payment** | 44 | 35 | 18 | 3 | NC |
| Inclusion of false extra cost to a claim** | 42 | 39 | 17 | 2 | NC |
| Bribes from the building contractor to win maintenance contracts | 53 | 19 | 23 | 5 | NC |

Notes: C = Common; NC = not common. * = see text. ** = see text.

Source: Adapted from the CIOB Corruption in the UK survey.

to a meal at the end of a job. It is easy, however, to overstep the mark in terms of the size of the gift or method/frequency of entertainment so that a client's representative might make promises for additional work or a contractor receiving a gift from a supplier will be influenced to negotiate with that supplier or to overlook poor work, etc. This might not be the intention, but the outcome in the UK would be most definitely defined as bribery and corruption under the Prevention of Corruption Act if the two actions – giving of gift and subsequent business favour – were formally connected as they could easily be. The most important thing in this situation is transparency for the receipt of gifts and entertainment and approval at a senior level that this did not contravene the company's ethical policy or rules.

The term 'quid pro quo' means that there is a specific understanding or agreed terms for the size of the gift and its meaning in terms of compensation and not illegal influence for unfair gain. Rose-Ackerman (1999)[33] describes four conditions with relation to defining whether a gift is a bribe or a tip is a price in Table 10.2. The bribe emerges from paying an agent on your behalf to give you compensation or favour which is not known to the principal, where the principal is the one providing the service you seek and the agent is an enabler of the service. A tip is a small undefined payment (no quid pro quo) to the agent, which for better clarity should also be declared before being taken by the agent.

In many cases companies have indicated clear guidelines to their employees, but in others it is a matter of precedent or sometimes negligence in leaving employees to decide where they were influenced or not. The latter is unsafe and may lead to concealment and hazardous actions. Transparency is the safest way to stop bribery-induced fraud and unfair advantage. The International Association of Accountants has suggested a matrix for its members.

This model is useful because it deals with different levels of normality and gives a knowledge-based rather than a prescriptive system which can be adjusted to different cultures. It can be applied to any gift or entertainment offer received or made in deciding an appropriate ethical level. In offering gifts, Cunningham (2003)[34] suggests that there is an obligation on the giver to also assess whether it is undue reward.

*Table 10.2* The determination of a bribe, gift, tip or price

|  | Quid pro quo | No explicit quid pro quo |
| --- | --- | --- |
| Payment to principal | Market price | Gift |
| Payment to an agent | Bribe | Tip |

## Case study 10.4 Deciding on status of action

1   If a client sends back a diary or a bottle of wine saying that they are unable to accept the gift, will you land you up in prison or lose your job?

Tip: Is this the client's morals and organisation's policy? Are the gifts intended to be influential?

2   Asking or accepting free services after hours from a friendly bricklayer working on your project to build your extension is definitely an influence on their judgement and clearly is a request that the subcontractor will feel they can't refuse because of your dominant role in their livelihood.

3   Is using excess materials ordered for a project a corrupt action if they are not likely to be used anywhere else on the project?

Tip: Why are the excess materials there? In the case of a mistake whose materials are they? Should you need to make a formal request to a senior manager? Do you need to record it? If so why?

4   Taking *bona fida* waste material form the skip or using equipment which is written off is unlikely to be a problem. However, if you as a manager wrote the equipment off or declared the material unusable and later gained from its use, then a senior approval process must be used. The key principle is transparency.

This borderline between gifts or a bribe is a very important one and is often perceived differently both by the giver and the receiver. However, it is clear that the size and the expectations of a gift are often in conflict and it is very helpful for companies to give some guidelines in both giving and receiving gifts to their employees. In the international situation, this will be a steep learning curve for those working out of their own culture and seeking to make a moral stand without upsetting the sensibilities of the client and business colleagues around them. Clearly from a moral point of view it is better to be safe than sorry and upset people. From a business point of view, there is a lot of pressure particularly on young professionals to get business in very competitive conditions and this may turn over their preferred values either because of direct pressure or because of the need to please or prove their worth. Those 'longer in the tooth' are in danger of deadening their conscience and justifying the means for the ends and move across borders that escalate gifts into bribery. This is a very fine line and practitioners are advised to have an external mentor who they use as a check and balance. This is the recommended case for the Chartered Institute of Management Accountants.

---

### Case study 10.5 Lecturer–student relationship

1   A lecturer in their relationship with students is unlikely to accept any gifts before they have completed the assessment and examination of all work that student has or will submit.

2   If they have a conflict of interest, such as a relationship with the student, they may ask someone else to mark or double mark the work or request strict anonymity by the student. The external examiner is a good judge here who also needs to know about any conflicts of interest.

---

## Competitive bidding

In the case of open competitive bidding, the client has an obligation to use the cheapest contractor. They may not have been advised on the pitfalls of a 'fixed price' contract allowed through the traditional method of bills of quantities or drawings and specification, where changes and unspecified gaps in the scopes of works can often lead to contractor claims.

In terms of value for money they would also want to reduce their running and maintenance costs, the fees for the various services they receive from their consultants, the cost of funding for money borrowed and any tax breaks they may get to offset the capital cost. This can only be assessed by having a balanced scorecard assessment on different published criteria. Regrettably inexperienced clients are often not properly informed of the difficulties of not having a single point of contact to negotiate with the contractor, the costs and loss of reputation if there are health and safety incidents caused by poor management and resources, and the tendency for some contractors to find loopholes in the specification and drawings provided and to make claims to claw back losses.

If many are bidding, it creates waste through many hours of unproductive bidding costs from losing contractors. A client who engages contractors to bid for single stage design and construction will incur particularly heavy costs to contractors who will have to factor in their bidding costs to their degree of success – a one in ten success rate means that ten sets of bidding costs must be paid for in each successful contract.

### Ethics

The client may be tempted to use a Dutch auction[35] with the lowest contractor to drive the tender price down, thus destroying the integrity of a competitive bid and the contractor may be tempted to offer further services to turn the client's head from the quality of the first. The integrity and trust of a level playing field have been lost. This is only partly justified if the client pays for bidding costs.

Under Stansbury's (2003)[36] definition, it is clear that both a Dutch auction and the pre-planned intention to tender low and clawback are corruption. However, an inexperienced client may also see themselves disadvantaged in a negotiated situation where they might feel vulnerable to expedience by others at their expense.

Under Rose-Ackerman's conditions for corruption, open tendering creates some frustration which could lead to bribes being offered. This triggers the ethical argument for selective or negotiated tenders. A selective competitive bid involves first pre-qualifying contractors on the basis of the ability and experience for a particular project – effectively a two-stage tender. Conversely this creates conditions for earlier bribe offers in order to get onto tender lists. Codes have been produced to make sure that there is a common and familiar standard.

## Codes of practice

A Code of Tendering seeks to provide guidance and is compulsory for some public bodies. Such a code has been prepared by the New South Wales State Government[37] in Australia, which is based on ten ethical principles that covers open, invited and negotiated tenders for public projects. The ten principles refer to honesty and fairness, avoiding collusion, intention to proceed, equal conditions of tendering, clear requirements, open evaluation criteria available at time of tender, confidentiality and declaring conflicts of interest. These are applied at all levels of the supply chain in this particular code and therefore close one of the loopholes. Poor tendering controls is seen by TI (2003)[38] as one of the key areas for bribery and corruption in the procurement of construction services.

Alternatively a client may be keen to do business with a particular contractor or design company of known compatibility and has the right to make a decision to negotiate or partner with a single or group of contractors. Public clients need to justify negotiation, for example the EU Public Procurement Directives,[39] see negotiation as potentially uncompetitive and non-accountable to taxpayers where a negotiated price has unproven best value. It is also not seen as a level playing field, but may overcome the problems of lowest price 'by stealth' described above.

## The European Union (EU) Public Procurement Directives (1996)

The principles of the EU Procurement Directives are to provide equal access of qualified companies across Europe to bid competitively for larger public projects and services. A lower threshold was applied to purchasing above which the Regulations apply. This is approximately £3.6m for public works and £144,371 (2006) for services and utilities. This allows for smaller contracts to deal with smaller and more urgent projects locally and on a quicker time scale. The Regulations call for an advertisement in the *EU Journal* (OJEC), minimum tender periods and a right to know the reasons for rejection of a bid. They also call for systems which make the criteria for selection absolutely clear and particularly reasons that do not select the lowest cost. Some projects can get exemptions or a degree of exemption to justify negotiated final prices. This is the case in the award of PFI and PPP contracts which can gain value from negotiating with a preferred tenderer after the initial advertisement and competitive bid criteria to get on a short list have been satisfied. The European Commission continues to monitor this type of tendering process. There are exclusions in the Regulations to restrict procurement for public works which are connected with national security. Unfortunately arms type contracts, which may fall into this category, are some of the most prone to corrupt agreements.

The updated Regulations now allow electronic tendering and have merged the original three sectors into one simplified Regulation for services, works supplies and utilities. Authorities also have to publish evaluation criteria and weightings which enhances their transparency.[40] The construction sector mainly comes under works supply. The telecommunications sector has also been interestingly deregulated, as coming to private competitive maturity.

### Ethics of procurement

Ethically the principle of transparency and integrity in public contracts is underpinned by trust supported by certain common-sense precautions, which do not limit the flexibility and promote value for money. Public procurement is fraught with danger of bribery and corruption as explained above. With reports indicating the delicate balance of public procurement with bribery and corruption such as TI's, there is a need to be seen to be fair in public procurement, but not to lose value for money.

### Case study 10.6 Bid for Council project

I am Director of Gubbins Builders and my estimator puts in a sealed bid for a project for a school for 'Costing Money Council'. The Council's chief quantity surveyor whom I have worked with before, tells me that I am the lowest price by far and if I check the bid and make sure that I had not made any mistakes I could still raise the bid a few thousand and still be the lowest and get the job. He rationalised his action by saying 'You won't think the Council is trying to get something for nothing' and 'You and I are old friends, now, what would the world be if we couldn't help each other out – we both know you are the quality job and we can work something out in the valuations.' I resubmit my bid at a few thousands higher and . . .

## Questions

I hope that you would agree that enough has been said to make this a clear corruption case, but can you explain what it is the QS and /or the builder have done wrong?

1   What is the likely detailed offence that they will individually or severally be charged with? Why? (Tip: look at the Stansbury TI summary above.)
2   Is spotting a mistake and not telling the contractor wrong?
3   Is there any legal recourse for the contractor if the mistakes are found after signing the contract?

## Solution

The QS has shared confidential information with the contractor on what would contravene the code of practice for selective tendering which the council would have to comply with. A public official should not suggest a level of correct pricing to contractors. The OFT under the Competition Act (1999) would consider that this is giving unfair advantage to one tenderer and would be able to build a case for collusion and deception of the other tenderers who were told this was a sealed bid. The contractor should not be allowed to alter his bid on the basis of knowledge of the other bids and could be fined. They do not need to prove a gain has taken place, just evidence for collusion with the council or other bidders. If no other information is forthcoming, all contractors could be fined.

There is a corrupt payment implied in two ways.

1   The contractor gains thousands of pounds more than they planned.
2   By implication, the QS gets a cut of the money saved, both of which are money fraudulently obtained.

This would imply a charge of deception and passive bribery and fraudulent gain by the Director who has concealed this information from others. An estimator who has changed the bid on request may also be considered to be fraudulent if s/he has information or suspicions why. As for the QS, a charge of bribery and embezzlement of funds, as the money goes to a private bank account after the revised bid adjusted for a payment to the QS. This stopped his employer from receiving a better value for the contract.

### Fair terms

In the case of a mistake made by the contractor, most codes allow a scrutiny of the rates and if a mathematical error has been made (not a low rate unless it appears to be a typing error) then they are given a chance to stand by the tender or withdraw. This is a fair stand as:

1   The QS can calculate the size of their error, but do not know where they stand in relationship to others.
2   A contractor would be working under intolerable temptation to make other spurious claims if they had made a sizeable error(s) that disadvantaged them by a significant percentage. If it was an error upwards, then the Council is also likely to gain from a correction later after acceptance.

If higher rates for work to be done and paid for earlier in the contract are included, this is called front loading so that contractor cash flow can be improved. If the QS believes this is deliberate when he evaluates the tender, then he is also at liberty to invalidate a bid under most tendering codes of practice. To negotiate at this stage would be unethical to the others and, if accepted, the client should be told the implications. There is a marginal recourse to fraudulent accounting of front loading by the contractor, but the usual way would be to disqualify the bid.

The above case study shows that there are a number of choices a contractor has ethically. The guiding principles are transparency by requiring a clear specification and publishing the deciding criteria and equality (level playing field) by publishing any clarifications to all and not entering into further negotiations with a single or favoured group of contractors. A further preferred bidder(s) round should only be played when the tender adjudication is sure that selected bidders have met conditions to be preferred.

## Competition and collusion

Fair private competition in the EU is based on the concept of open access to competition and a level playing field for competently matched suppliers chosen by the clients (Article 81/82). In their turn the competitors ensure they do not in any way restrict the competition process so that prices may be artificially raised through processes which are described below.

## Prosecuting collusion

In the UK, the Office for Fair Trading (OFT) is the body responsible for policing the Competition Act 1998 which regards collusion and sharing of information as a contravention of competition. OFT have a duty to ensure that markets work properly and remain competitive. A competitive market is described as one where there is equal chance to a range of bidders (open or pre-selected) to win work on equal terms, and covers bid competitions for public or private work and for any size of contract. If there is a sense in which competition is restrained beyond the apparent position by the suppliers, then this is considered to be deceptive and/or fraudulent in its effect. The resulting penalty is a fine through the OFT based on a formula they have devised to tax unfair gain to the participating contractors. As these gains are deemed to spread across a continuing market, they are based on a percentage of turnover of the company, but if a company also had a clearly independent windows division, then this may not be considered in turnover. Thus roofing has been defined as a market. Roofing has been discussed in more detail in the case study in Part II, Chapter 14. OFT regard four contraventions which spoil a competitive market at bid stage which are:

- *Cover bidding* – this is used when a contractor feels unable to put a bid in due to over-commitment, but believes that if they do *not* submit a bid it will count against them next time the client makes a similar enquiry. It involves contacting another contractor on the tender list to get a 'cover price' so that they agree a price which is above the other contractor.
- *Bid suppressing* – an agreement among suppliers to restrict the number of bids or to withdraw bids so that the odds of the remaining bids are higher.
- *Bid rotation* – an agreement for a pre-selected supplier in a given market submits the lowest bid on a rotating basis. Here competition between participating bidders is ruled out as other will make sure that they bid higher.
- *Cartel* – regarded as a formal agreement among a few dominant players (oligarchy) to introduce monopoly conditions for forcing prices higher in a sustained attempt to manipulate the market. This is regarded as a criminal offence under the Enterprise Act and a high level deception and cover-up which can severely distort market. It often includes attempts to close the market to new entrants by underhand methods.

The common factor (RICS 2006)[41] is 'that a contract price is determined by the suppliers and not by the market'. The OFT seeks to ensure that the market is working properly, which means there is nothing distorting competition among suppliers in that market. It does not have to prove that the final price was not competitive, but simply that there was an intention by suppliers to distort or suppress the market or to be involved in non-competitive bids. They have the power under the Competition Act 1998 to fine any guilty companies up to 10 per cent of their world-wide turnover on production of evidence. This is very much in the spirit of Article 81/82 of European competition law. The fine OFT decides, is on the basis of

- global turnover;
- mitigation or severity, e.g. the direct involvement of a director is considered more serious;
- leniency – a 100 per cent rule has been offered by the OFT to the first company in the market to admit involvement and to offer information as to the extent of the collusion. Further partial leniency is offered to other competitors who can give additional information before they come under further investigation.

This feature of penalising collusion among bidders on the basis that it distorts fair competition with penalties directly connected to the trading position of the offending organisation is levelled by the investigating agency, OFT, and not the courts. This has great potential to deter offenders because it is quick and a direct penalty against unfair profit gain. It also incentivises reluctant parties who weakly participate because of the imperative of market forces to expose uncompetitive practices. In addition, in the case of serious long-term collusion, such as major cartels, a company can be prosecuted criminally under the Enterprise Act 2002 and the case prosecuted by the Serious Fraud Office through the courts. Penalties in this case can include imprisonment of directors and banning from director positions as well as company fines.

Ethically this definition of *corruptness as a distortion of the marketplace*, purposely by all or not, is a transparent way of defining corruption through an economic argument that value is lost in the system with possible gain for a few. This is a utilitarian argument for the benefit of a wider audience.

The principle of investigating the marketplace and recognising a collective culpability in the marketplace for those with intent or wishing to ignore the law, seems more appropriate and has proved effective. It is easier and cheaper than trying to prove beyond reasonable doubt that a client has suffered harm at the hands of a supplier and is a powerful way to deter and challenge the collective might of the 'corruptive' intent. It also challenges the loosening morals of successive collusive actions which have become normal practice and lack of transparency, justified by business confidentiality. A classic example of justification is the use of a cover price as seen in the CIOB survey (2006). Through the leniency clause, those pulled in by the majority who would prefer a Kantian ethic, are given a whistle-blowing incentive to get business processes above board so that end does not justify the means with no controls.

## Integrity pacts

In the pre-qualification and bidding stages we have seen there are a number of ways in which a bidder can gain advantage or in which a project owner can ensure a win from their favoured bidder. Another approach for inducing transparency is to try and create a voluntary pact to create a level playing field of integrity, where client and contractor alike agree specifically to report any corrupt practices that emerge to a third party. This makes whistle-blowing easier to do because it is agreed as a primary method by all bidding parties. It needs to be a commitment throughout the successive layers of the supply chain to ensure that first-level suppliers are not ignoring or practising corruption at lower levels of supply. By itself this has been successful in limited situations where an influential client has been fully committed to this such as the Indian Oil and Natural Gas Corporation (ONGC) which has signed agreements with 500 of its suppliers[42] and quite a few government departments world-wide including the Indian Ministry of Defence who have insisted on these provisions.

It should be noted that in all cases where this pact has been used, a neutral authority has been nominated so that confidence in an investigation fair to both sides permits solving the problem and maintaining essential trust by both parties. The parties agree to a degree of trust that is not overruled by more senior partners. The integrity pacts are project based and signed as such by senior project personnel. They will fail if not taken seriously and this usually means the client initiates this in the contract. In the case of public corruption which may reach a high level, the public agency must be able to circumnavigate political intervention and report directly to a parliamentary committee.

TI has devised such a pact[43] and offer themselves as an independent arbiter for reporting the corruption so that neither party has the sole basis for determining a complaint. This has worked well in some situations but has not taken off as expected. There is also a similar tool used by the IMF in their own contract conditions that has worked well in the international contracts they control because of their leverage to withdraw funding. It has not been used as an instrument to secure convictions.

## A hybrid anti-corruption model

Fewings and Henjewele (2008)[44] have considered a competition value integrity model (CIVM) to bring together the machinery for an integrity pact *with* an agency for policing competition so that penalising corrupt activity in the fraught area of international procurement is easier.

Project anti-corruption measures are needed in the procurement of the project when an anti-corruption pact signed by the client is sent out to bidders, making whistle-blowing for corruption both easier and an obligation during the execution stage when training, monitoring and enforcement against corruption take place. These measures are supported by the OECD Convention which requires home countries to monitor and prosecute international companies who bribe foreign officials. However, these measures can break down because prosecution is long-winded and witnesses disappear or are subject to long-term pressure and evidence is hard to collect. Large companies also initiate the bribes and are not easily hurt by long-term intentions which result in inter-government compromises in a wider perspective of aid agreements. Motivation to blow the whistle on corruption is still low because, on a single project, corruption is still lucrative to the briber and the bribed. Partnering so that project integrity pacts are transferable to other projects does not occur on public projects so good practice cannot be passed on as the case for a fully committed multi-project client such as ONGC.

To overcome these obstacles it is proposed that an anti-competition agency such as the OFT in the UK is introduced which provides an independent conduit for penalising multinational companies by their home countries. The fining procedure is based on a single agency on the basis of a known set of rules and to an extent that can levy quite hefty penalties if based on a percentage of the trading turnover of very large companies and act as a deterrent. The companies bidding for a series of industry-related development contracts could be subject to a single monitoring and fining process by the OECD or prosecuted by a single agreed home country anti-competition agency in an extension of the OECD Convention using a common level of fines.

This agency would also have a leniency clause for companies that did whistle-blow existing corruption so that they paid nil or less fines for their own involvement in the corruption at the bidding stage. This would motivate companies to admit to one or several corruption cases and would provide evidence to break cartels and rings of bribery and corruption within the supply side.

An investigation would also spark off the integrity pact so that the independent assessor could investigate the client and political corruption within the receiving government and present evidence arising from the company investigation. Other privileges such as export guarantees could also be withdrawn from companies by individual home governments under the existing UNDP Convention (1998) as an additional deterrent.

## Case study 10.7 BP America anti-trust

In the case of BP America, the company overcharged consumers for propane supply by restricting supply and causing higher wholesale prices. The USA Justice Department[45] charged them fines of £372m which is an example of how anti-competition penalties can apply to the largest of corporations. These can be followed by unlimited class actions by groups of individuals and corporation world-wide reputation suffers. CIVM might be helpful in this situation.

This model would bring together voluntary resolve for transparency and improvement of procurement conditions through the pact with a network of organisations with powers to penalise anti-competitive practice and the use of bribes. This together with other wider international measures against money laundering and the outlawing of bribes so that companies cannot get tax breaks would have the potential to contribute to breaking the culture of bribery and corruption that surrounds the awarding and execution of large contracts.

### Complexity of construction

TI regards construction projects as complex with many participants and contracts down the supply chain with subsequent work concealing previous work completed. There are also complex client financing arrangements to aid cash flow for the client and interim payment valuations which represent part payment for work done which may change in scope. There is a culture of commercial confidentiality in the agreements between suppliers. None of this aids the transparency of accounts either in the financing of the project or in the payment systems during execution so full details of due payments are hard to audit effectively. Public officials may be comparatively independent in their authority to accept work and can extort additional payments through subcontractors in the context of the financial complexity above.

Further progress obviously depends upon helping weaker government anti-corruption mechanisms with measures that can be backed up by stronger economies to overcome the international violations by multinational corporations which have turnovers and influence much larger than the developing countries they work in. These companies have a greater moral responsibility to do fair business.

## Corruption in property deals

There are a number of temptations to do fraudulent dealings in property and these revolve around the under-valuation of property for which there is an interested buyer who is prepared to pay out some of the gains made by resale of a property which was bought below market price (briber led). Or the retaining of select properties to sell at an average rate to a favoured buyer (bribee led). The unauthorised lending of money on terms below market rate in order to make privileged purchases and the false transfer of ownership of property are two others.

A lot of property fraud is also connected with valuing property highly to obtain larger mortgages. In some cases these properties are hugely overvalued. A case involving overvaluation has emerged where estate agents have been under investigation for overvaluing three commercial properties at £16m in order to get a dishonestly large loan for their client.[46] Cases have been reported with the changing of land register details which were displayed online. This resulted in the service being withdrawn. A further case has involved sending fraudulent documents to the land registry claiming a change of ownership.[47] The use of bribes to get agents to carry out these services is also common so that the fraud itself can be kept at arm's length.

The matter of trust is an interesting one as Gambetta (1993), in Rose-Ackerman,[48] refers to the way in which the public lost trust in the Italian government agency to manage property transfers safely because of corruption and under-capacity causing delays which led to vendors and buyers employing third parties to protect their interests. Because of the Mafia connection which emerged for protection, he distinguishes between trust in a reliable process such as a legal agreement and trust in close personal ties, the latter of which he sees as open to corruption, because it is not trust based on competence and objectivity, but on favouritism. This Mafia trust has been connected with the emerging situation in Russia, i.e. that because of government inability to protect legal interests through legal law and order agencies, other illegal agencies come into being. Three case studies follow to illustrate different aspects of property collusion.

## Case study 10.8 Selling scam[49]

The Focus Housing Association case involved the selling of 140 properties owned by a private vendor to Focus Housing Association for inflated prices. This took place between 1991 and 1995 in the West Midlands, UK, and involved payments to housing association officers who approved the price over and above the market value. Some £22,000 was given out in bribes to the officers for properties worth £3.5 million as the vendor shared some of the gain he made over slightly inflated prices. £16,000 was received by the director of the property department and £5,500 to a manager in the same department. The case came to court in 2000 and convictions were made, jailing all three for 18 months (vendor), 12 months (director) and 9 months (manager) respectively. They also had to pay court costs totalling £20,500 between them. The case was referred to the West Midlands police by senior management in Focus who had suspicions and the evidence looked at 12 counts of bribery shared between the HA officers. Interestingly the two senior proponents pleaded not guilty until they realised the evidence was overwhelming. The manager pleaded guilty straightaway to five counts. Subsequently it has been a tactic by the Competition Agency to give leniency to the first admitting party to flush out the others.

### Questions

1   What is the economic harm of overvaluing houses?

2   What is the difference between ordinary inflation of house values caused by demand and excessive profits?

3   Who lost out on the activities above and was value of housing provision undermined?

## Case study 10.9  Borrowing scam[50]

This concerned a case of a property developer borrowing money from a bank with collusion between a director of a development company called Trebane, and its associated construction company called Construct Reason, and a senior manager of a bank called Samuel Hill. Skingly received corrupt payments in order to give favourable borrowing terms to the developer/contractor. Over a period 1988–1992, payments were made totalling more than £650,000 by the developer director to the bank official, in return for larger than normal commercial loans and for full payment of contractor claims. This made the work very profitable and the loans were more affordable.

Samuel Hill was later taken over by Lloyds TSB who suspected corrupt dealings and informed the Serious Fraud Office (SFO). In the court case in 1999, the director and the manager both ended up in jail for 9 months and had to pay costs of £450,000 between them. Interestingly, the partner of the director was not charged with bribery. The charges were made under the Prevention of Corruption Act.

### Questions

1   Is favourable lending to borrowers always bad? Substantiate your answer.

2   What should guide the size of the penalty in order to deter or to reform future corrupt activity?

3   Is imprisonment or a fine the greatest deterrent? If so why?

4   What sort of checks and balances could be imposed by the bank to prevent fraud?

## Case 10.10  Urban planning scam in Marabella[51]

In this case, the Marabella City Council was dissolved and a management committee installed by the Spanish government in the face of serious allegations that the Head of Urban Planning had illegally built up a £1.7 billion fortune from bribes and backhanders offered to him over 15 years in office in approving 600 illegal developments in the city. Other members of the city council have also been arrested for being involved either in ignoring or gaining advantage or benefit. The head of planning is believed to have amassed a fortune of up to 10 per cent of the value of the developments approved and is also alleged to have turned down developments, bought the land himself and changed the city planning rules so that he can develop the land himself. The rot also seems to have spread to other departments as the police chief has also been arrested on charges of corruption in receiving bribes from a businessman for the right to run the city's illegal parking, tow away scheme.

### Questions

What are the ethical issues involved in the planning case under

1   A utilitarian approach (the good of the many is met by the least bad to the few)?
2   A Kantian approach (the means is important in spite of the end)?

If the terms of the UK 1998 Corruption Bill, what are the charges that could be made?

1   To whom would these charges be directed?
2   Is it likely that rot spreads if corruption is not stopped? Why?

The above case studies show a range of situations which question the integrity of property and planning deals and contribute to a breakdown of confidence not only of the company or authority in question, but spread a cloud over the whole profession. Corruption in property transactions reduces the scope and resources for new development and reduces client value.

## Money laundering

This particular aspect of corruption is the hiding away of money that has been illegally obtained so that its destination is hidden. In its simplest form this means moving the money offshore to a protected bank account where there is an agreement not to disclose the customer's details. Under the new anti-laundering agreements internationally, it is now difficult for such banks to refuse access to a crime agency who have distinct evidence of a specific account, as used to be the case in Switzerland. However, more subtle electronic transfer which moves money from one account to another very quickly until it reaches a less controlled banking system is less easy to track and to investigate if the individual bank is unknown. This has led to the Money Laundering Regulations which means that businesses are legally required to report any suspicions of money laundering with clients and to carry out a risk assessment of new clients.[52] Signs of money laundering might be the overpayment of a contractor in order to receive a clean cheque back for banking.

The Proceeds of Crime Act 2002 also requires awareness and risk assessment to recover any property that is associated with crime including cash. A risk assessment and training of staff are requirements to spot any likelihood of property being stolen and this might apply to shady contractors bringing on stolen plant or materials.

In the Philippines, Marcos was accused of using reform to make it easier to corrupt systems and get the money into a safe offshore bank account[53] and Benazir Bhutto was initially convicted for corruption over alleged receipt of payments from capital projects into Swiss bank accounts[54] during her time as prime minister in Pakistan (1988–1990, 1993–1996), but she was cleared by the Pakistan Supreme Court. Money laundering to Swiss bank accounts had been the key accusation.

International anti-money laundering efforts are guided by agreements through the UN Office on Drugs and Crime (2007)[55] who run an anti-money laundering unit.

## Achieving reform of corrupt value

It is generally recognised that corruption will always retain some role in the economic sector as it is never possible to eliminate inequalities completely in complex state systems as indicated by the four Rose-Ackerman conditions. Procurement corruption could also be a subset of fierce competition as clients can demand more specifications for less reward. Human nature requires more certainty and building up relationships which secure more certainty and reliability can sometimes become too comfortable. The equation of need and greed can also override 'decent' behaviour.

However, reform has been connected with recognising basic causes and trying to undermine its base. The context of the political system is important as indicated in state capture. In a democratic system with informed voters and a reasonable chance of voting a government in or out, politicians need to declare interests such as directorships, shares and payments made so that fiscal transparency is maintained in the government and individual accounts. A system of parliamentary committee needs to exist so that government ministers can be questioned with impunity and confidential information accessed. An independent judiciary which is not connected to any political colour or party needs to adjudicate against legislative measures and over-rule government behaviour if necessary, without fear of coercion and with juror protection. A free press is an essential instrument for sniffing out indiscretion and bias, though it can also bias outcomes by strident and spurious leading opinions. It is also important to ensure that party campaign funds are collected diversely and not dependent on a faction of party policy or a few wealthy benefactors who are also lobbyists and could influence party policy away from the democratically voted members.

## Conclusion

This chapter has only scratched the surface of an area of corruption which is passionately written about by academics in many texts, but there is strong agreement about the harmful economic effect of corruption and bribery even in relatively mild forms, due to its pervasive effect in leaking resources away from productive use. This happens particularly in construction which has the worst record of any industry and reduces investment in the built environment. This chapter has investigated the strong role that corruption plays in construction in developed and developing countries, albeit in a different form. Although state structures to deal with corruption are relatively strong in developed countries, there is a tendency for new forms of corruption to arise through the distortion of free competitive processes. Many traditional trading practices in construction are outlawed by the TI, amounting to fraudulent, deceptive and dishonest activity, which over the years has become the norm, particularly in the procurement of new work. In developing countries, a lot of work is needed to develop fledgling state administrative structures and to produce further transparency into the way that government performs.

In its international forms, corruption has received much attention from development agencies who have tried to create a level playing field to define and outlaw corruption and bribery in the area of procuring construction projects financed or offspinning from international aid. The OECD has recognised that there needs to be a focus on the bribe givers who have often been tacitly ignored by donor countries and who are tempted to see the aid country as a huge fillip to their export balance of trade and have ignored the cumulating corrupting effect that bribes have had in encouraging officials and politicians in developing countries to steal value from their economies. In addition, some large companies have justified an alternative standard of ethical engagement and have weakly stated that they have to abide by a standard they themselves have corrupted with bribes which leak value out of the system. The OECD agreement requires home countries to police the multinational actions and signatories have agreed to criminally prosecute and deny further export guarantees to those who contravene the higher ethical standards.

The competitive climate for construction tendering is a harsh one with low margins and a recurring economic cycle of booms and crashes. Companies have to make sudden cutbacks with dramatic loss of demand and work to tight margins in order to win tenders against other work hungry contractors. This can encourage clawback of loss by spurious claims. In addition, it is noted that some collusion is exposed in the UK, but has been tackled quite effectively by the use of a competition agency to set fines in the competitive market. If these powers could be extended internationally under international agreements with other countries this would be an additional weapon in the fight against crime.

## Appendix: Corruption and bribe payers indices

Corruption is defined by the Transparency International[56] as mainly to do with the acceptance of reward by public officers for private gain. The main ethical problem with corruption is that poverty is likely to be perpetuated as money that is targeted for public benefit such as the building of schools is uncontrollably diverted away from this purpose. Tables 10.3 and 10.4 indicate that some of the poorest countries are also the most corrupt. When this happens in developing countries, there are other related interactions.

- Aid is likely to be diverted from those countries perceived to be corrupt.
- Where less money than expected is received for the public projects that the governments are committed to, then the country becomes indebted and this causes another spiral of waste through interest payments.

This vulnerability of developing countries to bribery creates a vicious circle which is a brake on development which affects investment and especially the construction industry. The most consistent corruption is measured in the Corruption Perceptions Index which measures the public experience of bribery and corruption, and the political parties are consistently cited across the board as significantly corrupt. The perception of the levels of corruption in those badly hit is generally between 10 per cent and 20 per cent of GDP expended.

Apart from political parties, different manifestations of bribery affect different parts of the world and vary between government legislature, media, judiciary and police corruption. For example,

- In Africa, the perception is that customs and police bribery prevail.
- In Latin America, it is the legislature.
- In Western Europe, the media and political parties.

The implication for this breakdown is that public monies are perceived to leak to political parties and this may well affect the quality and development of the built environment. Sectors perceived to have less corruption are education, the military, NGOs and religious bodies. Business and the private sector come fourth in seriousness after public services, but it is considered worse by those living in the Western and Central Europe. Much of this corruption is perpetuated in developing companies by the propensity of multinational countries to play dirty away from home and to offer bribes in order to gain access to services. Their dilemma is that other companies are offering bribes and they need to match these bribes and add more in order to get a foot in the door. This has created an escalation in expectations by local officials accustomed to small service payments and by politicians who are playing the higher stakes. This results in a runaway loss of value to the economy.

## Bribe Payers Index

TI compiled a survey of developing countries and asked them what propensity twenty-one developed countries had for paying bribes and produced the Bribe Payers Index (BPI). This shows that Australia is top with 8.5 out of 10, with the UK at 6.9 (Table 10.3). China, Korea, South Korea, Italy and Hong Kong scored below 5 and the USA, France and Japan just above 5.[57] Russia is bottom at 3.2.

Domestic companies were the lowest scorers in comparison with the perception of what foreign companies were doing in that country as locals were rating them higher. This propensity of developed countries to drop their standards when working in developing countries has led to an agreement in OECD countries[58] to penalise and withdraw benefits to their own companies working abroad who pay bribes. Osbourne and Clarke refer to the problems of export guarantee credit corruption and the emphasis on making sure that money guaranteed in this area is not diluted by the payment of bribes.

## The Corruption Perceptions Index

The Corruption Perceptions Index (CPI) is collated through combining survey material from a range of analytical sources by Transparency International. It is quite robust. It draws on 16 polls from 10 separate institutions.[59]

The top sixteen countries score above 8 in the index, with Iceland scoring the top at 9.7 and six others in Scandinavian countries, Singapore, Switzerland and

Table 10.3 TI's Bribery Payers Index, 2002

| Rank | Country | Score | Rank | Country | Score |
|---|---|---|---|---|---|
| 1 | Australia | 8.5 | 12 | France | 5.5 |
| 2 | Sweden | 8.4 | 13 | United States | 5.3 |
| 2 | Switzerland | 8.4 | 13 | Japan | 5.3 |
| 4 | Austria | 8.2 | 15 | Malaysia | 4.3 |
| 5 | Canada | 8.1 | 15 | Hong Kong | 4.3 |
| 6 | Netherlands | 7.8 | 17 | Italy | 4.1 |
| 6 | Belgium | 7.8 | 18 | South Korea | 3.9 |
| 8 | UK | 6.9 | 19 | Taiwan | 3.8 |
| 9 | Singapore | 6.3 | 20 | PR China | 3.5 |
| 9 | Germany | 6.3 | 21 | Russia | 3.2 |
| 11 | Spain | 5.8 | 22 | Domestic companies | 1.9 |

*Table 10.4* Countries in range 8–10 on the TI Corruption Perceptions Index, 2005

*TI 2005 Corruption Perceptions Index top sixteen*

| Country | Ranking | 2005 CPI score | Range | No of surveys |
|---|---|---|---|---|
| Iceland | 1 | 9.7 | 9.5–9.7 | 8 |
| Finland | 2 | 9.6 | 9.5–9.7 | 9 |
| New Zealand | 3 | 9.6 | 9.5–9.7 | 9 |
| Denmark | 4 | 9.5 | 9.3–9.6 | 10 |
| Singapore | 5 | 9.4 | 9.3–9.5 | 12 |
| Sweden | 6 | 9.2 | 9.0–9.3 | 10 |
| Switzerland | 7 | 9.1 | 8.9–9.2 | 9 |
| Norway | 8 | 8.9 | 8.5–9.1 | 9 |
| Australia | 9 | 8.8 | 8.4–9.1 | 13 |
| Austria | 10 | 8.7 | 8.4–9.0 | 9 |
| The Netherlands | 11 | 8.6 | 8.3–8.9 | 9 |
| United Kingdom | 12 | 8.6 | 8.3–8.8 | 11 |
| Luxembourg | 13 | 8.5 | 8.1–8.9 | 8 |
| Canada | 14 | 8.4 | 7.9–8.8 | 11 |
| Hong Kong | 15 | 8.3 | 7.7–8.7 | 12 |
| Germany | 16 | 8.2 | 7.9–8.5 | 10 |

*Table 10.5* Countries in range 0–2 on the TI Corruption Perceptions, 2005

*TI 2005 Corruption Perceptions Index bottom nine*

| Country | Rank | Index | Range | Surveys |
|---|---|---|---|---|
| Angola | 151 | 2.0 | 1.8–2.1 | 5 |
| Côte d'Ivoire | 152 | 1.9 | 1.7–2.1 | 4 |
| Equatorial Guinea | 152 | 1.9 | 1.6–2.1 | 3 |
| Nigeria | 152 | 1.9 | 1.7–2.0 | 9 |
| Haiti | 155 | 1.8 | 1.5–2.1 | 4 |
| Myanmar | 155 | 1.8 | 1.7–2.0 | 4 |
| Turkmenistan | 155 | 1.8 | 1.7–2.0 | 4 |
| Bangladesh | 158 | 1.7 | 1.4–2.0 | 7 |
| Chad | 158 | 1.7 | 1.3–2.1 | 6 |

New Zealand above 8.9. The UK is at number 12 with 8.6. The USA, France, Japan and Spain (just) fall in the bottom half of the 7–8 range, the highest Arab country is Oman at 6.3, Israel is 6.3, Botswana, the highest African country, is at 5.9, Italy is at 5.0, Brazil at 3.7, China at 3.2, India at 2.9 and Russia at 2.4. Below 2.0 are five African countries and Bangladesh, Myanmar (Burma), Haiti and Turkmenistan.

Out of these countries there has sometimes been improvement over the period 2000–2005. The UK, India and Botswana have stayed in almost the same position in this period. Italy, Australia and Nigeria have improved substantially, and South Africa has got worse. There will of course be fluctuations dependent on the economic circumstances and Arab countries were not included in the 2000 polls.

## Notes

1 De Graaf, G. (2007) 'Causes of Corruption: Towards a Contextual Theory of Corruption', *Public Administration Quarterly*, 31(1): 39–86.

2 European Council (1998) *Criminal and Civil Law Conventions on Corruption.* Luxembourg: European Council. http://www.conventions.coe.int/Treaty/en/Reports/Html/173.htm

3 Langseth, P. and Michael, B. (1998) 'Are Bribe Payments in Tanzania "Grease" or "Grit"?' *Crime Law and Social Change*, 29(2–3): 197–208.

4 Rose-Ackerman, S. (1999) *Corruption and Government: Causes, Consequences and Reforms.* Cambridge: Cambridge University Press.

5 Ibid.

6 Joint Committee (2000) *Draft Corruption Bill.* Online. Available http://www.parliament.the-stationery-office.co.uk/pa/jt200203/jtselect/jtcorr/157/15704.htm (accessed 12 December 2007).

7 Namibia's Zero Tolerance for Corruption Campaign (2007). Online, Available http://www.anticorruption.info/law_acts.htm (accessed 14 November 2007).

8 Rose-Ackerman, op. cit.

9 A small unit of the Economic Crime Department working with the Serious Fraud Office.

10 United Nations Office on Drugs and Crime (2003) Convention Against Corruption 58/4. 31 October.

11 OECD (1997) Convention for Combating Bribery against Foreign Public Officials in International Business Transactions. 17 December. Signed by the OECD Countries and Brazil, Argentina, Bulgaria, Chile, Slovak Republic.

12 Transparency International (2006) *Bribe Payers Index.* Online. www.transparency.org/policy_research/surveys_indices/bpi (accessed 13 August 2007).

13 IMF (2000) *Manual on Fiscal Transparency. Referring to the OECD Measures.* Available. Online http://www.imf.org/external/np/fad/trans/manual/sec03a.htm

14 Geddes, B. and Neto, A.R. (1992) 'Institutional Sources of Corruption in Brazil', *Third World Quarterly*, 13: 641–61.

15 Ayittey, G. (1992) *Africa Betrayed.* New York: St Martin's Press.

16 The Corruption Perceptions Index is updated yearly. TI ranks 180 countries and these are scored 1–10 using up to 14 different sources.

17 Transparency International (2007) *Corruption Perceptions Index.* Online. Available www.transparency.org/policy_research/surveys_indices/cpi (accessed 10 March 2006).

18 Grey, C. Hellman, J. and Ryterman, R. (2004) 'Corruption in Enterprise, State Interactions in Europe and Central Asia 1999–2002', *Anti Corruption in Transition Series No 2.* Washington, DC: World Bank.

19 Steves, F. and Rousso, A. (2003) *Anti-corruption Programmes in Post Communist Transition Countries and Changes in the Business Environment 1999–2002.* EBRD. Online. Available, http://www.ebrd.com/pubs/econo/wp0085.htm

20 Rose-Ackerman, op. cit.

21 Transparency International (2005) *Global Report on Construction. Construction Review.* March.

22 Fewings, P. and Henjewele, C. (2008) 'International Tendering Practice: Towards a Construction Industry Competition Integrity Value Pact,' CIB *W107 Symposium 'Construction Procurement, Ethics and Technology'*, 18–20 January (Ed. T.M. Lewis). Port of Spain, Trinidad and Tobago.

23 Preventing and Combating Corruption Bureau (2007) 'Corruption Allegation in the Contract between Tanesco and RDC', 12 May online www.freemedia.co.tz/daima/2007/5/12/index.php (accessed 14 August 2007).

24 Fan, L. Ho, C. and Vincent, N. G. (2001) 'A Study of Quantity Surveyors' Ethical Behaviour', *Construction Management and Economics*, 19: 19–36.

25 Vee, C. and Skitmore, M. (2003) 'Professional Ethics in the Construction Industry', *Journal of Engineering, Construction and Architectural Management*, 12: 601–10.

26 Pearl, R., Bowan, P., Makanjee, N., Akintoye, A. and Evans, K. (2004) 'Professional Ethics in the South African Construction Industry – A Pilot Study', in A.C. Sidwell (ed.) *Proceedings of the International RICS Foundation Cobra Conference*, Brisbane, Australia, 4–8 July 2005, ISBN 1–74107–101–1, 60–71.

27 Stansbury, N. (2003) *Anti-Corruption Initiative in the Construction and Engineering Industry: Introductory Report*. September. London: TI.

28 Gillam, K. (2006) 'Bad for Business', *Construction Manager*, February. p. 12.

29 Chartered Institute of Building (2006) *Corruption in the UK Construction Industry*. Survey. Ascot: CIOB.

30 Stansbury, op. cit.

31 Ibid.

32 CIOB, op. cit.

33 Rose-Ackerman, op. cit.

34 Cunningham, M. (2003) 'Present Danger', *Construction Manager*. November–December. http://www.construction-manager.co.uk

35 A Dutch auction is one where an informal unannounced negotiation takes place with a contractor which the client/MC wishes to use in order to reduce their higher bid to the level of the lowest bid, so that they can justify choice and reduce tender costs. This is a direct distortion of the 'level playing field' principle of the competitive tender.

36 Stansbury op. cit.

37 NSW Construction Policy Steering Committee (1996) *Code of Tendering*. Department of Public Works and Services, New South Wales.

38 TI, op. cit.

39 EU *(1992–2000) The Public Procurement Directives*. The original ones are implemented in the UK as The Public Works Supply Regulations 1992 (SI 3279), Public Supply Contracts Regulations 1995 (SI 201), Utilities Contracts Regulations 1996 (SI 2009). Updated to be one in 2006.

40 Osbourne and Clarke (2005) *Public Procurement Update*. Winter. Indicating expected amendments for January 2006.

41 Pinsent and Mason (2007) *Tender Strategies, Anti Competitive Behaviour and the Consequences*. RICS. Online, Available: http://www.rics.org/Builtenvironment/Constructionprocurementandtendering/tenderstrategies_anticompetbehaviour_consquences.html (accessed 17 June 2007).

42 The Integrity Pact: TI India's efforts. Online. Available http://www.tiindia.in/content.asp?ta=The%20Integrity%20Pact&ma=Programs (accessed 21 August 2007).

43 Transparency International (2007) Template 5 of the Anti Corruption System (Construction Projects) (PACS) distributed with bid documents.

44 Fewings, and Henjewele, op. cit.

45 *Washington Post* (2007) 'BP Settles Case for £373m', 26 October. The £373m was a

mixture of two serious pollution cases and anti-rights fines, but the larger quantity of £250m was set aside against a cost of only £53m to the consumer. In total, the incident in March 2006 and the Texas explosion in 2005 and the Alaska incident also in 2006 have cost the company reputation dear in a lucrative market. On line http://www.washingtonpost.com/wp-dyn/content/article/2007/10/25/AR2007102502552. html (accessed 19 November 2007).

46 Hassell, N. (2007) 'Erinaceous Linked to "£10m Property Fraud",' *The Times*, 29 March.

47 Black, M. (2007) 'Solicitor Denies Property Fraud', *Bradford Telegraph and Argus*, 12 November

48 Gambetta, D. (1993) *The Sicilian Mafia*. Cambridge, MA: Harvard University Press.

49 Serious Fraud Office (2000) 'Three Jailed in Housing Association Fraud', press release. http://www.sfo.gov.uk/news/prout/pr_50.asp?seltxt=. (accessed 17 April 2007).

50 Serious Fraud Office (1999) 'Former Investment Loans Manager and Client Jailed for a £22 Million Collusion'. Online. Available http://www.sfo.gov.uk/news/prout/pr_25.asp?seltxt= 25 June.

51 *The Times* (2006) 'Spanish Resort Council Banned over £1.7bn Corruption', 6 April. Online. Available http://business.timesonline.co.uk/tol/business/law/public_law/article702066.ece (accessed 12 November 2007).

52 The Money Laundering Regulations 2003 (Statutory Instrument 2003, Number 3075) http://www.legislation.hmso.gov.uk/si/si2003/20033075.htm (accessed 22 December 2007).

53 US State Department, Manila (1983) 'Creeping State Capitalism'. Online. Available http://www.gwu.edu/~nsarchiv/nsa/publications/philippines/phdoc2.html (accessed 12 December 2007).

54 BBC News (2007) 'Benazir Bhutto's Extraordinary Career', 18 October. Online. Available http://news.bbc.co.uk/1/hi/world/south_asia/2228796.stm (accessed 22 December 2007).

55 UN Office on Drugs and Crime (2007) *An Overview of the UN Conventions and Other International Standards concerning Anti Money Laundering and Countering the Financing of Terrorism*. Vienna: UN.

56 Transparency International (2005) *Media Pack for the CPI*. Online. Available http://www.transparency.org/policy_research/surveys_indices/cpi/2005/media_pack (accessed 29 August 2007)

57 Transparency International (TI) (2002) *Bribe Payers Index of 21 Developed Countries*. op. cit.

58 OECD (1997) Convention for Combatting Bribery against Foreign Public Officials in International Business Transactions. 17 December. Signed by the OECD Countries and Brazil, Argentina, Bulgaria, Chile, Slovak Republic.

59 Transparency International (2005) *Corruption Perceptions Index of 151 countries* getting at least 3 independent poll results.

# Chapter 11

# Delivering ethical improvement through contractual good faith

## Jim Mason

The interaction between ethical principles and legal rights and duties was noted in a paper given by Uff in 2003[1] as 'extending the definition of acceptable professional conduct'. The paper makes the argument that an ethical approach may hold the key to a number of the problems which have beset the construction industry for many years and may lead to a fairer and ultimately more prosperous future.

At the time, Uff was involved in the formulation and dissemination of an ethical code for construction professionals produced by the Society of Construction Law.[2] This chapter seeks to establish another method of promoting ethical behaviour through recent innovations in contractual practice.

The direction and implementation of each stage of the construction process are overseen and managed by professionals. Depending on the nature of the work, they will be engineers, architects, surveyors, lawyers, construction and project managers. Each of these professionals operates under an existing ethical code and the same professionals bring this ethical code with them when they operate in the field of construction law. Notwithstanding this, a widely adopted uniform code does not currently exist and the sheer variety of personnel involved on a typical construction project and the diverse nature of their tasks make such a code difficult to contemplate.

## The construction context

That there is a place for such a code appears to be beyond question. In his paper, Uff[3] identifies a number of individual issues including: whether it can be ethically justifiable to allow contracts to be let on inadequate ground investigation data or in circumstances where a major variation to the works will be inevitable or where grossly under-priced contracts should be let at all. He notes that in the area of tendering, the rule appears to be that anything goes. Unconscionable dealings of tendering contractors can be well matched by the practices of employers. An ethical code would seek to prevent cheating in the same way that duties are owed under the criminal law to avoid or prevent bribery or corruption.

It is often the case that when placing risk through the terms of the contracts and subcontracts the relevant risks are simply transferred to the level of sub-contractors and sub-consultants. Projects then run into difficulty as a result of the parties having taken on contractual obligations which are more onerous than anticipated in terms of time and money. The solution here is to suggest that there is an ethical burden to ensure that the contract draftsmen are properly informed as to the feasibility of the tasks that are being created for others and lawyers owe a reciprocal duty to avoid the creation of risk which cannot be practically and economically borne.

At the construction stage ethical considerations apply in giving warning of avoidable disaster. There are a number of cases where the duty to warn has been considered. However, the point is that the rulings of the court on the existence of legal duties cannot be taken as defining the extent of an ethical duty in such circumstances.

Much of the construction process is still operated and controlled by professionals who are often appointed to carry out an 'independent' certifying role where, despite being engaged by the client, the certifier's function involves 'holding the balance between the client and his contractor' as was identified in the case of Sutcliffe v Thackrah.[4] A common complaint of contractors is that the certifier failed to act in an ethical manner being reluctant to 'concede' extensions of time or to sanction variation orders.

In the third project stage, construction industry professionals will be involved at all levels in the settling of the final account including the formulation and resolution of claims and counterclaims. This also covers adjudication and arbitration should either of the parties so elect. This gives rise to a series of ethical issues concerning the way in which facts are analysed and expertise utilised to support claims.

## The wider context

Many people do not realise that actions which they regard as 'part of the game' are actually criminal activities. Corruption not only includes bribery, but also fraudulent practices such as tender collusion, claims fraud and deliberate supply of sub-standard products or incorrect quantities. Corruption on UK projects leads to waste, defective products, inefficiency and unnecessary litigation. According to TI's Project Director Neill Stansbury,[5] the construction process forms the UK's core commercial activity. Construction in the UK is, however, economically wasteful and excessively costly. This unnecessarily enhanced cost is further increased as a result of the large number of claims and disputes that construction generates. Particular complaints are: inefficient working practices, poor planning, inadequate identification of the scope of required work, unsatisfactory design, detailing and specification and low standards of workmanship. Complaints are compounded by a lack of trust, lack of co-operation and information sharing between participants at every level of the design, planning and construction chain. This is then exacerbated by the unsatisfactory way that many specialists are hired with inter-locking contractual and commercial relationships.

## Partnering and good faith

In the past ten years the introduction of partnering to the UK construction industry has sought to simplify contractual and commercial relationships. Partnering promotes a co-operative approach to contract management with a view to improving performance and reducing disputes. The relationship between a contractor and a client in a partnering contract contains firm elements of trust and reliance. In so far as partnering is delivered through the medium of contracts, those contracts more often than not contain an obligation that the parties act in good faith to facilitate delivery of those aims. The very existence of the principle of partnering implies that the participants will act ethically in seeking to achieve the stated objectives. The premise of this chapter is that by encouraging the concretising of the duty of good faith in construction contracts an improvement will be discernible in terms of improved ethical standards within the industry. The abuses of contractual positions referred to above can be alleviated by the adoption of good faith as the cornerstone of contractual relations.

Under the existing law, provisions of good faith in construction contracts are a long way from being able to deliver improvements in ethical behaviour. The perception is that these type of terms of contract are not taken particularly seriously at the present time.

The background to this is that partnering contracts pose a problem for contract advisors containing as they do 'hard' and 'soft' obligations. While all conditions of contract are equal, some, to misquote George Orwell, are more equal than others. Clients can be advised and terms drafted stipulating hard obligations such as payment and quality standards. But what of the soft obligations – and in particular the duty of good faith – what is to be made of them? As one leading commentator put it: 'We in England find it difficult to adopt a general concept of good faith ... we do not know quite what it means.'[6] The resulting situation is that 'soft' obligations are often overlooked and not given any particular importance. This sentiment was picked up by a report expressing the consensus of construction lawyers as being that duties of good faith are not likely to be newly recognised in law by reason of their introduction into partnering contracts.[7]

This consensus of opinion invites the question whether this is what the users of construction contracts want. Parties having taken the trouble of entering into a partnering contract may feel disappointed to learn that their voluntarily assumed mutual obligations are not enforceable. For the benefits of good faith provisions to be felt, there must be a 'concretising' of the duty of good faith by judiciary and/or parliament to deliver what the parties have chosen for themselves.

## The newer contract forms

By far and away the most popular forms of contract are those which make no mention of partnering obligations.[8] The dominance of the JCT lump sum and design and build forms remains intact. However, the growing trend is to use contracts which move away from formal legal 'black letter' contracts to contracts fulfilling a different role which includes seeing the contract as a management tool and a stimulus for collaboration. The challenge for these newer contract forms is to capture this new role while providing sufficient contractual certainty in the event that disputes arise.

The link between contracts, partnering and good faith was initially made by organisations such as Associated General Contractors of America making statements such as: 'Partnering is recognition that every contract includes an implied covenant of good faith.'[9] These connections are relatively straightforward in the United States, a legal system that recognises the duty of good faith in contracting. The principles of partnering are congruent with the doctrines of trust, open communication, shared objectives and keeping disputes to a minimum. Making the connections in the English context is more challenging given the absence of the general duty of good faith. In its absence it is the partnering contracts themselves which fill the gap.

In the fourteen years since the Latham Report, partnering contracts have become significantly more sophisticated in terms of the wording of partnering obligations and the conduct expected. The duty to act in good faith is a common thread.

There are variations on the exact imposition of the duty to act in good faith in partnering contracts. A distinction can be drawn between those which are intended to regulate the parties' behaviour through the contractual terms and conditions (binding) and those which place a non-contractual partnering framework on top of another contract (non-binding). The latter have been described as seeking to influence rather than mandate certain behaviour.[10]

The parties to the JCT Non-Binding Partnering Charter agree to 'act in good faith; in an open and trusting manner, in a co-operative way and in a way to avoid disputes by adopting a no blame culture'. The binding multi-party PPC 2000 requires that the parties 'agree to work together and individually in the spirit of trust, fairness and mutual co-operation'. The NEC ×12 Partnering Option calls the parties 'partners', and requires that partnering team members shall 'work together to achieve each other's objectives'.

The latest contract to enter the fray is the JCT Be Collaborative Constructing Excellence Form. The contract goes further than the other partnering contracts in introducing an over-riding principle which includes a duty of good faith and stipulates that this principle takes precedence over all other terms.

This contract completes the transition of good faith-type provisions from being somewhere on the under-card of contractual terms to being the main event. A significant proportion of the standard forms of contracts now available to the construction industry expressly impose an increasingly onerous duty on the parties to act in good faith. This chapter will briefly review the history of the duty of good faith before examining the reasons why the consensus of rejection of the legal significance of this development exists.

## The duty of good faith

The attraction for contract draftsmen to use the phrase 'the parties owe each other a duty of good faith' is understandable. The phrase resonates with the reader who has an instinctive grasp of what it is the contract is trying to do. This resonance is due in part, to the long history and high esteem in which the duty is steeped.

The concept of good faith has great normative appeal. It is the aspiration of every mature legal system to be able to do justice and do it according to law.[11] The duty of good faith is a means of delivery.

Good faith has an ancient philosophical lineage and is referred to in the writings of Aristotle and Aquinas.[12] They were concerned with the problems of buying/selling and faced the dilemma of how to achieve fairness while not stifling enterprise in commerce. This dilemma is still an issue today and its successful resolution is a major challenge for those seeking to (re-)establish a duty of good faith.

The ancient concept of good faith in a revived form went around Europe, England and United States like wildfire at the end of the eighteenth century. Lord Mansfield described the principle of good faith in 1766 as the governing principle applicable to all contracts and dealings.[13]

The duty of good faith subsequently fell into disuse in England in favour of encroaching statute law and the emphasis on the promotion of trade. Emphasis shifted onto contractual certainty in contracting instead.[14] Contractual certainty has remained the cornerstone of standard form construction contracts since their inception at the start of the twentieth century. Procurement and contracting in the twenty-first century, however, are different. The role of the contract is changing and the re-emergence of the duty of good faith is an important element in this development. The advantages of recognising the legal enforceability of the duty have been presented as safeguarding the expectations of contracting parties[15] by respecting and promoting the spirit of their agreement instead of insisting upon the observance of the literal wording of the contract:

- regulating self-interested dealings;
- reducing costs and promoting economic efficiency;

- filling unforeseen contractual gaps;
- providing a sound theoretical basis to unite what would otherwise appear to be merely a series of disparate rights and obligations.

To this list must be added the benefit of improving on some of the unethical behaviour highlighted at the start of this chapter.

The support for introducing the duty of good faith among industry commentators has to date not been overwhelming. Academic studies in this area tend towards mild encouragement for the judiciary or parliament to take action and introduce a general duty.[16]

Making the case for the imposition of a general duty of good faith is as challenging as attempting a definition. Despite its beguiling simplicity it has proved to be an elusive term. The attempts to define good faith, or at best replace it with equally vague and nebulous terms. The danger, as one commentator put it, is that any definition would 'either spiral into the Charybdis of vacuous generality or collide with the Scylla[17] of restrictive specificity'.[18]

The difficulty of defining 'good faith' is not necessarily a problem for partnering contracts which tend to evoke the spirit rather than the letter of the law. However, progress has been made in defining the term, particularly by the Australian judiciary. The parallels here are striking – a common law jurisdiction grappling with the issue of how best to 'concretise' the duty of good faith.

The Australian Judge Paul Finn made the following useful contribution towards definition in the common law tradition:

> Good faith occupies the middle ground between the principle of unconscionability and fiduciary obligations. Good faith, while permitting a party to act self-interestedly nonetheless qualifies this by positively requiring that party, in his decision and action, to have regard to the legitimate interests therein of the other.[19]

Thus far, the English Courts have denied themselves the opportunity to engage in this shaping of the meaning of good faith in the modern construction context despite its historical relevance, its resonance with the public and even in light of other recent stimuli to its introduction.

## Other stimuli towards the introduction of a duty of good faith

As mentioned above, English law made a choice to promote trade through contractual certainty rather than through widely drawn concepts. In Europe, the duty of good faith has flourished to the extent that its existence or otherwise in contract law is one of the major divisions between the Civilian and Common Law systems.[20] The great continental civil codes all contain some explicit provision to the effect that contracts must be performed and interpreted in accordance with the requirements of good faith. For example, Article 1134 of the French Code Civil and Section 242 of the German Code.

In France, the rather vague concept of good faith or 'bonne foi' has been given clarity and definition by judicial decisions, which cumulatively have produced a number of 'rules' relating to the performance of contractual obligations and, possibly more importantly, to the obligations of parties before a formal contractual relationship is entered into. For example, good faith is the legal basis for the rules relating to the French doctrine of abuse of rights ('l'abus de droit'). This is where the court adds a further qualification to the specific express contractual obligations to prevent the purpose of the contract being thwarted by a manifestly unfair attempt to rely on a contractual right.

The development of the doctrine in France has also been determined by the nature of the contract being considered. The courts have developed different types of duties based on the general obligation of good faith that are specific to certain categories of contract. In the context of engineering and construction contracts, there is authority that the developer must provide all relevant data that are necessary to the proper completion of the project by the engineer.

German law has adopted an even more positive approach to the doctrine of good faith than the French. Good faith creates positive extra-contractual obligations and is used as a justification to facilitate performance of the contract. The doctrine is contained in sections 157 and 242 of the German civil code, the Burgerliches Gesetzbuch, which provides that:

> S157 – Contracts shall be interpreted according to the requirements of good faith.
>
> S242 – The debtor is bound to perform according to the requirements of good faith.

The wider statutory basis extending the application of the doctrine from performance to the definition of contractual obligations explains the different approach of the German courts, which has been used to create a positive duty of co-operation from one party to the other. For example, in one case where German long-term contracts were adversely affected by inflation after the First World War, it was held that the principle of good faith allowed the judge to re-allocate contractually agreed risk pursuant to Section 157.

In summary of this point, the experience in France and Germany has been that through the duty of good faith the German and French Courts have had the freedom to develop its doctrines without incurring the reproach of pure judicial decision law-making. This has enabled the identification and solution of problems which the existing rules do not or seem unable to reach. Whether or not either of these models could be successfully adopted in England and Wales is not a question that can easily be answered. A 'bolting on' by domestic law-makers of a French or German-type duty is extremely unlikely in any event. The move contemplated in this chapter is much more modest in scope: where the parties have expressly contracted in good faith, there ought to be a detectable legal meaning to give some weight to their undertaking.

It is unsurprising, given the establishment of the good faith doctrine in continental legal systems, to discover the duty is enshrined within European law. For example, the Unfair Terms in Consumer Contracts Directive 1993 may strike down consumer contracts if they are contrary to the requirements of good faith. The Commercial Agents Directive 1986 also makes reference to good faith.

Moves towards the harmonisation of European Contract law by the European Contract Commission stopped short of outright commitment to the duty of good faith but did state that regard is to be had to the observance of good faith in international trade.

Nor is good faith a concept unknown to English Law. The obvious example is in insurance contracts which are subject to a duty of utmost good faith owed by the assured to disclose material facts and refrain from making untrue statements while negotiating the contract.[21] The duty of good faith is also apparent in areas of law where there is a special relationship such as family arrangements and partnerships.

A pattern is discernible towards the re-emergence of the duty of good faith in English law. Despite this encroachment (or possibly because of it) suspicion and hostility abound, in the words of one commentator: '[The duty of good faith] is a vague concept of fairness which makes judicial decisions unpredictable.'[22]

Another argument against the imposition of a general duty of good faith is the preference given to *ad hoc* solutions in response to demonstrated problems of unfairness. In other words, good faith outcomes are already being achieved through other means. Examples of these outcomes have been given[23] as the contractor's duty to progress the works regularly and diligently and the employer's duty not to obstruct and to co-operate. However, *ad hoc* solutions can lead to unsatisfactory results. Contract draftsmen have given the judiciary a unique opportunity to create new law based on the key concept of good faith. This chapter now examines judicial attitudes in this area.

## Judicial hostility?

The grounds for the seeming hostility (with one notable exception) of the judiciary to the concept of good faith has already been stated – suspicion of broad concepts. The approach is, to paraphrase Lord Bingham in Interfoto Picture Library v Stilleto Visual Programme Ltd[24] to avoid any commitment to over-riding principle in favour of piecemeal solutions in response to demonstrated problems of unfairness.

The judgment of Lord Ackner in the case of Walford v Miles[25] sums up the prevailing sentiment: 'the duty to carry on negotiations in good faith is inherently repugnant to the adversarial position of the parties involved . . . how is the court to police such an "agreement"?'

From time to time the courts have, at least, entertained submissions about the more general application for the duty of good faith.[26] A trilogy of cases[27] in the Court of Appeal suggested a move towards a more general principle. Lord Bingham was at the time dropping heavy hints such as: 'we would, were it material', imply a term but the door seemed to be more firmly closed on the introduction of a general duty of good faith by His Honour Judge Seymour: 'the development of the law in this direction would, it seems to me, be fraught with difficulty . . . I should not be prepared to venture into these treacherous waters.'[28]

There has been one case from which encouragement towards concretising the duty of good faith can be ascertained. The case of Birse Construction Limited v St David Limited[29] featured a non-binding partnering charter and the judgment specifically highlighted that the parties had entered into a partnering arrangement.

His Honour Judge Humphrey Lloyd recognised that the terms of the partnering charter were important in providing the standards of conduct of the parties. Although such terms may not have been otherwise legally binding, the charter was taken seriously as a declaration of assurance. In short, the parties were not allowed to interpret their relationship in a manner which would have been inconsistent with their stated intention to deal with each other collaboratively.

It is possible to discern support from this judgment for the parties' expressed desire to operate in good faith in their dealings with one another. This support fulfils the role of meeting the expectations of the contract users. Increasing numbers of contract draftsmen have been bold enough to include good faith provisions in their contracts. The contracts have been welcomed by their users. If they find themselves in difficulties, then the users have a reasonable expectation to be bound by their promises to one another. The challenge for the judiciary is to decide on the appropriate level of support to be given to the more prescriptive and onerous terms of contract now employed in the latest construction contracts.

## How best to deliver what the parties want?

It is beyond this chapter to provide a blueprint of how a general duty of good faith might operate. One commentator has pointed out that if good faith is to be of any practical utility, it needs to provide a few clearly understandable action-guiding principles of conduct.[30] The small print solution of listing every possible potential misconduct on the part of any party is not suitable given the complexity of construction contracts and the move away from voluminous forms. One approach would be to allow the judge/arbitrator/adjudicator a wide discretion so that they might 'concretise' the duty in line with the principles of conduct as they see fit or in line with experiences in other jurisdictions.

Good faith in negotiations could mean an inquiry into the reasons for breaking off negotiations. Examples of bad faith might include negotiating without serious intention to contract, non-disclosure of known defects, abusing a superior bargaining position, arbitrarily disputing facts and adopting weaselling interpretations of contracts and willingly failing to mitigate your own and other parties' losses and abusing a privilege to terminate contractual arrangements.

The effect of the court recognising the duty of good faith as a hard obligation has been likened to recognising the general duty of care in negligence or the principle of undue enrichment.[31] As a result the principle may remain relatively latent or continue to be stated in extremely general terms without doing too much damage to the important virtues of certainty and predictability in the law. The principle could also provide a basis on which existing rules can be criticised and reformed.

The alternative way of introducing a duty of good faith is to set down guidelines in a statute. A statutory obligation to act in good faith was recommended by Latham as a measure which would lead to the improvement of the performance of the construction industry. The government of the time chose not to move in this direction. The time may have come to revisit this decision.

## Conclusion

Good faith has been described as 'repugnant to the adversarial position of the parties'. The duty is surely not so repugnant to an industry currently characterised and actively pursuing an agenda not of adversarial relations but of collaboration. The industry is also making moves towards an ethical agenda. It would appear to make clear sense to further promote collaboration and ethical behaviour through the medium of good faith provisions. In this regard the benefits that could be felt across the industry would be considerable and potentially lead to completely new framework in which construction activity would be characterised by increased professionalism, integrity and fair dealings.

The industry would benefit from some clear messages from the judiciary as to the enforceability of their collaborative arrangements. The positive stance taken in the Birse v St David case is encouraging in terms of direction but further concretising of the exact meaning of such obligations on the particular facts of any case would be helpful. Re-ordering the structure of construction contracts by introducing the sound theoretical basis presented by the duty of good faith is an achievable and laudable aim. The expression of this underlying principle with its uncluttered simplicity may serve to bring clarity to the dense contractual conditions for which the industry is renowned.

The benefits of concretising the duty of good faith would be felt right across the board, not least in the improvement of the standard of acceptable ethical behaviour among the disparate set of people working in today's construction industry.

## Notes

1 Uff, J. (2005) 'Ethics in Construction Law – Two Years On', *Society of Construction Law* paper, available at scl.org.uk
2 Available at scl.org.uk/ethics
3 Uff, op. cit.
4 Sutcliffe v Thackrah [1974] AC 727, HL.
5 Stansbury, N. (2005) 'Unethical Behaviour and Criminal Act', *Society of Construction Law* paper, available at scl.org.uk
6 Goode, R. (1999) 'The Concept of Good Faith in English Law', Saggi, Conferenze e Seminari. [http://www.cnr.it/CRDCS/goode.htm] 1992
7 Honey, R. and Mort, J. (2004) 'Partnering Contracts for UK Building Projects: Practical Considerations', *Construction Law Journal*, 20(7): 361–79.
8 RICS (2004) *Contracts in Use Survey*, available at: http://www.rics.org/Built environment/Building-contractforms/Contracts+in+use+report.htm. 2007 use of contracts survey which indicates a growing use of NEC.
9 Heal A. (1999) 'Construction Partnering: Good Faith in Theory and Practice', *Construction Law Journal*, 15: 167–98.
10 Miner, M. (2004) 'Time to Contract in Good Faith?', *Construction Law* 15(2): 20–2.
11 Sim, D. (2001) available at: http://www.cisg.law.pace.edu/cisg/biblio/sim1.html
12 Jansen, C. and Harrison, R. (1999) 'Good Faith in Construction Law', *Construction Law Journal*, 15(3): 346–73.
13 Carter v Boehm [1766] 3 Burr 1905.
14 Groves, K. (1999) 'The Doctrine of Good Faith in Four Legal Systems', *Construction Law Journal*, 15(3): 265–87.
15 Goode, op. cit.
16 Miner, M. (2004) 'Awaiting developments through the common law is likely to be slow; the time for appropriate legislation may now have come', *Construction Law Journal*, 15(2): 20–2; B. Colledge (1999) 'Future explicit recognition of the concept (of good faith) is not inconceivable and would appear to demand only a re-definition rather than a sea change in judicial analysis', *Construction law Journal*, 15(3): 288–99.
17 Charybdis and Sylla were famous in Greek mythology. The meaning of the phrase is 'between a rock and a hard place'.
18 Stansbury, op. cit.
19 Finn, P. D. (1989) 'The Fiduciary Principle', in T. G. Youdan (ed.) *Equity, Fiduciaries and Trusts*, vol. 1. Toronto: Carswell, p. 4.
20 Macqueen, H. (1999) 'Good Faith in the Scots Law of Contract: An Undisclosed Principle?' in A.D.M. Forte (ed.) *Good Faith in Contract and Property Law*. Oxford: Hart Publishing, pp. 5–37.
21 Owen, D., Birds, J. and Leigh-Jones N. (2004) *Macgillivray on Insurance Law*. London: Sweet & Maxwell.
22 Goode, op. cit.
23 Colledge, B. (1999) 'Good Faith in Construction Contracts – the Hidden Agenda', *Construction Law Journal*, 15(3): 288–99.
24 [1989] 1 QB 433.

25 [1992] 1 All E.R. 453.
26 Jefford, N. (2005) 'Soft Obligations in Construction Law: Duties of Good Faith and Co-operation', Keating Chambers seminar, 12 May.
27 Philips Electronic Grand Public SA v BSB [1995] EMLR 472; Balfour Beatty v DLR [1996] 78 BLR 42; Timeload Limited v British Telecom [1995] EMLR 459.
28 Hadley Design Associates v Lord Mayor and Citizens of the City of Westminster [2003] EWHC 1617 [TCC].
29 [1999] BLR 194.
30 Honey and Mort, op. cit.
31 Groves, op. cit.

# Part II

# Case studies of good practice

Part II has been compiled with the knowledge that the discussion of ethics in the built environment is not complete without the application of ethics.

The case studies have been culled from examples of good practice in the UK with the exception of one which indicates the case of companies which have been prosecuted for unethical actions. They were chosen because they illustrated some of the issues that were being discussed in Part I. They set the scene and the context and either illustrate the decision-making process or they give an outline of some of the issues in ethics such as:

- influence of policy;
- influence of ethical approach;
- the role of trust and integrity and the level of control;
- how communications worked;
- management of relationships;
- dealing with external stakeholders;
- making decisions and handling risk;
- frustrations and achievements.

These issues are critical to a working application of ethics in good practice and it is interesting to see how many of them were in position for good practice to occur. The information was collected through interviewing senior managers who were personally involved with the day-to-day workings of the project or the situation described.

## The case studies

Chapter 12 is on CSR applied to contractors and developers in practice. This case compares and contrasts the CSR reports of a contractor and developer who have experience of over five years. It looks at their policies and compares the economic, social and environmental reporting characteristics and the standards that are used to record their targets and progress.

Chapter 13 is on partnering trust and risk management. This case study looks at the innovative use of the collaborative JCT Constructing Excellence contract which encourages partners to assess the risks involved with each project and to allocate responsibility for that risk. It also assesses a cost and time value for that risk based on probability and impact and looks at ways of reducing that impact, which would save on the target cost put in the contract sum. The case study indicates how time, cost and quality were significantly improved over a series of projects and that good relationships helped to innovate ideas and make further savings.

Chapter 14 is on distorted competition in the roofing industry. This case study follows the distortion of competition which occurred in a particular project and how the OFT leniency policy unearthed the evidence and set the fines to penalise the offenders. The ongoing investigation is industry-wide and work done in one group action contributed to another prosecution.

Chapter 15 is on employment conditions at Heathrow T5. This case study describes the construction labour agreement used at Heathrow discussing good conditions and remuneration and improving working relationships between key contractors. It contrasts the more common situation of what has become known as false employment where many self-employed workers are being exploited.

Chapter 16 is on health and safety systems in a large PFI Hospital. This case study illustrates a real effort to move towards a culture of improvement which embraces the whole supply chain. It illustrates the current state of the art and is a system that has greatly reduced the accident rate and is working towards a zero accident goal. As a large contract there is room for lessons to be learned and reapplied, but it needs the co-ordination of a huge workforce.

Chapter 17 is the Stroud District Council planning case study. This case study contrasts the difference between the problems of planning being seen as a bureaucratic system and the ethic of inducing a partnership of trust and respect so that effective decisions are made in an acceptable time frame bringing on board external stakeholders.

Chapter 18 is on the use of training to establish small-scale organisations. This case study looks at the way in which training provision has been made accessible to smaller organisations by identifying a critical area of concern and extending this to the enhancing business management and enterprise. It also integrates the relationship between client, designer and contractor.

Chapter 19 is on manufacturing quality and trading relationships. This case study looks at the pursuit of continuous improvement in maximising quality and output and considers the difficulty of developing trust in working relationships which work further up the supply chain with both main contractor and client.

Chapter 20 is on educational partnership and sustainable contracting. This case study examines social sustainability in practice between a university and a contractor during the completion of a contract. It considers how mutual objectives for education and corporate social responsibility can come together.

Chapter 21 discusses trust and relationships on a mega property development. This case study considers the importance of the various aspects of trust during the progress of the development such as prior relationships, integrity, reliability and loyalty. It particularly examines the stakeholder management process.

As these case studies have exemplified leading practice, they have shown a particular concern for reaching beyond minimum levels of compliance and dealing with people ethically with integrity, respect and dignity. There are some barriers which have been indicated in the case studies which in most cases have been overcome.

# Chapter 12

# Comparison of CSR between a developer and a contractor

## Introduction

This case study investigates the motivations and aspirations of two leading UK-based companies who are well known for their commitment to a sustainable and ethical approach. Both have produced corporate social responsibility reports for a number of years. A contractor and a housing and regeneration developer were chosen to see how their policies compared and contrasted in reporting and implementing corporate social responsibility reporting (CSR). Information was collected through in-depth interviews and through looking at publicly displayed CSR reports and website information.

## The contractor

One of those chosen was a contractor offering building and civil engineering contracting, design and facilities management services. The contractor engaged mainly in collaborative-type contracting where negotiated continuing partnerships were available. They were also involved in joint ventures such as PFI procurement where involvement in managing real estate for the client was longer term. In this organisation the environmental manager responsible for co-ordinating CSR reported directly to the board and had developed a range of supportive training materials, diagrams and employee handbooks. The CSR report was easily available on the shareholder website. The programme was associated with getting project managers and employees on board who could also motivate others to work with stakeholders.

## The developer

The other was a developer specialising in residential and mixed development schemes mainly sold on completion. In particular, mixed development schemes offered challenges to find sustainable solutions and were built on inner city brownfield sites. They were involved in detailed negotiations with a wide range of stakeholders on their city centre sites and had adopted regeneration as an aim for their sustainability approach. Some of their work was let out in these schemes to other contractors, but they managed the developments overall on a programme basis with development and implementation inputs at director level. The CSR was co-ordinated by a dedicated 'environmental manager' reporting direct to the board. It was clear that they needed to be attractive to potential clients for the mixed use and to comply with strict city requirements for sustainability in the community as a whole in high profile regeneration at the heart of city. They also needed to sell attractive residential city living direct to home buyers at rates to recover the regeneration costs and often within strict time scales and cost constraints.

Information collected was restricted to their reporting processes, the nature of their ethical guidelines, employee and management recruitment, retention, buy in and training, sustainability targets, stakeholder relationships, and community leading practice.

Significant differences in approach were found between these two sub-sectors in the construction and property industry. Some differences in their approach are indicated in Table 12.1. Information was extracted from in-depth interviews with the environmental managers and also from published CSR reports.

These differences could be due to the nature of the sub-sector (housing vs contracting) or may be connected to a different policy approach. The assumption is that policies have developed in response to the prevailing stakeholder pressures. Overall, Table 12.1 indicates the developer's focus on the product and the contractor's focus on the people delivering the *product*, which means that external trust is built up by the developer through the quality of their product and by the contractor through the quality of their *service*.

The contractor shows their ethical responsibility by a particular emphasis on building direct relationships with the client/stakeholders and on training ethically focused staff. It is important that ethical relationships are built up which lead to responsible decisions. The contractor has used a set of targets which are monitored and verified, but act more to focus improvements in management and employee performance. This focus on the service makes the openness of the ethical guidelines more important and significant effort is put on reputation, leadership and communicating company values to all staff. Staff are encouraged even to practise sustainable values at home.

The developer shows an ethical and social responsibility by the proven attainment of the quality and sustainability of the product by reducing its net impact on the environment and ensuring that there is also some social cohesion in the design and use of such products. For them, it is important that socially responsible systems are operating that give products which result from more explicit measurement of the after-effects of the product on users and other stakeholders. This leads to a different emphasis in the format and content of the report. There is an emphasis by the developer to report specific KPI audited achievements which will help to build the confidence of the customer in the product.

*Table 12.1* Corporate differences

| Factor | Housing developer (2006) | Major contractor (2006) |
|---|---|---|
| Characteristics | Public company, £600m turnover, 850 employees | Public company, £2000m turnover, 18,000 employees |
| Report type | Social and environmental report with targets fully audited GRI and WWF reporting standards | Sustainability report, independently verified not audited, GRI reporting standards mainly |
| Policy focus and reputation | Strong emphasis on KPI achievements. Product focus. Formal GRI tables | Strong emphasis on KPI targets and leadership. Graphical models for easy communication |
| Importance of listings | BiTC* middle rating to show performance against other members of the sector | BiTC* verified, 1% club member to show compliance |
| Ethical focus | High quality customer environment, equal opportunities and rewarding career, sustainable and efficient business | Openness, collaboration, mutual dependency, professional delivery, innovation and sustainable profitability |
| Sustainability focus | Competitive industry comparisons e.g. 75% brown field development. Natural step | People performance, training people in sustainability. Natural step |
| Performance | Emphasis on eco-home ratings | Emphasis on retaining people |
| Retention and training of staff | Full potential, mutual benefit of business and individual's welfare and career | Attracting 'green' recruits, raising retention, differentiate sustainability |
| Stakeholder relationships | Leading home buyers awards. Planning gain generous | Sustainable approach. Supplier charters |
| Attracting customers | High customer and user satisfaction | High employee satisfaction leading to good customer relations |
| Community engagement | Community donation | Community projects |
| Profit sharing | 0.5% profits to relevant charities | 1% profits to community |
| Health and Safety | Safe and healthy products and safety targets | Health and safety zero targets |
| Employee relations and management style | Contribution to sustainable standards testing. Training in eco-homes index | Lifelong training, satisfaction surveys and management communication plan |

*Source:* Adapted from Fewings (2006) Table 3.[1]

*Note:* *Business in the Community.

Other customer aspects are also differentiated in the same way. Good ethical relationships with suppliers are considered by both to be the basis of profitable operations, but the developer has a much more standardised approach while the contractor has a need to develop effective supply chain management to develop confidence and enhance quality. Both have a strong sustainability focus and both have used the Natural Step philosophy to inform their policy.

The community aspects seem to be differentiated by the developer making *contributions* to the community as a compensation for developmental impacts and the contractor *working with* the community on community projects. Both perceive an overarching priority for safe and healthy solutions. Ethical and sustainable behaviour is sensitively judged by the community and reputations can easily be lost by not handling complaints and concerns promptly and effectively. Clients have also stepped in with requirements to protect their own name.

## Conclusion

The two interviewees were very passionate in their approach to CSR and believed that sustainable business was good business. Their was a sense in which they were still developing their systems and both had started off as a result of the specific commitment of the chief executive who was well known in their championing of sustainable issues. Money had been put upfront by each company in the belief that it would be returned with interest.

The different emphasis in the reports reflect the difference between the economic sectors. In housing and regeneration very prescriptive targets for sustainability by the government have been concentrated on environment issues where progress in the residential project needs to be proven and hard figures are given. Private housing buyers are only generally pushing and city partnerships require fairly predictable social and community inputs. In the contracting sector it is client-driven and there is a unique formula for each project or programme in order to deliver a sustainable service with quality and life cycle costs as an important aspect together with managing the unique set of stakeholders.

## Note

1  Fewings, P. (2006) 'The Application of Professional and Ethical Codes in the Construction Industry: A Managerial View', *International Journal of Technology, Knowledge and Society*, 2(7): 141–50.

# Chapter 13

# Partnering trust and risk management

## Introduction

VOSA is the government agency that is responsible for the official testing of HGV vehicles throughout the British Isles. In 2003, it started a modernisation programme for its test stations which afforded an opportunity for a strategic partnering contract which was considered a more efficient and ethical way of delivering the programme and to introduce improvements that would involve the whole team.

## Transparency and risk

The contract that was chosen was the Built Environment Collaborative Contract (BECC). The contract philosophy is to have the same contract form for all parties from consultant to specialist contractor and it affords the opportunity to get early involvement of all parties. One of the key aspects of BECC is that it requires the aggregation of an agreed risk allocation with financial quantification based on the agreement of all parties involved. The risk cost is separately allocated from the costed bill of quantities so that the prices paid for services are transparent and accounts are presented on an open book basis. This provides for both cost savings where a risk does not occur and for extras to be added without the encumbrance of obsolete risk premiums being added to the rates of work. Overheads and profits are ring fenced for the contractor. A pain–gain incentive scheme is also operated under the contract conditions so that when the agreed guaranteed price is exceeded and the scope is unchanged, a penalty will be imposed on the contractor. Any saving on the guaranteed cost through value gains is shared between the client and the contractor. In traditional contract forms risk is allocated and costed from the perspective of one party only and may not be efficiently managed or costed by that party as the late procurement of the contractor allows no collaboration before the price is fixed (Figure 13.1).

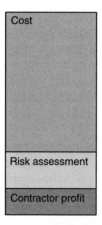

*Figure 13.1* Target sum make-up.

## Value for money

The pain–gain, open books clause in the BECC gives an incentive for a win–win situation and helps to eliminate the suspicion that additional cost is less than that charged. It also gives the contractor a leverage to suggest value for money improvements. Since a contractor only loses the profit margin if they go over the guaranteed sum, it also provides an incentive for efficiency. The practice of this theory was tested by VOSA by commissioning a third party QS to calculate the estimated time and cost for a traditional procurement of a building of the same specification with the actual costs under the BECC contract. These turned out as shown in Table 13.1.

It is interesting to note that the cost-saving proportion appears to be less than the earlier collaborative projects. However, the price per metre ratio of successive similar test stations modernisations has been reduced steadily from Aberdeen (4 March 2005) at £635 to Grantham (6 October 2006) at £550. This

*Table 13.1* Comparison of traditional and BECC costs and time

| Procurement | Aberdeen 4 March 2003 | | Northampton 5 August 2005 | |
|---|---|---|---|---|
| | Cost (£) | Time (weeks) | Cost (£) | Time (weeks) |
| Traditional procurement | 1,861,275 | 28 | 1,791,914 | 28 |
| BECC cost | 1,616,559 | 24 | 1,591,748 | 24 |
| Cost/time savings from partnering | 244,716 | 4 | 200,166 | 4 |
| Percentage saving | 13% | 14% | 11% | 14% |

represents an average value improvement of 4.8 per cent[1] year on year and a total improvement of 13.4 per cent over three years. These figures are encouraging though they might not all be due to the conditions of contract. Inflation could be added to enhance this improvement. Value for money is a difficult concept to define as value does not just cover the capital cost, but is also relevant to the life cycle costs which include running and maintenance costs. Value for money brainstorming events were arranged at the beginning of a project to look at the possibility of adding value and/or reducing cost by innovation. This also involved the input of the facility management team and the manager of each centre as well as the design and production teams.

## Trust and working together

The client has been very much involved in developing the project team and has employed a project manager and used a single contractor and consultant over the whole modernisation programme. This has allowed for synergy efficiencies to cut out the learning curve for the project specification and conditions of contract and to build up trust between the parties. There is also a spirit of competition between the site teams engendered by the publication of key performance indicators, known as SETS (Simple Efficiency Target System), similar to the ones in Table 13.1.

Trust was recognised as an essential part of an efficient and ethical approach to the programme. A lot of work was put into the programme through the use of workshops and seminars to reinforce key points and to build up teamwork and co-operation across the whole supply team. Some of these events would be fun events to help people to build up confidence and trust in each other and get to know the other processes they were not a direct party to. The integrity of the client and management was also tested by putting them in close quarters with the workforce on the ground. The client adopted a high profile with frequent site visits and adopted an informal dress code to help build up a more egalitarian culture. The real interaction though was through direct involvement with workers and supervisors by inspection, with penalty or reward for the quality of the product. Individuals and work teams were made accountable for their work. An additional KPI was created for quality described below.

A few concepts were dreamed up to illustrate and make concepts come alive such as the 'nagging wife' represented by the client who had an additional inspection of the finish quality each time on site, 'Mum and Dad' management which viewed the whole team as a family and sought to get to know the workforce as individuals, with positive listening, encouragement and healthy 'sibling' rivalry.

## Quality, integrity and achievement

The quality levels achieved on early projects were not up to the expected client standards and systems were put in place to improve the culture and attitude

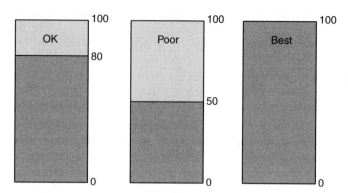

*Figure 13.2* Sparkleometer comparison.

towards defects. This particularly applied to finishing details which were highly visible. JD introduced the concept of the sparkle factor which measured finished quality and rated elements on a scale of 1–10. These scores produced an aggregate score over different elements and an overall score known as the sparkle factor out of 100 visually presented on a sparkleometer (Figure 13.2). The scheme proved very successful in raising standards in a fun way. It also encouraged teamwork as subsequent finishes also depended on taking over good quality superstructure. It encouraged discussion and contractors were able to score the quality of design drawings once a month at the site meetings. Design needed to be effective functionally and buildable.

Design information needed to be timely and accurate. This balanced the fact that site quality was measured by design consultants. There were two other non-standard KPIs:

- to reduce construction waste by 50 per cent;
- to reduce whole life costs by 50 per cent.

The sparkle factor score measured on different parts of the building such as the test hall, the office, external works and handover was recorded for several different projects so that parallel projects could compare their quality factor.

## Change management

Inevitably there were changes to be made in scope and design development and this was measured and controlled on change information forms (CIFs). These forms were a way of pricing and approving change spontaneously so its impact could be assessed before client go ahead was given. They provided a checklist of no more than 20 changes per contract and a running record of the state of the change orders was shown to encourage progress on the status of the approval. This position in relation to the overall target sum could be controlled. A colour coding was used to highlight red CIFs which were the late ones that now needed urgent attention.

## Risk management

The risk allocation schedule is an innovative way of managing risk. In this schedule incorporated in the contract risks were identified and assessed for any time or cost impact. These amounts were factored into the target cost sum (Figure 13.1). If a risk is only partly, or not used, it later represents a saving to the client. If it is exceeded, then it represents a cost to the risk holder who may be the client or the supplier. This gives incentive to those responsible for the risk to manage it properly. As the residual risk value is agreed by all parties at the outset, this seems a fair process and should ensure that the risk ends up in the hands of those best able to manage it – or it should do if the best value is to be obtained for the client. The value of any risk can be recorded at progress meetings until it has a final value and can be deemed 'risk over'. If risks are reviewed regularly, then it helps to provide early warning of any tendency for the risk to go 'over budget' so that corrective action can be taken. Many risk registers lie dormant as a paper exercise at the beginning of the contract and are worse than useless.

The risk register was a way of managing active risks and recording their costs. It was managed by a risk manager and financial quantification based on impact and frequency extracted from the risk allocation schedule in the contract for time and cost, but updated during the contract on the basis of risk over (green) and risk alive (orange). Risks had mitigation action plans that were the prime responsibility of a member of the project team who then also sought to minimise the impact and the frequency of the risk and the impact on the budget. A risk was deemed 'risk over' when the element of the work was completed and the full cost accounted for and monitored on the financial plan. A risk is quantified with a financial allocation and risk is separately quantified to cost (see Figure 13.1).

## Sustainability and whole life costs

The programme had two objectives, to reduce whole life costs and waste by 50 per cent. In order to achieve this and also work towards a BREEAM excellent rating and meet local planning requirements, a positive sustainable design policy was required. Three initiatives are listed below:

- The use of a turbine on two of the later projects was commissioned to provide site electricity and sell excess back into the national grid. A value for money exercise on this indicated a payback of 15 years if electricity was to go up slightly more than the price of inflation.
- The use of passive stack ventilation to eliminate the use of air conditioning

in the offices. This had a payback of two years and was used on all the projects.

- The use of roof lights to introduce more day lighting in the test halls and cut down on the use of artificial lighting which would supplement day lighting only where necessary and in the winter. This was calculated to have a payback of 5.7 years.

Sustainable development is a well-worn phrase to try and get a better use of scarce resources on an economic basis, but also to maintain a present and future quality of life that is acceptable to the community and employees. The discussion of an ethical way, say amongst designers, will produce contradictions between capital cost, quality of employee and community life and imperatives to reduce energy use and the effects of climate change. In the case of VOSA, they have a choice to be merely compliant to the latest Building Regulations, for, say, insulation, or to look at future possibilities which would also benefit the economics of their business plan for providing effective and efficient vehicle testing.

A turbine which takes 15 years to repay is a long-term economic solution as well as an environmental one to reduce global warming. It may, however, be noisy and aesthetically unacceptable which may dent its community benefit or be seen as an acceptable price to pay. Its whole life costs will also depend on maintenance cost escalation, replacement costs and the impact of future government policy on the electricity cost compared with inflation as a whole – better or worse. Under the precautionary principle (action to head off harm which might happen) some good may be done for society.

## Site efficiencies

This was a programme of workshops to be able to share learning and to encourage ideas from the whole workforce. It consisted of inter-site workshops such as the use of sustainable drainage systems. Meetings between the client project manager and the site operatives (omitting as far as possible the site management team) yield valuable ideas from the 'coal face'. The adoption of these ideas strengthened trust and mutual respect between all parties. It was also an efficient way of discovering and testing the general applicability of efficiencies on one project and driving the cost down on projects following on.

## Ethics

### Trust and transparency

The attraction of this case study is the way in which an open and transparent approach was achieved which was proven to also benefit the commercial realities of the client and reward the innovation of the contractors and consultants in a reasonably equal way.

Ethically trust and transparency are well documented and they are likely to be valued by some as a commercial benefit and so give some leverage to a wider range of stakeholders. A minimum level of trust is necessary to do business. Some of that trust is based on the robustness of the project planning process, such as effective risk management. In this case there is a clear commitment to risk planning and management.

A professional has a duty to act to engender trust. A client would do so to nurture their reputation and in a belief that they will attract less wasteful risk premiums. A contractor is likely to benefit from future work only if the bidding system selects on a value for money basis and they are able to deliver efficiently. In other words, an agreement to act in this way will only work where all parties agree and have a committed collaborative covenant which survives adverse situations and is based on good future planning.

In the case of VOSA, they have driven the trust and collaboration as a result of a bad experience with using other non-collaborative contracts which resulted in minimum co-operation. They budgeted for substantial input by the management/consultancy team who had to win and sustain the commitment of the design and production teams. It used the synergies of past experience well to drive further efficiency and effectiveness and was mainly justified by increasing overall value and not minimum capital cost. This was possible due to the obligation of the public client to be exemplary in this area. Also through the vision, enthusiasm and synergy of the senior management team and the flexibility for the operational team to research and decide the procurement method and contract type.

The ethic or 'religion' of collaboration was backed up by several complementary ideas which helped people think they were working together such as, the competitive rivalry of the sparkleometer, the 'nagging wife', the 'Mum and Dad' management, the culture of management and worker equality and the encouragement for ownership of ideas. The family analogy allows different roles and different levels of experience and authority. Collaboration and especially transparency heighten sensitivities to lack of communication or honesty which requires continuous management awareness. Members of the team or new members grafted into 'the family' may need space to develop their own ideas.

## Sustainable development

The precautionary ethics[2] of sustainability was driven by the two practical objectives set up in the strategic business case to reduce the whole life costs and the waste generated in construction by 50 per cent in both cases. These had been fully approved at senior management level and used to justify time spent on generating innovative sustainable solutions. Together they had strong economic leverage, but caused all parties to generate ways of reducing use of current and future resource.

Precautionary ethics is often seen as a drag on resources and progress. To counter this, there is a need to maintain the momentum with continuing new ideas and to combine the ideas for increasing value for money, which is also a feature of the workshops. The application of sustainable solutions in itself injects innovation risk into the project and final solutions may need to be developed throughout the delivery process (design and construction). Where untested solutions are *prescribed* in the tender process, it will be hard to project final costs and so contractor early involvement gives the potential for a full value management process and for an iterative development process so that quality can be better assured.

---

### Questions

- What alternative ethics could have been employed to the transparent and precautionary approach?
- How could you make this work for VOSA?
- How would a traditional approach effect the delivery and ethics of sustainable buildings?

---

### *Alternative scenario to transparency and sustainable development*

The use of a traditional JCT contract is an architect-led process and the client delegates the development and project management process to the architect. In the past, this has suited many clients who wish to concentrate on their core business and do not have to employ additional skills. In these contracts it is likely that a fully priced 'lump sum' bill will be available. This can be used as a basis for pricing future costs and for valuing work done on a physical measurement basis. Change is priced in arrears and the client takes the risk for design development only nominally being involved in the design approval process. The client only sees the 'virtual' bill of quantities figures. In short, the client is excluded from the real process which suits a client who wishes to remain at arm's length.

However, when delays and budget overruns have occurred, the client finds it difficult to identify where things have gone wrong and may have to pay extra money or wait longer for a product for no extra value. The business situation may also be dynamic requiring client changes to the facility which may be costly if they are identified in the implementation stage. The abortive costs are not understood or measured leading to a sense of aggrievement and loss of value and control.

In the case of VOSA, they became disenchanted with the lack of co-operation of the contractors with their objectives. These would have to be communicated to the contractors in pre-selection interviews *after* the design was completed and risk allocation determined. They would have no leverage under a non-collaborative contract to ensure value management. Innovation introduced by consultants may not be economical to the methods of the lowest priced contractor chosen and the contractors themselves would have no incentive for innovation and would require prescriptive specifications. Risk premiums would be combined with the bill of quantities rates and there would be no opportunity to eliminate these where risk was successfully managed.

Traditionally, a business case is fairly short term, dealing with the medium term 2–5 years. Dominance is assigned to the shareholder stakeholder interests and the economic bottom line is concerned with sustaining profit levels to attract investment. The complication of whole life costs is difficult and may also be considered to be a complication to the accounts. Historical accounting methods do not allow long-term value added, i.e. the maintenance costs in the future are only roughly indicated by depreciation and assume that the business environment is stable. Results depend on the competitive market-place which is controlled by public demand.

For a public organisation, this is easier to justify if it is supported by existing policy, but budgets are often tight, annualised and on an incremental basis, so do not allow additional capital expenditure without integrating the revenue and capital budgets. Results are in the context of electoral promises to reduce public expenditure, though there is a certain amount of public sympathy for a supplementary charge for proven environmental improvements, e.g. the ring fencing of additional water rates to improve flood defence. In the UK, all government agencies are required to take the lead on sustainable solutions, but evidence suggests that many departments fall below the standards set. It is very difficult to justify extra capital cost for long-term indirect (non-sales) gain, e.g. a turbine.

In the case of VOSA, the projects would be awarded with a design performance requirement and risk transfer to comply with government directives to reach a BREEAM excellent rating on 50 per cent of the buildings. Risk would be transferred for the outcome. The difficulty here would have been the lack of integration between the build programme and the facilities management to ensure environmentally efficient use.

Ethically this is closer to the utilitarian approach which differs from the precautionary ethic in its shorter term and less radical outlook. Outside consultants are likely to adopt a gradually compliant innovation rather than a step change shift in innovative practice. Government will also be guided much more by the prevailing public opinion in attaining environmental goals. In VOSA's case, this might have amounted to an upgrading of test stations according to existing and imminent regulations and targets.

## Conclusion

There is a necessity for the client to set clear objectives which suit their values. Once set there, the choice of a procurement method and the client's ethical responsibilities should be clearer.

## Acknowledgements

Interview with Joanna Davies (JD).

## Notes

1  Straight line average.
2  Theory. The utilitarian approach would be mainly short term to obtain the most good for the most people, which contradicts the sustainability argument. This might be a central tenant of a Kantian precautionary approach which might try and reduce the production of $CO_2$ in the world as a whole with some sacrifice and lack of amenity for most people. It also depends upon more multi lateral approaches so that savings and production limits are not taken up opportunistically by others profiting from the shortfall to popular demand. Williamson, P. and Radford, M. (2000) *Building, Global Warming and Ethics*. London: Routledge.

# Chapter 14

# Roofing contractors collusion case study

## Introduction

The OFT is a body responsible for policing the Competition Act 1998 which regards collusion as a contravention of competition and OFT have a duty to ensure that markets work properly and remain competitive. They consider there are three contraventions which spoil a competitive market at bid stage which are described in Chapter 10:

- cover bidding;
- bid suppressing;
- bid rotation.

The OFT seeks to ensure that the market is working properly, which means there is nothing distorting competition among suppliers in that market. It does not have to prove that the final price was not competitive, but simply that there was an intention by suppliers to distort or suppress the market or to be involved in non competitive bids.

## Corporate penalty

They have the power under the Competition Act 1998 to fine any guilty companies up to 10 per cent of their worldwide turnover. The turnover calculation may be mitigated to distinguish between the turnover of different types of products, but not in different locations. The fine OFT decides is on the basis of mitigation or severity, e.g. the direct involvement of a director. A special 100 per cent leniency rule has been offered by the OFT to the first company in the market to admit involvement and to offer information as to the extent of the collusion. Further partial leniency is offered to other competitors who can give additional information *before* they come under further investigation. The OFT fix higher fines on a greater awareness of the director and also on the number of times they have contravened.

In competition, the critical OFT test is whether the market is prevented from working properly. This means that competition is distorted in some way and clients are misled into thinking that the prices and services they receive are more competitive. The three cases above both prevent wider competition and create a falsely high price or reduced choice of contractor. Because this is a crime against the market, it is likely that several contractors have been involved and so the actions are a group action.

## Individual penalty

The consequences of being caught are far-ranging and individuals could lose their professional status or jobs, be fined or face jail. A director who has specific intent can be charged with systematic activity to further a cartel and be subject to a jail sentence under the Enterprise Act and debarred from directorship for up to 15 years. Pleading ignorance is not a good defence. Directors are expected to keep on top of practices which are being used in their own companies and to exercise reasonable control within the law.

In the case of a formal cartel, the OFT can also pass the case to the Serious Fraud Office (SFO) who may investigate and prosecute for criminal activity under the later Enterprise Act for what is called the 'cartel offence'. This offence is a deliberate fixing of higher prices where there is an oligopoly involving a few influential contractors who agree to carve up the market between them so that they take it in turn to bid.

## The group cases

The case in hand is the investigation by the OFT of thirteen roofing contractors in the English and Scottish market for infringements of the Competition Act 1998 for activities pertaining to the year 2000. All were believed to be involved in collusive cartel activity. The outcome was 100 per cent leniency for one company, Cladding and Roofing, for offering evidence first about the collusion and six other companies were also offered partial exemption for additional pertinent information and co-operation. Firms had a varying amount of fines depending upon their turnover, the number of infringements, their degree of co-operation when formally investigated, their ability to pay, i.e. proportion of profit, and the extent to which it represented director activity. The fines went up to £852,000 for one company with seventeen infringements reduced by 40 per cent for co-operation under the leniency rule to £511,000. Another company also had a non-leniency fine of £526,000 for one infringement.

In a follow-up West Midlands case,[1] which had a similar outcome for a range of companies convicted or owning up there was an appeal by a company called Price to the Competition Tribunal against the offence which was turned down and a fine of £18,000 which represented a 2.2 per cent of turnover. This was later reduced to £9,000.

### The Bull Ring scenario

To give a flavour of the complexity of the OFT enquiry it is worth looking at the Birmingham Bull Ring contract for asphalt roofing and waterproofing of decking which was finally let by the main contractor to Contractor B on 1 September 2001. The initial tender on 1 April showed a set of tenders[2] that indicated approximate prices for the asphalt roofing contract

| | | |
|---|---|---|
| Contractor A | £823,000 | fully priced BOQ |
| Contractor B | £1,011,000 | fully priced BOQ |
| Contractor C | £1,246,000 | fully priced Bill of quantities, but using exactly the same items as Contractor B |
| Contractor D | £1,476,500 | lump sum price, no rates |

The main contractor X did not immediately choose and allegedly offered a further contract to the two lowest contractors A and B so that a choice could be made on the combined contract (this is contested by A who is the lowest at present). In the event, only two tenders were received from A and B for the waterproofing of the decking. New combined figures meant that A was now cheaper on the combined figures by only £5,000 because their second price was much higher. However, X chose Contractor B because they believed that they would do a better job as they were also suppliers of the special polymer modified asphalt. Contractor A and D also received a payment from Contractor B for tender costs.

OFT have concluded in their report after representations from all the parties, that there was a breach of the Chapter 1 prohibitions for collusion for three roofing contractors, as Contractors C and D had cover prices in the first tender for Contractor B. In the second combined tender, including additional work for deck waterproofing, there was a further breach as the payment to Contractor A was illegal. It was designed to stop Contractor A doing the first contract at the original lower price. On this count, Contractor A was counted by OFT to have colluded in the second combined tender and received a payment from them. There was also a restricted competition for the second combined tender excluding Contractors C and D. Contractor B was guilty because they had made the payments and had won the contract after colluding with the three other contractors and been party to additional information from the main contractor in the second contract. In the event, all of these contractors have been fined and all have at least one other infringement in other contracts.

On the basis of the evidence given and that Contractor B has 17 infringements in three markets and has an £8.5 million turnover and co-operated, Contractor A is four times the turnover and did not come for leniency but had two infringements in two markets with suspected director involvement. Company D is large but came forward first to the OFT and was given 100 per cent

*Table 14.1* Comparison of Bull Ring bids

| Tender | Contractor A | Contractor B | Contractor C | Contractor D |
|---|---|---|---|---|
| Asphalt tender 01/04/01 | No collusion. Lowest price | Gave price to C&D. 2nd lowest | Cover price tender. Next highest price | Cover price tender. Highest price |
| Combined tender 01/09/01 | Alleged collusion? Lowest combined price | Collusion with A. Payment to A and D for tender costs | Not on list | Not on list |

leniency on all counts and Company C is quite small, but co-operated and received partial leniency (Table 14.1).

## Discussion

There appeared to be no allowance as to how much turnover was involved directly in the 'collusive' market or to whether there was a proven distortion of the competition and lost value by the client, e.g. the project did not have a go-ahead. The decision to fine on the spot in this way has taken the market by surprise and OFT does not have to go to court although it has had to defend its decision on appeal.

The choice to give leniency to flush out information and to widen the prosecution has been successful and has also set a precedent for the more general investigation into construction activity by larger main contractors. These contractors have been written to, asking them if they would like to declare any wrong doing and receive some leniency for co-operation with the enquiry to seek out their involvement and further organisations involved. This leniency will be reduced from companies who have come forward without letters of warning. In this latest investigation several well-known companies are involved with large turnovers and potentially very large fines.

## Question

Morally, who should be fined the most and for what reasons? State assumptions.

## Answer

The fine for Contractor B was based on the number of their infringements, the size of their turnover and the fact that a managing director was involved knowingly in the collusion and their were three markets involved in Scotland, the North East and the Midlands. This added up to nearly 20 per cent of turnover on OFT formulae. The maximum 10 per cent of their turnover, i.e. £852,000. On top of this there was a leniency granted for coming forward, reducing the fine by 40 per cent, giving £511,000.

## Notes

1  OFT (2006) 'OFT Fines Roofing Contractors in England and Scotland', 23 February. Press release. Online, available at http://www.oft.gov.uk/news/press/2006/34–06 (accessed September 12 2007). [The OFT has in fact made three CA98 chapter prosecutions, two for roofing and one for double glazing; West Midlands (2004); the North East (2005); Scotland (2005).]
2  OFT (2006) CE982006 'Collusive Tendering for Flat Roof and Car Park Surfacing Contracts in England and Scotland', Public Register CR98. Online: http://www.oft.gov.uk/shared_oft/ca98_public_register/decisions/flatroof.pdf    (accessed September 10 2007).

# Chapter 15

# The Heathrow T5 major projects agreement vs false employment

## Introduction

Jump (2002)[1] expressed the aspiration that 'a career in the building and construction trades' should be one 'which can be pursued with just rewards and with safety, security and dignity from day one . . . to retirement'. This is not an unusual aspiration for a job but is being threatened by what has become known as false or bogus self-employment and has occurred because of a misuse of the self-employed status, often for the purpose of avoiding tax and costs of the legal responsibilities of direct employment. The OECD (2000) has also identified this as more prevalent in some countries where taxation systems and/or labour policies encourage its development.[2]

## Self-employment in the UK construction industry

In the UK construction industry, as we saw in Chapter 5, many contractors do not employ direct labour and choose instead to subcontract it out. The Construction Industry Scheme (CIS) for taxation was introduced by the Inland Revenue originally to try and make it easier for employers to make gross payments to workers who were genuinely self-employed and therefore would deduct the tax and national insurance contribution (NIC) themselves in accordance with total payments received in any one tax year. Under the law, the status of workers who work for one contractor, are expected to work the same hours as direct employees, and are provided all equipment, are now interpreted as a direct employee for the purposes of tax and other schemes such as CIS 215.[3]

Unfortunately this system had been abused either by workers in falsely not declaring payments. Employers had also treated workers they used full-time as self-employed, paying them a gross wage with no other payments or tax. This almost certainly encourages short-term behaviour by the self-employed. Workers who fail to save are depriving themselves of pensions and the means to take time off if they are sick. The long-term effects of 'non-direct' and interrupted employment is that these workers lose their rights to notice, holidays, redundancy, pensions and training.

In the UK construction industry, there are an estimated 361,000 self-employed,[4] as many of these will not produce returns, they will not appear on the official statistics. Self-employment in UK construction is acute at 40 per cent, having fallen a little since the recession in the early 1990's.[5] Construction as an industry is very fragmented with many small companies (93 per cent) with fewer than eight employees. These companies also find it difficult to have in-house pension schemes, generous holidays and to employ apprentices, making the problem worse for the long-term health of the industry's workforce. Many of these companies give a good employment experience for direct employees who have the benefit of a close knit working environment and can lobby direct improved employment benefits, but they also have the more acute problem to cope with fluctuating workload and need access to a roving workforce off the books to cope with the peaks of workload.

However, the key worry is that the problem of self-employment is widespread in the larger subcontractors who choose to have a skeleton direct workforce well below the fluctuation 'low' and perpetuate false self-employment and under-cutting in a cut throat tender market where low margins and high turnover have become acceptable. It is clear from Harvey's research which sampled labour forces in large subcontractors and on mainly large London projects that 85 per cent of the labour force could be defined as falsely self-employed and *not* complying with the four tests of independence for self-employed tax status.

If this is the case, the ethics of open competitive tendering of those who are tendering labour below a market rate on the basis of false self-employment are low. It means they are likely not to be able to guarantee the quality of the outputs of their labour on a long-term basis and may be cutting back on apprenticeship training which in one form or another is so fundamental to sustainable future quality provision and capacity.

## Training

On a large scale, self-employment undermines the apprenticeship scheme which can only work where an organisation allocates experienced workers to teach others. If they do not employ them, then it is difficult to ensure that skills are passed on. It is clear that a short college-based scheme which is unable to get apprentices direct employment for the practical side is also less likely to be successful. It is estimated that less that 20 per cent of this type of entrant actually stay in the industry, inducing hidden costs into the training system and disillusionment for many and a damaging shortage of labour. The loss of the three-year apprenticeship is also a factor for the output quality in the industry. Typically the average length of employment is 34 months with 70 per cent having an average of nine successive jobs.[6] This has reduced to between 75–80 per cent in a later survey in 2001.[7]

## Health and safety

An additional ethical problem comes in the level of safety awareness and training that self-employed workers have time to arrange, because they are not perceived as the direct responsibility of their clients in terms of their basic and updating training. Again lower tight margin competitive bids will lead to contractors being tempted to do less training for all workers. Larger contractors may require induction and a CSCS card but these cards are not a reflection of the basic cultural awareness and behaviour of these workers if they are treated as temporary or casual status. Minimum compliance is unlikely to improve an already poor industry health and safety record. Money for health and safety training is needed from external agencies such as the CITB.

## Schemes for false employment

There are three main ways a self-employed (SE) worker may be paid. The main scheme for 'false' employment are operated through a CIS 4 arrangement which allows employers to sign up a self-employment contract specifically excluding benefits and deducting tax prior to payment. Typically the worker will be working on successive contracts and will be expected in many cases to work the employers' hours for direct workers and may receive some holiday shadow pay within the weekly rate. They will not be entitled to paid leave of absence for sickness, injury or holiday or for any other reason as the pay will be deemed to have included this allowance. The rate is determined by the employer and there is very rarely an ability of the SE worker to employ a substitute in their absence.

Agency workers are employed by an agency and so are an SE status, but may not receive continuous employment. They should receive a proportionate amount of holiday, pension and NI contributions during their employ. Paid leave of absence is applicable upon the qualifying period. However, because of the likely temporary contract of the agency to the project, they are unlikely to achieve a qualifying period of employment for some unpaid leave such as maternity/paternity redundancy, pension and will also not get basic training or apprenticed through the agency. This is, therefore, not a solution to the quality and dignity of the labour, even though pay may be good and, in this case, the contractor pays for the agency services.

Payrolling companies exist as an administrative mechanism to pay a 'self-employed' labour force. It is likely in this situation that a holiday period or a factor for holiday pay has been agreed and that there is a legal responsibility for the payroll company to deduct tax. But it expedites a loophole in the new April 2007 cancellation of CIS 4 status. The employer still side-steps the responsibilities towards statutory paid leave, training and pension contributions which are all deemed to be paid. It is also popular with the worker who is happy with a good 'take home' pay. However, long-term provision is unlikely to be satisfactory.

## Heathrow Terminal 5 (T5)

T5 employed 8,000 workers at its peak and 60,000 will have been associated with T5 by its finish. It has a £4.3bn turnover of work over 7 years due to open phase 1 in 2008 and phase 2 in 2010. As an exceptionally major project there is an estimated 40 million person hours of work on the site. As a major contract, the relationships are complex and there are 16 main contracts and around 60 first tier suppliers.

### The agreement and culture

To generate a fair labour situation T5 has a model agreement which is legally binding on BAA and its key suppliers, in that BAA accepts all the risk for the contract on the basis that it subcontracts the other parties. The principles are:

- focus on managing out the cause of problems;
- work in truly integrated teams given that it is an uncertain environment;
- focus on proactively managing risk rather than avoiding litigation.[8]

BAA has also identified the need to induce loyalty to the project and they have worked hard to get an integrated culture where the identity of the individuals is subsumed into a T5 culture. They have had some success in associating the permanent workers on the site with the success of the project, first, and the success of the organisations they work with, second (as a subset of the first). This loyalty hierarchy has helped to defuse a confrontational approach and together with the risk burden lifted has helped to encourage a blame-free environment. The spirit of the contract is expected to be passed down the chain by the first tier suppliers.

The evidence that is appearing so far suggests that the agreement was successful in that it was running to time and budget and good relationships have been engendered between the key players and these have been successfully managed further down the supply chain. As part of this arrangement it was thought essential that an employment agreement was negotiated between the unions and the main contractors to agree a full direct labour agreement. This arrangement is backed up by the client BAA, who take some of the risk of pay and employment cost escalation under the risk allocation arrangement in the contract.

### Employment opportunities

It was essential to make sure that there was sufficient labour and also that the local economy benefited from the employment created. To support this, some leading edge welfare facilities have been made available.

For example, Laing O'Rourke have produced an hourly paid, notably, *employees* handbook[9] that puts emphasis on its cultural change programme which it calls, Management and Operation with Vision and Efficiency (MOVE). It picks out communication, learning opportunities and safe working as critical to its cultural change. The commitment for safe working is to reduce accidents to 'one in a million' and to have a zero tolerance for unsafe behaviour or a substandard approach to health and safety. It encourages everyone to be prepared to challenge 'existing beliefs' and not 'walk by' so that there is a concerted action on improving practice.

There have been some headline stoppages on this site because of the electricians who are able to command a significant wage. The site has not been completely free from accidents, but in March 2007 the site managed no accidents in a million person hours for the eighth time. The mechanical and electrical contractors use the Major Projects Agreement (MPA). One interesting aspect of this is the commitment to integrated teams hand picked to work together, independently competent and inclusive of apprentices where possible. Efficiency was supported by performance payments which were shared throughout the project. When the MPA agreement was independently audited, 88 per cent of the respondents in the Amicus union and other subcontractors thought that the agreement had improved the project and integrated team working had improved performance and productivity. With sickness and industrial relations it matched other major projects in spite of the high profile electrician dispute.

## Training

The Heathrow Employment Forum is dedicated to the improvement of local employment. Some 70 per cent of the labour used on the site comes from within a 75km radius of the site.[10] BAA Heathrow and Construction Learning and Skills Council (LSC) have set up the T5 Construction Training Centre.[11] The centre is run by Carillion Training and supplies 80 apprentices for the project. These modern apprenticeships are part of a total Carillion programme of 1500 apprenticeships per year where the apprentices go through a two-year programme and come out with NVQ level 2 trade apprenticeships, but may also proceed to level 3 supervisory status where they show their worth. The take up of the apprenticeships is assured where there is a long-term project nearby and is preferred to the unattached apprenticeship. Another scheme to supply fourteen advanced apprenticeships for the mechanical and electrical services companies through sponsorship by the London West LSC with outside training by the West Acton College, Surrey, and Berks LCS support schemes. BAA itself puts in about £150,000/year into training and education initiatives.

## Discussion

The T5 scheme is unusual in the industry because of the special risk agreement with the employer and because of its size and length which means it is more unionised and has great influence. However, it does show that there are possibilities to work with the unions in an ethical way to fight the malaise in the industry which is destroying the quality of the workforce and creating unethical employment arrangements commonly called 'false employment'.

## Acknowledgement

Interview with Jerry Swain, UCATT London South East Regional Secretary.

## Notes

1  Jump, J. (2002) *175 Years of Building Trades Unionism*. London: UCATT, p. 49.
2  OECD (2000) 'Employment Outlook', p. 156. Quoted in Harvey, M. (2001) *Undermining Construction: The Corrosive Effects of False Self Employment*. London: The Institute of Employment Rights.
3  From April 2007, the CIS 215 scheme has been stopped by the UK government as one measure to discourage false self-employment and close the loopholes which allow employers to avoid employer responsibilities, and employers are required to identify the tax status of each SE person (subcontractor), make the deduction and pass it on to the HMRC. Those that require gross payment or reduced reductions must be registered with HMRC before gross payments may be made. This has led to the start-up of payroll firms who carry out this task and make the monthly updates.
4  Harvey, op. cit., p. 23.
5  Central Statistical Office (2000) *Labour Market Trends*. London: HMSO.
6  Harvey, M. (1995) *Towards the Insecurity Society: The Tax Trap of Self-Employment*. London: Institute of Employment Rights Survey. Two surveys covering workers and employers on 10 large sites in depth covering 3,500 workers and 55 follow-up representative workers respectively. These sites included infrastructure sites that traditionally carry more direct labour.
7  Harvey (2001) Survey with UCATT officials covering c5,000 employees on 22 construction sites.
8  BAA (2007) 'Fact Sheet and Case study'. Online, available at http://www.baa.com/portal/page/Terminal5%5EBuilding+Terminal+5%5EFact+sheets+and+case+studies%5EThe+Terminal+5+agreement/d0f8f00ee5003110VgnVCM20000039821c0a/448c6a4c7f1b0010VgnVCM200000357e120a/
9  Laing O'Rourke (2003) *Terminal 5, Hourly Paid Employees Handbook*. Laing O'Rourke plc, Dartford.
10  BAA Terminal T5, 'Fact Sheet and Case study'. Online, available at http://www.baa.com/portal/page/Terminal5%5EBuilding+Terminal+5%5EFact+sheets+and+case+studies%5ESetting+new+industry+standards/f2d75fe363cf2110VgnVCM20000039821c0a448c6a4c7f1b0010VgnVCM200000357e 120a
11  Learning and Skills Council (2007) 'Apprenticeship Case Studies'. Online, available at http://www.apprenticeships.org.uk/awards/launchevent/number10.htm (accessed 14 June 207).

# Chapter 16

# Health and safety systems in a large PFI hospital

## Introduction

The project is a £300m private finance initiative (PFI) project for a new hospital, designed and built for the City Health Care Trust on the site of the old, early twentieth-century hospital. Derby City General is a working hospital delivered in two phases in several blocks. The first phase consisted of temporary accommodation, demolition of old buildings, a new treatment centre and a new acute hospital of five storeys built on the site of part of the existing buildings. Phase two consists of a further five storey acute block, built on the site of the remainder of the demolished buildings. There is a complete refurbishment of the five storey gynaecological and obstetrics block. This block is joined to the new phase two acute block across the whole of one elevation. New access roads and traffic system are incorporated together with an extension of the energy centre. The hospital incorporates the facilities of the other city hospital.

Some 500 people work on site and they are employed mainly by package contractors. The site is spread out over a wide area with several gates for delivery. Entry is through a central security point with a health and safety induction cabin, where passes are issues for entry to site. Small site plans inclusive of site rules and emergency number are also issued to all visitors and new workers on the site. There is a large management team for Skanska which is located in site offices behind the main area of the new development work. There is a dedicated health and safety manager for the principal contractor (Skanska). This manager is responsible to the project director and liaises with the other package contractor contractors to co-ordinate formal method statements and risk assessment. The project has run with four reportable incidents over the life of the project. This meant that nearly one million hours of work was completed without a RIDDOR incident.

## Health and safety policy and ethics

Health and safety policy is not just on a basis of legal compliance, but policies and standards are rationalised for the project on the basis of a moral humanitarian responsibility and are justified to management as good business economically and commercially. Staff and operatives are encouraged to report incidents and incentives are given and prizes exist for major contributions. The health and safety manager also sees this as contributing to more productive and quality work. Where this is proven, extra expenditure on resources for health and safety is justified over and above the minimum contractual obligations. Senior management become involved in order to review and determine the effectiveness of policy. This is done by three monthly company director visits involving inspections of the workplace as well as the system documents. A no blame culture is encouraged and directors inspect in order to be proactive and to come up with ideas themselves. A health and safety control document exists for application of policy and implementation of health and safety procedures. There was no evidence of an integrated policy with sustainability or quality policies.

The order of ethical application for making health and safety decisions is:

- to reduce harmful consequences and increase happiness;
- to give parity and equality of information, training and discipline;
- to induce moral responsibility towards company, family and friends;
- to give dignity and respect to an individual's personal ethical values;
- to promote justice and compliance.

### How are decisions taken?

Meetings are held weekly with project operatives, contractor supervisors and Skanska management. These take place at a supervisor level chaired by the health and safety manager and review the feedback forms which are strongly encouraged from operatives and contractor management for improvements to the system. Up to 140 near miss forms (see below) are processed per month. Each one of these has to be actioned or acknowledged for timed action and accountability in 24 hours. The health and safety committee meets to discuss strategic issues and any changes to policy each month with the management team and worker and contractor representation called 'safety champions'. This committee takes strategic, site-wide action on a lessons learnt basis. VOICE (views of the operative in the construction environment) – is the fortnightly safety forum for operatives with a voluntary representation of ideas from the ground up. It is run by operatives for operatives with management in attendance. £20 is offered for any ideas received and prizes are given out each month for examples of good practice which are sustained and where applicable passed onto others.

In this sense *everyone* has some accountability and a role to play in decision-making.

## What level of control is practised?

Having said this, the principal contractor runs a strict warnings and disciplinary action system for contravention of site rules and unsafe behaviour. A verbal warning is first given for unsafe or dangerous behaviour. A yellow card is issued for incidents where a first verbal warning was unheeded. This means that the worker has to be re-inducted in the presence of their supervisor and a £250 administration fee is charged to the company. If the incident was particularly dangerous, then workers may be dismissed immediately from site. A relevant example of this is an operative who dismounted from a cherry picker directly onto a sloping tiled roof with no edge protection. This action is considered unacceptable and highly dangerous and directly counters safety measures put in place such as protection for working at height. Persistent violation of safety practice will also incur removal from site following a yellow card. The health and safety manager keeps a particular eye on older experienced workers who have become immune to changing to today's more stringent safety regime. The health and safety manager has powers to stop any work that he feels creates significant imminent danger and uses these powers sparingly where improvement notices can be safely used to ensure compliance in sufficient time. Trade contractors are expected to price for a maximum first line supervision to operative ratio of 1:12.

The culture here appears to be one of minimal tolerance for non-compliance with legal requirements and methods/rules that have been clearly agreed and consulted upon. This is in contrast to the self-management culture encouraged at the planning stage.

## What sorts of skills are needed?

There is a need for relevant, updated skills and awareness of dangers in the context of the workplace. This is provided by a minimum requirement for Institute of Safety and Health (IOSH) basic management training for all supervisors which is regularly updated and tested. All workers, management and visitors who enter the site need to have induction training and to be qualified through a relevant Construction Skills Certification Scheme (CSCS) card. Current hazards are indicated on bulletin boards at the entrance each day. Tool box talks are used to raise awareness and to provide opportunities to focus on particular health and safety measures with the opportunity to explain why they are in place and identify the dangers and the impact to other workers. The problem with the talks is that they need to be delivered effectively and measures to guard against symbolic compliance are important. Talks are based about immediacy and relevance. Supervisors and trainers are encouraged to

• deliver interactively to give ownership to ideas for prevention of harm;
• maintain a no-blame culture;

- drive home personal moral obligations by concentrating on the personal impacts of outcomes on the home and family and fellow workers.

One of the key cultural issues upholding the dignity and respect of the worker is involving them in the risk assessment process and rewarding them for improved methodology. The VOICE group consultation may also produce updating of skills and improved methodology in the light of working experience and feedback.

This project is big enough to employ an occupational health advisor. His/her role is to provide preventative medicine and health education. S/he has a role in delivering health-based tool box talks relevant to current working and it has been found that workers respond better when they know why and how something causes a health danger. Simple issues like sun burn, eating habits, smoking and dermatology are recurrent themes. More technical issues like lead working, working on existing asbestos products, noise and vibration effects can be used to espouse safe working to mitigate longer-term effects, particularly when they are connected to imminent activities.

A package pre-start meeting is used to draw together a health and safety approach to the delivery of the work and identify corporately with the supply chain managers, managing contractor estimators/QS and package/quality managers so that health and safety information is passed on and adequate resources discussed with reference to training, feasibility and methodology for delivery. The presence of the designer has not yet been required to discuss the buildability issues, but this is critical to complete the loop of responsibility. The skills of the designer to ensure post-occupancy safety and access are not discussed in this forum, but may be relevant later. Some degree of involvement by the facilities management team on the PFI joint venture team or client team may be focused where particularly relevant. Training on the health and safety file is given for the facilities management team on handovers.

Legal knowledge is carefully explained and impacts discussed, but skills are justified on the basis of rationale and common sense and not blind compliance. Risk assessments are aired at the meeting. There is also a 'near miss' database which is used to inform risk assessment Adjustments to the assessment and method statements are made and it is important to communicate changes to the workforce. In order to justify skills training to middle managers, there is a need to make an economic argument to support the release of resources and management time.

## Management involvement

There is a policy for all management staff to carry out group audit reviews on a weekly basis and to make on-the-spot reports to deal with safety incidents and recommend further improvements to prevention and better systems. These are at an operational level, but include all management categories on the basis that they raise their awareness of risks, impacts and any resource shortages required in proper implementation of policy. The culture needs to be a no blame culture with a corporate responsibility for action and not a blatant risk deferral approach to lower levels of the supply chain. There are questions here about how equal awareness and skills may be obtained by the second and third level supply chain management, as these are on the ground to monitor improvements and responsible for adjusting and communicating revised method statements.

One of the key issues is the co-ordination at the interface of different trade contractors where it is well known that a sequence of events by different, but co-located contractors is more complex to predict and cause dangerous situations. Different contractors' work may make another activity unsafe. There is a need to make imaginative implementation of the CDM (2007) and other specialist regulations to ensure a proper co-ordination of the workplace and not depend on individual package contractor method statements. For example, the stripping out of asbestos or working with it in existing buildings needs to be communicated to all passing through, so that they are not exposed and are aware of the emergency procedures. If information has not been passed through the client and designers about existing hazards, then dangers can occur.

Materials that spontaneously combust on contact with hot work carried elsewhere or as an afterthought need co-ordinating. In these instances the principal contractor holds weekly meetings with supervisors to discuss the impact of adjacent work and changes to design and programming in a given area. These require trust to be built up and commitment for attendance at the meeting with a clear line of accountability for the co-ordination function. Certain activities may be going on overhead unexpectedly endangering workers or, worse, the public. Central systems of co-ordination include the segregation of pedestrian and plant, delivery control, identification of delivery and vertical transportation access points, welfare access during breaks. One such solution is the 'permanent' construction of a segregated surfaced pedestrian route with handrails for access to work places and a controlled cross-over with traffic routes. This is a response to good practice and a traffic incident on site. It also helps to control, but not to eliminate poor habits. Behaviour which ignores the routes is an extremely serious offence, putting individuals and their friends in danger, as happened on the site when the path was ignored.

### Frustrations

There are inevitably elements of frustration in the system, which make it difficult to achieve targets. These are described by the safety manager as:

- lapses of support by senior management in moral and financial resources, especially in changeover periods when a good working system is threatened;

- lack of support by harassed middle managers unprepared to consider the wider picture;
- blatant and uninformed dangerous activities which are pursued in spite of warnings, especially the complacency of older experienced workers who are particularly targeted.

These frustrations are not unusual and may be overcome, but can be demoralising where a strong case has been made for additional resources. There is a need to persist in appealing to the relevant business justification for better management and not to manage a one-size-fits-all approach to different individuals.

## Relationships

Those managing health and safety try hard to build up relationships with the first level package supervisors who are key motivators of the level two supply chain. These are formalised at the weekly review meeting, but continuous contact in between helps to engender a productive, on the spot, two-way problem solving for health and safety problems taking simple actions out of a bureaucratic loop and supporting the improvement cycle. The intention is to build up trust and respect so that health and safety becomes second nature. Health and safety is delegated down to spread the effect and to induce formal discipline procedures direct to operatives if possible. The culture here is one of understanding and no blame so that both sides can work together to stop further accidents.

The health and safety manager is well known by many of the operatives on site on first name terms. Relationships are encouraged by the formal health and safety documents and talk about behavioural safety and there are measures in place to do this through health and safety training and tool box talks. Building relationships is more difficult because of the short-term nature of the attendance of specialist trades operatives. The main reward for good practice is praise, which balances the discipline system.

## Achievements

The site has been able to restrict accidents to four serious ones with nearly a million person hours without a reportable accident using the culture they have built up. The same culture also produces feedback of 140 unsafe incidents per month which are reported by any operatives. They post a written postcard to the principal contractor in boxes provided. Each of these is either dealt with on the spot (simple remedial works) or, in the case of a near miss, it is investigated fully. The site has won a gold Considerate Constructors award – a score of over 40 out of 44 points. Some 28 prizes were awarded for innovative ideas for health and safety in one month.

## Conclusion

Ethically, it is easy to just argue the existence of a code or a set of standards – clearly, ethical practice consists in ensuring implementation. As a whole, Skanska Integrated Projects (SIP), which is responsible for the large PFI projects, has an integrated system which is designed to reduce harmful consequences and provide parity of action. The most innovative system is the VOICE committee for operatives, which is well attended and tries to generate ideas from the operatives in a no blame atmosphere. Ideas may also refine actions in response to the introduction of new requirements. The system has been successful because management has made itself accountable to report progress on requests received at the meeting. The other innovation is for managers outside production to do group audits on a weekly basis. It has the benefit of involving a fresh set of ears and eyes and is carried out on all projects. It is hard to maintain the broadness of this approach, but it is supported by IOSH training for *all* managers which has to be updated at least yearly. The company directors walkaround audit is also claimed to have an impact like an HSE inspection and keeps directors in touch with all the projects, encouraging interaction of improvements between them.

As this is a very large site with inordinate pressures to work to tight targets, with a large geographical area and a lot of package contractors needing to co-ordinate with each other, there are a lot of things that could go wrong and it is difficult for one person to effectively co-ordinate it. This makes efficient and reliable delegation a keystone of the policy. If personnel move, relationships of trust have to be rebuilt. It is also clear that new staff and operatives need to 'catch the fire' and not depend on an arm's length induction video. When the health and safety manager's message is passed on by workers to each other, then a cultural revolution is in the making.

## Acknowledgement

Interview with Steve Iddon.

# Chapter 17

# Stroud District Council planning case study

## Introduction

The planning permission for Stroud College in Gloucestershire (SCG) (Figure 17.1) was granted on 14 June 2005[1] and included the building of a twenty-first-century facility to replace the outdated 1970s buildings which presently occupied a large site. Learning and Skills Council granted a 47 per cent grant for the £18m building and land was sold off to a housing developer to build 149 dwellings to pay for the shortfall. It will include a number of energy-efficient and environmental features as part of the planning approval. Existing trees and

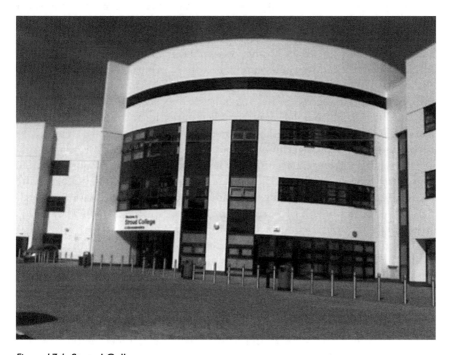

*Figure 17.1* Stroud College.

wildlife habitats will be substantially retained in the scheme which is on the edge of prime parkland. The housing scheme will also include substantial Section 106 agreements (S106) to improve transport provision, offsite open space and a proportion of affordable housing on a brownfield site[2] with a frontage along the main access road.

## Development control policy frustrations and ethics

Nationally prescribed targets for the determination of planning applications are 8 weeks (domestic and minor applications) and 13 weeks for major schemes. Hitting these tight targets is incentivised by the provision of performance related funding called the planning delivery grant. Perversely this grant creates the option to turn down applications without negotiation where the negotiation would take time for determination beyond the 8 or 13 week target. A subsequent application will be classified as a separate application for the purposes of the delivery grant. This satisfies targets, but builds waste and frustration into the system for the applicant and ultimately the council who both replicate work and aggravate the delay period for what might be a basically sound application with a technical hitch. Where this is done with a low application approval rate and the lack of proper pre application advice, there is an effective breakdown in the develpment control system. In a council's defence, the time for approval of 13 weeks is very difficult to deliver given the complexity of the approval and checking systems and the need to synchronise with a member approval meetings within the approval period. Action such as the request for additional details giving little time for response, is an example of the impact of target-based incentives and represents a more cynical use of the system at the expense of an unsuspecting public. Other mechanisms used by developers include the automatic use of the appeal system as soon as planning is turned down, which means that decisions are taken out of local control for the wrong reasons. This again reflects the frustration of a developer who might have been turned down on a technicality, but feels better served by a central non-democratic appeal system even though it might take longer. Proposed rules to charge for appeals are designed to ensure the completion of the dialogue with the local authority, but ethically they should also be associated with a commitment of local authorities to work with developers logically on a 'how to', win–win basis, rather than being a confrontational basis.

Officers are also being given greater delegated powers so that they are able to deal with less controversial schemes strictly within the PPS guidance and without recourse to a member committee. Authorities can and do ask for a range of discretional requirements beyond strict policy expectations which puts them ahead of the game in areas like renewable energy, transport plans and affordable housing. As discretional standards they can create an unequal 'post-code' lottery for requirements and drive worthy development away from the area. Also there can be uncertainty about what standards should be budgeted for at the inception stage, because authority requirements are not planned into the future and significant developments are delayed for long periods, during which these standards change.

## Achievements

A sensible solution to rebuild confidence would be for a greater transparency for future requirements so that planning authorities work closely with potential applicants and that changes to requirements are planned into a sensible implementation time frame to allow for more stable planning. This might encourage developers to work with authorities in a more committed way and build up more credibility and trust in the system. Section 106 agreements are those agreements that allow councils to mitigate adverse impacts on the communties they affect. Historically these agreements have been cited for enabling developers to effectively buy planning permission through lucrative offers. These agreements are controversial and a more equitable and consistent approach has benn discussed for several years,but remains to be enacted.

## Stroud District Council (SDC)

### Control and policy

Stroud District Council is commended for its Beacon Status and the way that it carries out the planning role. The improvement reports check out the ethics of councils in the areas of integrity, codes of conduct for the members and officers, the effect of policies on its planning decisions, the perception of the public towards that and to ensure that scrutiny is free from party influence.

The Improvement Report (2002)[3] for Stroud District Council indicates that the integrity of SDC is 'generally very good', there is a code of conduct for members and officers on the development control committee and they have received training in their planning duties. However, in their findings, they say that the public perception of planning is poor in Stroud District because there is a belief that there is no openness and no public participation and that it was suggested that council members do not understand the value and limitations of the S106 agreements and the committee does not have a strategic context because there is currently 'no local plan'. They make recommendations for more training to deal with S106 and to ensure that scrutiny and planning and DC meetings are free from party influence with perhaps a revision to the code of conduct for members.

With the 2004 changes in the planning law and the delay in approving the local framework, there is now a very tenuous basis for supporting evidence as the old local plan (2004) runs out of currency (two years) and the precedents based on old superseded planning policy guidance (PPG) make it more difficult to support new decisions when challenged. It also means that more work in justifying decisions from first principles is required and the resources associated with this have not increased. This is leading to the charge of unfairness and an increase in the use of planning appeal across the board and the delay associated with this.

### Achievements

In this context it is interesting to note that the council now has a well-advertised practice to ensure a much greater degree of pre-application consultation with officers to ensure that when applications are made, they can meet the tight target dates required by the government to make decisions on applications in 8 weeks and 13 weeks respectively for small and large schemes. Now the structure of the Council ensures that the determination of an application is not biasedwhen the Council itself is the applicantor promoting a particular scheme. This stops a conflict of interest and the charge of favouritism towards council schemes by those who are seeking approval of opposing schemes. The balance of power between the officers and the members is that officers make recommendations on the basis of the central guidance in the Planning Policy Statements (PPS's) and the local Development Framework. and will also draw on precedent controlling previous decisions. With the 2004 changes in the planning system and potential delays in approving parts of the Local Development Framework, the robustness of the Local Plan becomes increasingly tenuous as the basis for determining application. This also means that more work in justifying decisions from first principles is required and the resources associated with this have not increased.

### Communications, stakeholders and relationships

Part of the culture then needs to be an even greater effort on communication to retain public confidence and to ensure fair time scales for simple planning decisions without undue pressure from dominant parties. There is also a commitment to spend time pre-application to agree a timetable for information required from applicants and to agree what is likely to be acceptable. After registering applications and technical examination there are several areas that need to go to different forums and this includes a non-elected LSP to engage a community, a three-week period for public objections and comments as well as keeping members informed without creating any conflicts of interest.

Those areas such as design appearance and function, sustainability and energy use reduction, conservation if relevant, transport and highways impact and disability accessibility are varied and impinge on each other. In the context of lots of applications, SDC endeavour to get applications grouped for presentation to different panels and specialist officers. Some of these panels including members may wish to impose S106 agreements. For any project that has a range of these issues, it is almost impossible to do this in the time available unless time is given to pre-application discussion, negotiation and planning.

### The Stroud College case

Stroud College is an example of how this process was made successful by the use of pre-planning and agreeing targets. In the initial stages after the concept design, the college designs were worked up with a national contractor who was appointed early on a design and build basis. This design was submitted for community consultation and a meeting was advised through a letter drop. The community welcomed the development of a new college to raise the profile of post-16 training in Stroud and to modernise facilities. The meeting allowed the developer to get feedback on community concerns and to see if there were improvements that would induce value into the design. Section 106 agreements were beginning to emerge. A housing developer had been found who was interested in developing the old site.

The method that worked for Stroud LPA in partnership with Stroud College is shown in Figure 17.2.

Discussions were ongoing with the DC officers in a partnership with Stroud College to establish compliance with the PPS requirements and with the local plan as it stood at that time, albeit not fully agreed. On top of this, a S106 agreement ensured that increased traffic onto the main road was controlled by

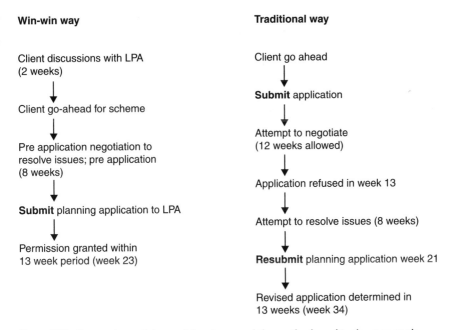

*Figure 17.2* Comparison of the traditional way and the method used in the case study.

the redesign of the junction and the establishment of a one-way system on site and a transport plan to reduce car use and encourage car sharing.

### Building trust and respect

In the case of Stroud College, there was an initial engagement between the head of planning and the project manager for the college to set up discussions. There were six pre-application meetings to deal with the principles and the detail of the application which took place between the contractor design team and SDC. This included detailed discussions over the design. There was a pre-member committee site panel visit and the college obtained the support of the members and the Parish Council prior to application. As a result, the decision was made in just 7 weeks after application (Figure 17.3).

There was also a significantly reduced amount of time by the officers, members and applicants as there was a pre-application time which was productively used and in the case of Stroud it was possible to get the planning permission in less time as well, so the total period was 17 weeks instead of 34 weeks and there was a whole lot less frustration.

Confidence and certainty were built in from the start, making the process less stressful. However, there was quite a lot of expectation for the project from Council with a requirement for developing the brownfield land that was vacated by the demolition of the old building with a proportion of affordable housing built in. Issues to gain compliance with sustainability and reducing energy use were negotiated so that there was a trade-off in the design between the energy use, social, economic and climate change impacts.

It is controversial that some of the land was sold off to developers to pay for the construction of a new college. This application for a 149-house residential scheme took place in parallel with the redevelopment of the college because of the inter-dependency of the two projects. Six pre-application discussion meetings took place between the Developers, Stroud College and SDC and there was an agreement for the Developer to provide 30 per cent affordable housing, the

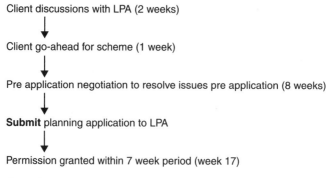

Client discussions with LPA (2 weeks)

↓

Client go-ahead for scheme (1 week)

↓

Pre application negotiation to resolve issues pre application (8 weeks)

↓

**Submit** planning application to LPA

↓

Permission granted within 7 week period (week 17)

*Figure 17.3* SDC reduced cycle time for college application, win–win way.

new road junction which was upgraded for additional road traffic and open space contributions. After the submission of the application, the approval was given in 12 weeks including the signing of the Section 106 agreement.

## Conclusion

The ethical issues in this case study revolve around the loss of public land by the college in order to fund the project, the suitable nature of the residential development so as not to degrade the quality of the site or to congest the road system, and the proportionality of the requirements of the Section 106 agreement which have to be paid for by the buyers of the new homes instead of being spread more widely across the community. There is a strong ethical argument for partnership to reduce the uncertainty for a scheme in principle and the need to make this scheme more environmentally acceptable in terms of global impact and to preserve open space and affordability. Best value to compare value for money for this scheme against another could be an equally important ethical issue, but it is not the main point of this case study. The community stakeholder consultations were also important to increase the profile of the business case for SDG and to ensure better transparency for the sale of public land for housing. The consultation also needed to take on board views and refinements offered from the meetings so that views were not ignored on the basis that the meeting was not representative.

## Notes

1 SCG (2005) 'Government Grant Gives College the Green Light'. Archived news, 27 July. Online, available at http://www.stroud.ac.uk/news/news_archive.html (accessed 27 September 2007).
2 SCG (2005) 'Planning Approval Granted for Stroud College Regeneration Scheme'. Archived News, 20 June. Online, available at http://www.stroud.ac.uk/news/news_ archive.html (accessed 27 September 2007).
3 Stroud District Council (2002) *Ethics, Standards and Conduct: Chapter of Improvement Report*, 20/22 November.

# Chapter 18

# The use of training to establish small-scale organisations in construction

## Introduction

This case looks at the development of a training framework for small and particularly micro organisations (≤ 5 employees). These organisations make up 92 per cent of the construction industry and represent a sector where there is a significant rise in the proportion of accidents and lack of awareness of employment legislation. The work that these contractors do is mainly refurbishment and extension work. These types of projects are generally small and are often overlooked by conventional courses and can fall outside the coverage of the CDM regulations. It is also likely that their proprietors are very busy and would find it hard or strange to attend conventional awareness and health and safety (H&S) training which is also expensive proportionately for the overhead of a micro organisation. They are sometimes unaware of the complexity of the legislation and their obligation to relate to basic good practice such as an H&S policy and the proper compilation and monitoring of risk assessments and method statements. The HSE has recognised the danger of poor H&S planning and needed help to target micro organisations to raise awareness and to ensure compliance.

## Enabling an ethical and strategic entrepreneurial outlook

Organisations need a more strategic approach to cope with the range of legislation that becomes applicable as they grow from self-employment to employing others. There is an ethical obligation to provide a quality of employment for others which matches up to standard levels of compliance and fulfils a psychological contract. These skills are taught in many business courses and also through the NVQ system, but both of these approaches are quite long-winded, needing money and time organised over extended periods. Funding is available from the Construction Industry Training Board (CITB) if firms are registered, but many micro firms cannot afford registration so nothing is done. Regular upgrading on changing legislation is another issue.

## The centrality of health and safety ethics

Many small companies have suffered from a lack of awareness and training in health and safety and there are restricted resources for doing so. Local authorities (LAs) now require evidence of knowledge and basic compliance with health and safety law and its implementation in construction projects. This needs to fulfil an organisation's moral obligation to its employees.

CHAS (The Contractor's Health and Safety Assessment Scheme)[1] is a scheme for the assessment organisations. It has been devised by a consortium of London LAs in an effort to reduce the duplication of health and safety assessment for each project, with particular attention to small and medium-sized organisations who proportionately have less resources for registration. A reduced fee is offered for micro organisations employing ≤ 5 people. The scheme, which has now spread geographically, is committed to feedback to those who need to resubmit. It also complements the CSCS card which certifies individual competence. This gives it a training and development angle to bring up the standard of health and safety performance.

### The emerging ethics of sustainability

Government White Papers and subsequent directives have led to a series of requirements for environmental and sustainable solutions which are mostly relevant to new build. However, the granting of planning permission and building regulations for extension and change of use projects makes it relevant to show how sustainability is taken into account in the design and use of the building. Householders will also have more questions about the house sale requirements of a home improvement pack which needs an energy assessment. Extensions and home improvement will provide opportunities for householders to reduce energy ratings and consumption and it is becoming more important for designer and builder awareness.

### The mantle of good employment practice

Employment practices need to keep up with developments and there is much legislation to digest including the new requirements of the CIS scheme. This is a revenue requirement to pay tax at source on all transactions and the HM Revenue and Customs (HMRC) now put responsibilities on contractors to verify the status of their subcontractors and the self-employed before payment. There is also a desire to collect more VAT by the Revenue on informal contracts. Alongside this there is a campaign by unions to ensure that employees of small organisations do not shirk their responsibilities for the longer-term costs of employment such as holidays, sick pay, maternity leave and pension provision to be in place through the NI contributions and stakeholder pensions. There are also obligations to pay redundancy after two years of service and to have fair schemes for discipline, grievance and dismissal. Other legislation tackles discrimination and the need to cope with rising claims to unfair or virtual dismissal. For a small employer they miss the importance of these changes and need a supportive environment to develop an ethical response.

## The training structure

Apprenticeships in construction and property cost money. Tenders are competitive and depend upon whether the opposition is also providing training opportunities and fair employment terms, so that tenders from responsible contractors are not priced out by those who cut corners. This depends to some extent on client attitudes in selecting contractors on a value rather than the lowest cost basis.

## The E scheme

The proposal for the 'Essential E' (Figure 18.1) is a four-pronged training programme for the development of a foundation of H&S with the associated development of business strategy, environmental practice and the individual training of workers in minimum competence to attain their CSCS cards.

It is a Constructing Excellence Demonstration Project and the Team Leaders for the demonstration are ISG Pearce.

### Decisions

The pilot scheme phase one was run by a director seconded by ISG Pearce and in conjunction with two local councils in order that support could be rallied and to ensure an integrated approach between the council objectives and the training given. Training agreements for the second phase are with local colleges and these are control more at arm's length to any central control. The

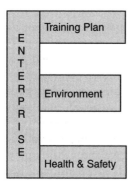

*Figure 18.1* The four strings of the 'Essential E' plan.

scheme relies on the charisma of the director and the leverage of the councils to insist on registration. The SMEs need registration to get work in some LAs and this is where the pilots were rolled out. There are further plans to launch this scheme to more authorities.

## The development of trust and relationships

With the inception of the idea, there was a challenge for this to reach the target audience of micro organisations and to fund it at a level that could be afforded and match the time that could be made available by proprietors. Without this match there would be no take-up of the training and no progress towards the vision of competent service and safety for SMEs.

Meetings were set up by working with local authorities who would benefit from better qualified and more efficient small organisations in their own work and in the registering of those who met standards in their CHAS schemes. (Private clients will have separate goals.)

The initial foundational H&S training was carried out over two days in the SHAC (Safety and Health Awareness Cabin) which consisted of a mobile Portacabin set up outside the builders merchant where contractors were often to be found first thing in the morning. This gave an informal venue designed to be familiar to those mainly site-based proprietors, so that they might feel comfortable. It was cheaper and could also be used as road show, getting to where the builders were. The event took place over one day. Funding was available from the Skills Alliance[2] for training and the project was piloted successfully in Plymouth and rolled out across the five main areas of Somerset working with Somerset Health and Safety and Building Control Departments. The key to success was using building control officers to advertise the events as these officers had regular two-weekly contact with each of the projects through their official capacity to inspect building structures. They agreed to encourage small builders to attend the training and also had some leverage through their desire for quality compliance. This programme rolled out across the five areas and attracted 80 of the target audience, to the training days.

## The development of skills

On the first day, builders were introduced to the CHAS schemes and given the tools to collect information for the health and safety accreditation such as a policy and a risk assessment of key activities. This was brought back the next day for assessment so that additional advice could be given and the organisation could prepare confidently and quickly for registration with very little trouble. About 50 per cent of the organisations get through assessment the first time and CHAS also gives written feedback to those who don't. If organisations are well short of a standard, they are encouraged to carry out some more major changes to the organisation before they come back.

The next stage of the programme was to back up the initial 'urgent' foundational training with the other aspects of business training in environment and enterprise. This programme covers a range of business aspects including accounts, employment, contract and insolvency law, costing and pricing. Agreements were made with a training provider to develop modules which could be delivered in a one-day package, and a certificate in practical business was offered. This meant that in both cases the effort was rewarded with a tangible recognition of competence.

This training had a further module to develop knowledge and understanding in environmental law and practice and compliance with recent government requirements such as the sustainability code and the Eco-Homes minimum standards. The final follow-up stage to the programme was to develop an ongoing training programme for the company.

## Culture

The training programmes indicated sensitivity to the existing reactive culture of micro organisations which meant that 'time equals money' and only immediate things that stared them in the face would be dealt with. It offered some positive added value to the organisations by saving them the cost of constant registration for LA work and gave them enhanced ability to advise clients on sustainable systems and to get workers onto closed CSCS sites. It could also offer some subsidy and bulk buy economy by bringing organisations together.

However, it was more likely to reach those who were meaning to do something, but had not had the time or the push to do so, but was unlikely to force compliance from those who still believed they could bury their heads in the sand and the problem would go away. The process of culture change still has some way to go, but at least if financial success by this method could be shown, then others might follow.

## Client participation and ethics

The second link in the project chain is the client. The project aimed to reach clients who after all would *make the decision* as to whether they would use cheapest price or best value in the long term. Many 'cheapest prices' actually cost more by the end of the project and may not meet the needs of clients or fail in some other way, e.g. a serious health and safety incident, inferior quality or avoidable delayed finish. Many of the clients of smaller contractors are one-off and inexperienced, but many do offer repeat work. These clients could be targeted, but needed to be convinced.

Working Well Together (WWT) organised a conference for clients to gain their support to look at the value of a contract. This inevitably targeted clients who were experienced or about to embark on a project. The aims of the conference were to raise awareness of the CHAS registration system, to indicate the harmful effects upon them of poorly executed, unsafe work and to highlight their own responsibilities and how they had increased in the Construction (Design and Management) Regulations 2007. The Regulations recognise the dominant role of the client in making decisions to use contractors who are properly resourced, but may have an initial higher cost. Under the Regulations it is a client's offence not to choose a competent and safe contractor and not to monitor safety plans properly.

### Design health and safety issues

The designers are the third link in the chain and have a strong influence on a small contractor's actions and also on the safety of the design in use. It is also possible for the HSE to sue the designer in the case of unsafe designs even if a health or safety incident does not occur. A separate programme for design health and safety training has emerged following dialogue between the Bristol-based designers and the WWT Health and Safety Campaign for the industry. The training is delivered via a group called Safety in Design (SiD). This is a series of modules sold to architectural and engineering practices to raise the knowledge and understanding of safe design. Awareness raising road shows have been set up to reach as many designers as possible close to their place of work and WWT have joined forces with Constructing Excellence and Construction Industry Research and Information Association (CIRIA) to deliver these seminars.

Modules for the SiD training cover hazardous activities like the removal of asbestos, the labelling and use of hazardous fluids and materials and the proper surveying of construction sites to identify contamination. They also draw attention to the maintenance activities which result from design such as access for cleaning and servicing and the ergonomic hazards for buildings in use.

## Reflection on the ethics of the project

All contracting organisations have a duty of care to provide a healthy and safe environment and there is a moral and contractual commitment to deliver competently the product promised. There is also a legal and moral duty by the client and the designer to provide the basis for safe manufacture, installation and use. These latter organisations might also be small and inexperienced. Micro organisations which begin to employ people will immediately take on more responsibility to comply with legislation. All the above can be described as law-abiding and is a duty not a choice, though many micro organisations are in danger of falling on the wrong side of regulations and ignorance is no excuse.

The requirements for ethical behaviour are to at least meet legal minimum standards, to meet contractual commitments and, importantly, to provide a service that gives confidence to the user both for the promises made and for the competent standard to which they are carried out. Competence should mean reliability and a degree of planning and checking is implied. The customer will also ethically look at track record, and be anxious to exclude incidents that affect their own reputation such as serious accidents during construction or maintenance problems for the safety and comfort of their staff.

Sustainability is another area where there is a choice of ethical behaviour and designers need to produce evidence of targets which help to reduce environmental impact such as energy use reduction, low carbon emissions and less use of scarce or unrenewable resources such as unmanaged use of hardwoods for development control. In use, this might translate into better access to public transport and more use of recycled materials. They can also go further and anticipate future standards and targets.

Employment conditions for employers are often measured by the retention rate and anonymous surveys by employers of employee satisfaction. Those who are self-employed or subcontracted may also be mistreated by a degree of commercialism in delayed payments, non-payment of retention and the wrong classification of employees as self-employed therefore avoiding pension and National Insurance payments which provide coverage for sickness, redundancy and pensions or full allowance for independent schemes. These are basic and rewards for service would represent a more proactive ethic.

## Conclusion

The project has sought to reduce the lack of compliance with emerging health and safety legislation and to improve the business skills of micro organisations including designers so that more proactive work is done by the contractors and designers and provide the chance to develop the basis of a competent service which is robust when crises arise and ethical. Through its wide-reaching actions, it has attempted to close the link with clients so that value-driven bids are given credence and pricing allows a competent contractor to do their work. If this link fails, it could drive out good practice by the client practice of choosing lowest price.

An iterative process of introducing good practice, adopting efficient good practice and driving up productivity could cancel out the price difference, but this needs time to be nurtured. Likewise, good practice contractors may find they can be choosy about their clients and insist on giving value.

## Acknowledgements

Thank you to ISG Pearce and Tom Harper.

## Notes

1 CHAS is a self-financing registration scheme for small builders to enhance their health and safety policy and management. The LAs run it on a budget to keep costs low and registration up.
2 The Skills Alliance was a partnership between the Qualifications and Curriculum Authority, the Trades Union Congress and the Department for Education and Skills to help develop more skills in industry.

# Chapter 19

# Manufacturing quality and trading relationships

## Introduction

Specialists contractors are normally contracted by the main contractor and are sometimes subject to practices which TI see as unfair, such as delayed payment structures, Dutch auctioning and monies withheld on retention for long periods. Paying when paid is outlawed under the Construction Housing and Regeneration Act, however contractors exploit a loophole in the law which allows contractors not to pay the supplier if the major contractor's customer defaults on payment because of certain exceptional reasons. Liquidated damages are often charged back-to-back in contract conditions and blame can be passed down the chain disproportionately. Subcontractors need to retain their business relationships and decide on their strategy.

This case study is based on an award-winning steelwork fabrication company, operating in the South West of England, which employs 90 people over two sites. It specializes in constructional steel frames in one factory and structural steel products such as towers, masts and quick fix bridge structures in the other. It is well located for access to the whole of the UK.

## Manufacturing

### Manufacturing control and achievements

The company has a modern manufacturing facility with CAD-CAM technology for direct cutting and drilling from the detailed design programmes (Figure 19.1). The detailed design process is worked from structural steel designs and automatically feeds instructions to the CAM saw and drill machines. The steel is delivered on a just-in-time basis in the order planned and is cleaned and is fed into the beam saw and drill line. The smaller fittings such as braces, baseplates and angles are prepared on a parallel line and sent to the fabrication shop with the structural members. In the fabrication shop the parts for welding are assembled and pass on for painting and direct loading onto the lorry to give an efficient sequential access to steel for erecting programme on site. A separate

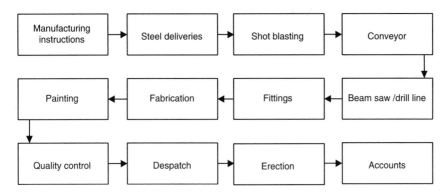

*Figure 19.1* Manufacturing process.
*Source:* William Haley Engineering.

team of steel erectors works off the sites. Quality is a key concern both in getting repeat work and in avoiding the costs of abortive work in the installation process. The company works on a continuous improvement lean construction basis to drive out waste in the system and although the factory is divided into sections, improvements are invited from all employees to apply to their own or other sections. This recognises the interrelationship in the factory that one problem will have direct or indirect influence on other parts of the system. In making changes output has increased 27 per cent and tonnes per employer has increased by 10 per cent and non-conformances have improved by 42 per cent.[1] This level of achievement is not so sustainable by the whole construction project.

### Ethical values

The main ethical approach is to build up integrity in relationships to the extent that promises given are promises kept, problems are shared and there is mutual respect to mainly work without reference to contract conditions. The director shares the business objectives with the workers openly.

Most steel and all bolts are ordered from one supplier and a single hauler is used so that relationships can be developed. This means that relationships grow closer as manufacturer, suppliers and haulers can develop more efficient ways of working together. The steelwork specialist gets excellent service and reliable deliveries to operate a just-in-time manufacturing process and the supplier gets continuity in orders.

### Worker relationships

The quality policy of the company is to be reliable and to achieve 995 out of 1,000 pieces of steel defect-free. The company also has a feed forward quality check on material and supply and checks each piece before working on it. They believe in training their workforce and also invites them to identify their weaknesses. The company's continuous improvement process depends upon workers being aware of the business objectives for improvement which are advertised around the factory and reinforced through workshop meetings and critically evaluating the work process. Workers are rewarded for innovation. On the outputs the company seeks to delight their customer by 'under-promising and over-delivering'. They see mistakes as 'an unplanned investment creating a learning opportunity to find an improvement which will prevent repetition of the mistake'.

### On-site assembly

The contractor stockpiles some finished steelwork in their yard only where they are manufacturing ahead of programme. It mainly works on a just-in-time basis to deliver steel immediately it is manufactured to site for immediate assembly in programme sequence. They depend upon working to programmes agreed well in time to order the steel and process to suit each project. The beams are marked at the end. This informs the erector whether the end points north or west and, at the same time, that the marked surface is the beam top, so steel erectors can see to sling and orient the beams correctly and the right way up.

## Building up trust with the contractor and client

A specialist contractor's work is tendered through the main contractor whose main concerns are reliability and price. Their expectations are that the steelwork contractor will give value for money and it is a strong possibility that the steelwork contractor was asked to price prior to the main tender for the construction of the whole building. In competition it is also possible that the main contractor will go back to the specialist contractors after tender in order to re-quote for the best possible value for money. Some contractors expect additional savings at this point either:

- in a value engineering exercise for the client;
- in sharpening their own margins as a reward for the job;
- in commitments to making savings to offset other overpriced packages.

The main contractor will be looking at precisely matching requirements to get the sharpest price. Exceeding the quality is not going to be paid for by the client. Although time is critical, specialist contractors are often asked to revise the agreed programme at the last minute which also disrupts their planned manufacturing programme. An early involvement by the specialist involves them in helping the structural engineer and architect complete the design.

## Relationships

The steel manufacturer will want to please the client and so service and the ability to deliver on time are also important. The expectations of the specialist contractor are that they will give a fair price for a fair job. With steel fabrication, making savings on the initial price can only be achieved by increasing productivity or by reducing margins. They will be comfortable with certain quality levels which they will not normally undercut as the consequences of failure are very high. If a structure fails or needs maintenance, it is expensive to put right and has health and safety implications. With some innovation it may be possible to counter some of the risk in the design process if leeway is given for the structural engineer to work with the specialist contractor at an earlier stage, and communications and relationships will be enhanced.

Strategic partnering is another possibility and has the potential to build up trust over time and to involve the contractor early. Active value management may identify savings, but this only sometimes will apply to steel. Other possibilities are the savings that can be made on time which may be substantial if there is factory capacity and site labour resources.

Another facet of trust is the willingness to share problems which may arise. This arises out of a no blame situation. Success here is that there should be a mutual respect and commitment to minimise mistakes with the same degree of competency and to learn from mistakes. If it is one-sided so that one party feels they are constantly bailing out the other, then trust is lost in favour of confrontation. The specialist contractor has an important role to play in getting the structural engineer and architect to finalise the design. By being proactive the specialist can create a 3D model of the whole structure and show where information is lacking, setting out needs and defining where clashes of steel and other hitches might occur. The specialist can define the detail problems and check that the solutions are practical and work. In this way the specialist can be a vital aid to the main contractor in getting the design finalised, because the structural steelwork is a critical path activity and will aid timely completion.

## Frustrations and risk

Most specialist contractors will have factored in a risk element based on their assessment of impact and likelihood of the risk, but will need to remain competitive to get work. Some of the fruits of productivity may be shared with the workforce. There is also risk of losing productivity if there is poor information supply or delays in construction forcing a less economic sequence, constant job switches or an under-capacity operation. In practice, a lack of respect or understanding for each other's problems allows these conflicts to escalate and in excess will impact on project time, cost and quality objectives and induce overall project penalties.

As the supplier is unlikely to be dominant in size in this sector, they are likely to have to try harder to please the main contractor and will be frustrated if their management of the contract does not respect the manufacturing and quality constraints they are working within. Repeat business or collaborative arrangements build up the likelihood of more robust relationships, but the dominance factor can cumulate where there is not an understanding of supplier issues. Inevitably trust does play a role in the relationship especially in terms of keeping to promises given on both sides, which underpin the tender figure and also govern reciprocal behavior. If this information is asked for ahead of time for no given reason, then the trust is also broken.

## Case study 19.1  Fitting the price

A contract for 500 tonnes of steel was awarded under a partnering arrangement. A price was given which was considered by the client to be 5 per cent over the budget. The steel supplier reduced the price to comply with their view of give and take and a belief that there would be a later benefit to making this concession. After this had been taken forward, the main contractor returned saying the price was still above the contractor budget. In the meantime the steel prices had gone up by £20/tonne and there was an acute situation for the steel supplier to either comply and to supply fabricated steel under price or to drop out of the contract. In the end the steel fabricator agreed to partly drop the price for the sake of the partnership and the two parties agreed to look at some cost-cutting measures.

### Questions

1   What would have been a fair outcome in this situation?
2   If steel prices go up during the negotiation, how can they be fairly resolved?

## Case study 19.2  Partnering commitment

A contract for 200 tonnes on another partnering contract consisted of a framework agreement of three specialist contractors with the main contractor. A commitment was made to the steel contractor to take their price because it was the lowest. In the event the developer requested that they should use another steel contractor that was not in the partnership as they had worked with them before. The contractor reneged on his promise to give the contract to the specialist partner on the basis that they should comply with the client's wishes.

## Questions

1   Is this acceptable ethically or should the main contractor have discussed their own commitments with the client?
2   Should the client have the right to interfere and overrule the contractor's right to choose their own partner?

## Case study 19.3  A case of wrong instructions

On a refurbishment contract the steel subcontractor was informed that the main contractor would check out the plumb and line of the steel. In the event the main contractor gave the wrong measurements and the steel went out of line and this was only found out at a later stage when the steel had been built in and a component did not fit. Because of the disruption and the problems caused, the contractor blamed the steelwork contractor because it was in the contract that they should plumb and line the steel. As a result the contractor refused to pay for the work done until it was put right – an element for the disruption. The steelwork contractor put a claim in for the monies unpaid as they had put the steel right and the contractor put in a counter-claim for disruption. The case went to adjudication.

## Questions

1   Was it the steelwork contractor's fault for not double checking the line and plumb as they were contracted to, or was it the main contractor's fault because they did not carry out the line and plumb job correctly or indeed at all?
2   Who do you think ethically ought to pay for this?

## Case study 19.4  Claim and counter-claim

As there was no liquidated damages agreed in the contract, the contractor put in for seven weeks of delay and claimed £60,000 per week. The steel contractor thought that this was unacceptable and they thought the adjudicator would adjust the weekly rate claimed, and put in for their own expenses at £5,000/week for the cost of delay. The adjudicator apportioned the responsibility for the delay at six weeks by the contractor and one week by the steelwork specialist. He accepted the weekly amount put in, but the counter-claim for the contractor against the steel contractor meant that they were out of pocket.

## Question

1   Given that a claim had been put in, what would the outcome above be fair?

## Conclusion

This case study shows the way in which a continuous improvement regime can yield very pleasing results and also gain a reputation. It also indicates the difficult conditions which a specialist contractor has to face in a commercial situation. It indicates that even in a partnership, trust is tested to a great extent and relationships need to adjust to a give and take situation and cannot be all one-sided. Ethically there is a particular concern that what has been promised or agreed as part of an agreement should come to fruition in the formal resolution of conflict. Often this is not the case and resolution involves a negotiated middle position where a reluctant understanding has taken place in order to retain relationships.

## Acknowledgement

Thanks to Bill Haley.

## Note

1 BERR /MAS (2007) 'Case study on William Haley Engineering'. Online, Available http://www.mas.dti.gov.uk/content/resources/categories/case-study/cs_William_Haley.html (accessed 3 January 2008).

# Chapter 20

# Educational partnership and sustainable contracting

## Introduction

At the beginning of September 2004, Carillion won a new design and build project worth £77 million for 2000 new student places and a new sports hall on the University of West England (UWE) Frenchay Campus due to start on site in December 2004. It was due to get planning permission in October 2004, but this was only granted with major design changes. Carillion were responsible for incorporating these design changes, keeping to the original start date and completing the building by the end of July 2006. These presented interesting challenges for design and construction management. The project was to use a system of prefabricated concrete floor and wall units together with toilet pods in order to meet the client's requirements for robustness and quality and to cut down the build time to 20 months. The building was going on beside the Faculty of Built Environment (FBE) and had potential for becoming a learning resource for a range of planning, construction and property students.

## The agreement

Carillion had a commitment to provide a good working design, but also wished to maximise the educational aspects of innovative building methods, high quality management and to meet sustainable targets both in the design and the construction. These could be formulated in a locally generated project sustainable action plan (SAP). Carillion therefore generated commitments to evidence their social responsibility in the community and wished to give experience to their staff and to put something back into the educational process. UWE, not limited to the Faculty of the Built Environment, wished to make use of the live project and the expert staff to further the educational value of site visits, research opportunities and industrial contact. They also wished to do all it could to create a safe and secure environment for its students and staff.

This provided an ideal platform for the development of an agreement initiated through the University project board which gave extended and developing educational opportunities over two academic years which were easy to access and were operated between both contractor, consultant and UWE staff for the mutual benefit of both. The agreement was made between Carillion's senior project director and the University project board chair. It was co-ordinated by the Built Environment Faculty. Several staff were involved to help identify useful collaborative activities and a programme was formally agreed.

This agreement committed the two parties to the programme of activities which could be sourced from the project and also rightly limited the access of students to project staff. The agreed programme included site visits, briefings for student projects, research enquiries and surveys, a proposed testing programme for the toilet pods and a conference for further education students interested in applying to the university. Access was also granted to take photographs and one of the classrooms overlooking the site was designated as a place for viewing progress and for tutors to identify the various technical processes in progress. The viability of the agreement depended upon all parties working within the programme and formally agreeing any changes to it.

## The programme

The programme was co-ordinated by direct reference to the Carillion business development manager. Direct reference to the Project Director was made for the more detailed arrangements affecting site staff and access to site. The project director owned, and was accountable for, the SAP and the Considerate Constructors Scheme that drove the business case for collaborating educationally with the University. This programme was wider, involving other organisations and visits and interface with other communities impacted by the project. The objectives[1] which were derived for the programme were as follows:

- to provide a safe and accessible live environment for student learning outside the classroom;
- to develop relationships for mutual understanding of commercial/educational goals using the project director's time effectively;
- to execute mutually beneficial research consultancy on technical aspects of the project;
- to demonstrate management processes first-hand to construction-related courses;
- to provide an input into the Considerate Constructors Scheme;
- to be a platform to further the Carillion corporate social responsibility and sustainability action plan.

The University and the project director were jointly held accountability for the educational programme. Table 20.1 looks at a range of the activities in this programme and evaluates their success against the objectives of both of the parties.

*Table 20.1* The Student Village collaborative activities programme

| Activity | How it met the objectives of both parties |
|---|---|
| *Student projects*<br>Used the live environment of the project in planning and construction and also accessed the documentation (programmes, health and safety plans, drawings and risk registers) and the expertise of the staff in briefings and site visits as a basis of focused tasks to plan and evaluate feasibility and strategy which could be assessed in context. | The FBE uses live projects in training construction students. This provided excellent accessible documents and context to further two years of projects. Some limited staff controlled access was given to the projects, but this did not get used by students in practice. The Carillion staff were excellent and 6 staff were involved in giving briefings, health and safety, showing round. Recordings and pre planning meant that staff were not hassled by students. One thousand students had effective contact with the project and nearly 200 worked on live projects with it. |
| *Research and dissertations*<br>Access for 2–3 notified students/staff to get access to workings and data from the site environment. A plan for testing the sound insulation of the toilet pods in position was agreed using the University field sound testing unit to be carried out under the supervision of a UWE specialist staff. | An MSc dissertation student assessed the '4 projects' collaborative network alongside other networks and in terms of its effectiveness and implementation. A discussion on the CSR policy as implemented at site level to support staff research. The research was discussed, but not finally implemented. |
| *Considerate contracting*<br>This contractor scheme needs to identify social and educational projects bringing together the contractor and the surrounding community. The whole collaborative project came under this heading for the contractor. | The framework developed on this project received a mark of excellent with a score of 36.5 and 37 out of a possible 40 for the two audits. This led to a gold award for Carillion under the Considerate Constructors Scheme. The comprehensiveness of the community involvement was clear for all to see. The University programme strongly contributed. |
| *'Don't walk by'*<br>Carillion instituted this campus wide health and safety campaign to encouraged the whole University community to join the site workers in reporting unsafe and unhealthy incidents occurring as a result of the impact of construction on a site of over 8000 students and staff. The calls were co-ordinated by the University House Services and passed on to Carillion. | This scheme was effectively used by the University community throughout the 20-month period on-site and comprehensive measures to ensure safe working within and outside the site boundaries were clear for all to see. The University was also active in advertising this campaign on its own website. |
| *WRAP*<br>Carillion organised a recycling seminar in the University premises to invite people right across the University community to see examples of recycling. | The seminar was attended by estates staff, university staff and the environmental officer and addressed by the project director and WRAP staff. |

*(Continued overleaf)*

*Table 20.1* Continued

| Activity | How it met the objectives of both parties |
|---|---|
| *Summer placement*<br>As a result of the relationship between the University and Carillion a summer placement was offered which allowed a mature female student to work on a Carillion site. | Carillion already employed a University student and previous graduates on this site, but the contact made a summer placement possible on another site. |
| Development of a DVD with progress photographs and technical site visits was initiated for the piling and groundwork stages. Material was made available for teaching purposes. Regular progress reports were posted on the University website by the Estates Department indicating the primary activities going on to inform the whole University community. | This resource was only partially used and access to the site for safety reasons was not spontaneous so as to be connected to events, but the initial material was effective. The observer window over the site was used by staff to further technology and planning seminars. |
| A newsletter was initiated for the residents of Caroll Court to inform them of expected intrusive events and as to what was to be done about noise levels. | This was a high quality online newsletter on a regular basis and was an effective dedicated communication to residents. |
| Conference for FE and UWE construction students to brief them of the various aspects of the project. This supported the nation wide construction promotion week. | This conference was well attended by students and created a shop window for the project with a wide range of contractors in the supply chain as well as designers taking place to raise and enthuse awareness for construction course students and potential applicants. |

The programme had lots of spin off in helping to smooth relationships between the University and the contractor. The major objective to help educational development suited both the University and Carillion. It also made live projects particularly accessible and relevant to students for two cohorts and promoted staff networking contracts with Carillion and some of its major contractors. Students understood the major planning processes and health and safety requirements by personal experience and observation.

## Conclusion

Many projects work in isolation to the surrounding in which they are placed. From the design point of view there are a number of planning constraints, but in operation it depends upon the professionalism, skills and ethical responsibility of the contractor to work well with the community around it and to enhance the experience for them during construction. This may be to do with the impact of noise, dust additional traffic congestion and dirty roads, but this project indicates an extra dimension which helps to use the latent educational value of a project in promoting construction and property development as a positive feature. This proactive approach helps to optimise construction intellectual value resources such as specialist expertise and raise awareness of the processes of construction for a new generation of interested citizens and in this case construction and property specialists.

This has recently become more of a feature of sustainability action plans and extends beyond the considerate contractor and basic courtesy to realising the uniqueness of opportunity for a wider band of public stakeholders. This project has illustrated a good degree of success in building up relationship networks and knowledge exchange. It falls short of training on the job, but reaches a much wider audience with a single project.

## Acknowledgement

Interview with Colin Rooney.

## Note

1 Fewings, P. and Rooney, C. (2006) *Sustainable Contracting and Education Development Planning: Case Studies of Good Practice*. Construction Knowledge Exchange.

# Chapter 21

# Trust and relationships in a mega property development

## Introduction

The Cabot Circus city centre expansion project is worth £500m and covers a 14-hectare area of the centre of Bristol. Initial construction started in the autumn of 2005 with enabling and infrastructure works following approval of reserved matters in May 2005. There are three distinct areas of the project:

- Area A is the main shopping precinct which is anchored by a large department store. The main square and part of the three precincts are covered by an innovative 'flying roof', but are not enclosed so access is available all hours and a range of external levels are created. It consists of 100,000m² of mixed use shopping and leisure called Cabot Circus, including a major 13-screen cinema.
- Area B is Quakers Friars, which creates a square with residential and specialist shops around the historical Friary building and includes the only high rise in the development with 100 residences in the tower block above a new branch of Harvey Nichols. Areas A and B have over 300 shops including 15 major stores.
- Area C includes the 2,600-place car park and also associated offices, affordable housing, a 280-student residential block and a 150-bed hotel. The main arterial ring road divides this phase from the shopping areas.

## Purpose of the development and stakeholder management

The aim of the Broadmead city centre expansion is to revitalise part of a rundown shopping area, by expanding it and creating a development which remains fresh and dynamic. Ownership of the area is encouraged by introducing affordable and high class residential occupation. Existing residential neighbours in St Paul's and St Jude's need to be kept happy. Bristol City Council (BCC) are a key stakeholder whose aim is for the scheme to be a gateway into the city and to lift the appearance of a rundown area so that shopping can be returned to the city centre and traffic conditions will be enhanced without cutting off the area from its neighbours. It is hoped that the new and existing residents will play a role in improving security and themselves benefiting from the uplift. The scheme has specifically rejected the idea of an enclosed mall planted surreptitiously as a closed development during the night time. As the development is partly surrounded by trunk roads there is a need to maintain crossings that are attractive and safe.

## Procurement and management

The master design has been carried out by Chapman and Taylor, Alec French and other architects who have been involved in the individual concepts for each of the twelve numbered blocks in areas A, B and C. McAlpine was procured as the main design and build contractor for areas A and B and employ their own architects, Benoy, as well as liaising with the client's concept architects. The car park and some of area C was let to Norwest Holst and their site is segregated on the north side of the ring road. Individual developers have taken over the sites for the student accommodation and the hotel. Nuttalls carried out the initial infrastructure and enabling works including the realignment of the ring road, the diversion of Cutlers Mill, services diversions and some demolition. Keltbray were responsible for demolition of parts of the old Broadmead Centre, the multi-storey car park and a major 19-storey 1970s office block. Enabling work cleared the site ready for major groundworks and building works and on substantial completion of these works Nuttalls were novated to the main design and build contractor to manage and co-ordinate remaining road works together with the external works. The client project manager was responsible for co-ordinating the different projects, traffic management and also ensuring that different units were let and fitted out. The shopping centre is to be run by the client. The project manager also deals with the external stakeholders and in particular the relationships with Bristol City Council.

The client is the Bristol Alliance, a joint venture between Hammerson UK Properties plc and Land Securities plc. The values are customer service, respect, integrity, excellence and innovation. The client, with their project manager, has procured a number of other services such as public art and fitting out services on behalf of some tenants. The project manager is the main arbiter of disputes and makes final decisions about quality, organising inspection and handover and approving strategic sequence and methodology for carrying out the works. The client also employs personnel to sell space in the development and to liaise in the planning process for the fitting out works and common usage areas internally and externally to the buildings. There are key issues to market and maintain an attractive location for shopping tenants. Residents either buy or rent their apartments and there are issues to organise a safe and secure environment. There is a need to set up management contracts to maintain common space which, in the case of major retail areas will be managed by the client. The terms and conditions of various management contracts must be co-ordinated. There are many areas where the project must manage the general public as it has a large impact on a strategic and busy area of the central area of Bristol.

## Objectives and stakeholder consultation

At the beginning of the project a consultation exercise was carried out and the following issues emerged which were:

- improve transport congestion and access for the shopping facility;
- ensure sufficient car parks;
- reduce the impact of deliveries and prioritise reduced lorry movement;
- ensure mixed use and safety by design;
- integrate the new development with St Paul's and St Jude's;
- create landmarks and a gateway to Bristol;
- strike a balance between weather protection and fresh air.[1]

The stakeholder consultation has been a huge success and resulted in Bristol City and the developers working closely together to achieve the goals. The consultation started long before the implementation stage and has continued with success throughout the construction stage. A name competition was held to name the three streets and involve Bristol residents.

In analysing the success of the design in achieving these objectives a finished product is really needed. However, at this point in time some of the features are reviewed in Table 21.1 as a communication and trust exercise.

Table 21.1 implies the importance of initial promises made for a development which is remembered and as a baseline for comparison with the finished product. The ownership of such goals is in the hands of the Joint Venture and Bristol City Council. This makes it more difficult to achieve unless there is a co-operative accountability for the communications made. But relationships can be maintained on the basis of updating and rationalising the objectives in a consultative manner.

## The development of trust during the construction stage

The construction project phase is run by the client project manager who reports to the project director. There is a culture of collaboration which has been built up over a series of projects so that trust has developed at a senior level in the organisation. McAlpine tendered competitively on a value for money basis against two other contractors and were chosen on a combination of price and quality and level of service that they were able to deliver. They operate on a JCT Design and Build Contract. As the client and company know each other well,

*Table 21.1*  Broadmead city centre expansion – does it achieve its goals?

| Consultation | Design evaluation |
| --- | --- |
| Improve traffic congestion and access for the shopping facility | There is a new radial circulation built into the Bristol city centre, but traffic flows sometimes congest without a flyover. It is not worse. |
| Ensure sufficient car parks | The 2600-space car park is probably no bigger than the three demolished though it is more convenient in its access to shopping. |
| Reduce the impact of deliveries and prioritise reduced lorry movement | Deliveries are directed to the basement in Area A which is a neat solution, but they are over ground in Quakers Friars and may be disruptive to residents if out of hours. |
| Ensure mixed use and safety by design | There is a good mix of residential, shopping and leisure, but remaining old Broadmead is not yet resolved in terms of safety and image. |
| Integrate the new development with St Paul's and St Jude's | The residents of St Paul's and St Jude's have parallel improvements and there are plenty of crossings to the centre, but the ring road 'bracelet' is still in place for them with less bridges across. |
| Create landmarks and a gateway to Bristol | The Tollgate landmark has been replaced by the Harvey Nichols tower which is more in proportion, but the gateway icon is quite narrow on Bond Street North. The wider aspect of Bond Street S. has brought the refreshing icon of the student hostel. |
| Strike a balance between weather protection and fresh air | The innovative covered square and precincts allow sustainable natural ventilation and some weather protection. It needs to remain secure. |

they have expectations of collaboration and have not signed up to a charter of partnership, nor do they have a partnership contract. There is not a specific project code of ethics which is applied by Bristol Alliance, but there is no need for a formal control of culture and behaviour, but there is a strong culture on the project which has been assumed from previously working together.

The McAlpine website states, 'We believe in the benefits of co-operative working and seek to apply the principles of partnering and alliancing in the course of our work.'[2] In their annual report they refer to the fact that

75 per cent of their annual turnover is design and build, giving the client a single point of contact. They also state in their report, '[high standards] are complemented by a co-operative business culture which promotes and sustains long-term relationships with clients and members of supply chains, leading to repeat business and continuous improvement in customer service.'[3]

All parties would like to promote a collaborative culture. McAlpine have three other retail developments running including other projects by Hammerson.[4] This makes smooth relationships with this client of major importance strategically.

## Integrity and transparency

Integrity is an expectation at Cabot Circus and is defined mainly in terms of professional behaviour, but does not mean that that there will not be a case where integrity is breached by some. Unwelcome behaviour would be defined in terms of contravention of professional codes of conduct with regard to honesty, confidentiality, not declaring conflict of interest and not keeping the client and other parties informed. On the project a culture of transparency is expected to apply. This means that any information which is reasonably asked for is made available and that on the client's behalf instructions are given which are rationally explained so that mutually beneficial solutions can be resolved together. This will extend to informal agreements to bear the cost of change on a balanced incident basis between client and contractor where this cost is not huge and where it does not represent a change of scope. This allows for flexibility in detailed design and conversely, for helping out where contractor-based problems are becoming costly or difficult to deliver. This has been found to be an efficient way to work and avoids recriminations and blame. However, it would not be possible where trust did not exist that the client and contractor would exploit the situation at the other party's expense. It also matches one of Pareto's principles that a solution to a negotiation (or series of) should not give a net gain to one party. Exploitation is described by the construction manager as acting dishonourably, so for example,

- the contractor asks for more than market rate when the client has no alternatives;
- the client undervalues and makes late payment for services completed.

The overall account of credit and cost is tracked informally, but trust would be spoilt by even a single case of perceived exploitation. The informality works because of tacit approval and support at the executive level of both organisations. It can threaten to break down where there is not a similar tacit approval at the operational level of the supply chain and costs are sent back up the chain without an understanding of credit/cost balance. These can be resolved when they pass to the senior levels of the first level of the supply chain on the basis that trust between first time organisations at a lower level will develop where they receive a credit first. An example of possible credit is the greater involvement of client engineers to bring a work back on programme. If this is later appreciated, then minor late design changes can also be absorbed by a reciprocating subcontractor. Trust can be built to the level of knowledge-based trust where both parties have come to understand each other's ways and relationships are not just contractual.

Figure 21.1 is an illustration of trust that has a bounded fairness, i.e. that certain changes are permissible if both sides agree the disruption is containable. It could work on a progressively bigger or longer balance period, but saves the transactional time and cost of double accounting. It depends on being 'honourable' and on all parties in the supply chain acting equitably. It also needs a champion/arbiter at a senior collaborative level in the primary organisations. In the new culture, trust exploitation would be to take without giving and walk away contractually. Formal meetings bringing together operational and design staff are expensive on time, but they play a part in the transparency process by allowing information to flow equally in the spirit of participation and consultation. In short, they are part of the trust building process, and give reluctant team members less temptation to hide. Collocation is another way of doing this so that different parts of the project team have access to each other.

## Checking

Checking remains in place by the client and monitoring by the contractor in the spirit of quality assurance. Quality assurance is less paper-based as all parties are in the same building and contractors work on the same floor with designers. Checks are necessary in proportion to the impact. For example, 100 per cent dimensional checking is done on all pile caps and bolts for the steel frames as the impact of getting the buildings in the wrong place will increase frustration and resentment. Other checks and procedures are set up where there are

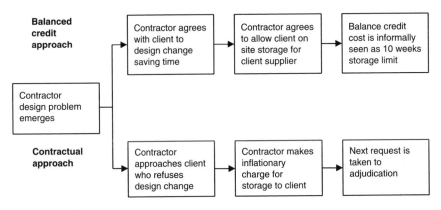

*Figure 21.1* Alternative balanced credit approach to build trust.

interfaces between different trades and contractors to ensure that contractors admit errors early so that any failures are not compounded. Interfaces between contracts, e.g. the construction of the bridge between the car park and Area B contracts; and between packages, e.g. fully rechecking steelwork before hand-over to cladding, are important. In health and safety, there is a concern to build up a culture of awareness and integrity by reporting unsafe incidents and behaviour. Unsafe practice may well be carried out by experienced workers who take their own short cuts and endanger others.

Financially there is a requirement to be transparent. The principle is not to exclude payments made, but to declare all arrangement fees and to audit the valuation of wok done in the spirit of project governance. From the contractor's expenditure point of view there is a particular concern to audit the expenditure of preliminary and prime cost sums. The number of cranes that are used is not an issue as this is a professional judgement and would only be questioned as part of a general review for savings or where there is patent evidence of under utilisation. Conversely the good of providing fair valuation and prompt payment is a basis of contractor and particularly smaller subcontractor trust. Contractors need to give clients forecasts of their cash flow to facilitate fairness.

The consultants are chosen on the basis of their professional behaviour and would not normally be double checked, but expected to self-assess their own actions. Where dishonourable behaviour is discovered, instant action would be taken as an exemplar to deter others to follow suit. Clients depend implicitly on the integrity of their professional team and this team will help to keep an eye on the contractor activities.

### Reliability

Reliability in this context is whether both parties can be depended on to do what they said. Reliability is a measure of trustworthiness and it is important for the truster to take action on a scale of trustworthiness for those they trust. Unreliability is epidemic in construction sites. On more complex activities it is easier to start or promise to deliver and for targets or quality to slip. Some issues will require more monitoring than others and the following conditions should be assessed by the project manager.

- Consider under what conditions trust can be relied upon and if reliability has already been breached.
- Assess the probability of reliability failing and causing embarrassment in delivering what was promised and agree accordingly.
- Consider the impact of the failure.
- Allow tolerance for trustworthiness to develop, especially where the relationship is new.

In the context of the Cabot Circus project, it is complex and not all parties are equally reliable or predictable. A known contractor is easier to deal with than one who is not. Things are likely to be planned more tightly than on a smaller project. Statistically more things can go wrong and only one needs to be unreliable to create a knock-on effect. At Cabot Circus, there is more interest by disaffected outside parties and they can draw adverse media interest in the case of things going wrong.

Heavy crane lifting was required for the pre-cast concrete beams over the substation near the car park involving the use of the largest mobile crane in the UK, delivered on 24 lorries. In this case reliability first time was crucial both to cost, safety, traffic disruption and maintaining city electricity supplies across the whole city. It was a unique exercise with no precedent and reliability could only be brought about by meticulous planning and the co-operation and co-ordination of a wide range of agencies. Ethically the client's and contractor's reputation was at risk and plans had to be proven to be viable. This event required an absolute trust by the client in a contractor they had not built up a long-term relationship with. The contractor had to have absolute trust in the designer and the electricity supplier in the contractor whom they knew even less than the client.

## Information

The construction manager sees information provision as a key aspect to ensure reliability and has systems to ensure routine information gets to relevant parties and a system to look for gaps. A wider net of information is provided in special circumstances to inform the community. The systems have mechanisms for dealing with unreliability – giving more outside help where it is available, putting in place procedures to provoke self-help or encouraging training programmes to build future reliability and ability.

## Planning and risk

Where reliability is unknown, then there is a need to identify weaknesses through the planning system by assessing risk of failure and putting in place mitigation as in any risk management process and ensuring competent responsible persons to monitor it. In the case of concrete floor pours cycles getting behind promised targets in the post-tensioned car park floors, the client's construction manager made his own engineers available to counter the resource shortage and to help to achieve a reliable cycle time. The action in a tight programme maintained trust and eliminated further risk. The extra cost to the client remained as a credit to him for future spending at the contractor's expense.

Planning is also flexible as critical programming events change. This was evidenced in the target date for the anchor department store being set back to relieve pressure on other parts of the contract when the company delayed their fitting out date. This is a result of making information available trustfully throughout the supply chain. Flexibility was also exercised when an improved cash flow was envisaged for the client by finding an office tenant earlier than envisaged. Fresh capital was released into the project to ease resource pressures to build contractor trust and time reliability, if not budget targets.

## Loyalty

Loyalty is a measure which allows a party to overlook mistakes on the basis of trust built up in the past. It is valuable as a way of developing a no blame culture, but also needs a mutual desire and respect to achieve the goal, otherwise it will be eroded and replaced by bitterness and a sense of betrayal of trust.

In building up loyalty in the Cabot Circus project, there is a culture of making people feel comfortable, where they do not take advantage of trust given and are therefore able to absorb some pressures and work harder. Internally there is a leadership by example that works hard. The hours for work on site have been increased from 8am–6pm to 6am–8pm due to time pressures and this has required commitment from supervisors and has been reciprocated by the management. Residents were consulted about the extra disruption and their suggestions used. In this case longer hours could be negotiated selectively to deal with quieter jobs and to keep external work away from the residential areas. However, these stakeholders have little power and may have to be content to go along with unpopular actions. In this case their loyalty is gained through future amenity such as extra money spent on projects of their choice.

## Relationships and communications

The role of communications in the building of trust has already been emphasised through the proper distribution of information and the opportunity for regular face-to-face meetings. On this project, however, there is also a collaborative network system which is used to share information. This has potential, but from the construction manager's point of view there are shortfalls in downloading for ease of reading information through the size of screen, through printing time and software that could help is not installed.

Collocation plays a part in cutting down on the number of meetings. Meetings have also been used as a way to keep stakeholders informed and to provide a forum to ask questions. Education widens the transparency process through the setting up of a project website and a public relations office and bringing the public, schools and universities to site to learn lessons from the project and express themselves.

## Conclusion

The project is big and complex and this case study can only hope to scratch the surface in the complexity of trust relationships and concentrate mainly on the client's trust relationships with others. The effectiveness of trust is dependent on the individual relationships which are developed in a project cycle, for example, between the client construction manager and the main contractor project manager. The scene is set for trust at an organisational level where this project illustrates the importance of previous work together and the commitment to collaboration at the highest level. It indicates trust is not a series of conditions but is a relationship built up over a period of time. This happened previously when they had already formed a successful partnership in the Birmingham Alliance and continued on other current contracts since. Trustworthiness and reliability were qualities that were previously assessed in the main contractor. These qualities are important not because of a perfect relationship but because it is a professional one which depends on a good degree of trust and engenders flexibility. It is seen that other organisations are encouraged to reach the same level of trust, but it requires more work and confidence building by the client's construction and project managers.

Integrity is at the heart of trust and is represented by acting professionally, honestly and with honour so that an approach is well planned and there is a respect for each other's competence even when things go wrong. This is at the core of trust and through the normative ethic of professional codes can provide a basis for this knowledge-based level of trust.

The role of loyalty is supported by the knowledge of probable future profitable work, but it also lies in the degree of fair play which is experienced and expected. A trust *premium* is possible by reducing a confrontational approach, but this does not have to be through a partnering contract as a standard design and build contract is used here. The time taken in keeping detailed change accounts is saved by flexibility and a contractual organisation lower down the supply chain can be influenced by the culture at the top of the chain.

An inclusive treatment of external stakeholders is taken very seriously through an initial consultation exercise which has been continued throughout, but the true test of success in this issue is the satisfaction of the community and the final customer in the finished product.

## Notes

1 Brown, R. (2007) Presentation to UWE Students, 26 Jan. 2007.
2 Sir Robert McAlpine (2007) 'Company Profile introduction'. Online. http://www.sir-robert-mcalpine.com/profile/index.html (accessed 27 September).
3 Sir Robert McAlpine (2006) Annual Report. Sir Robert McAlpine plc.
4 Hammersons (2007) 'Current Projects at Leicester and Aberdeen'. Online. http://www.hammerson.com/pages/113/Current+Development+Projects.stm (accessed 28 Sept. 2007).

# Chapter 22

# Making it work

One of the key things we have found about ethics is that there is a great variety of approaches in a postmodern world and it is necessary to get some sort of common language where people come together to work on the same projects. There needs to be a minimum societal or even cross-cultural norm. People need to feel respected and understand the decisions of others which affect their lives or impact upon the environment around them. The case studies have shown situations where this has been happening to some extent.

The influence of the built environment is of interest to all of us and its effect on the natural environment. Traditionally a level of agreement has been reached where people have come together in professional societies and have agreed a threshold of professional conduct. Some societies have produced additional ethical guidelines, because they recognise the greater need today for more guidelines in how to achieve the standards of behaviour which are looked at more critically by the public. Also that ethical behaviour is more than professional conduct. They also recognise that there is a pressure for more training because of the long-term effects on many people of many professional or expert decisions. Spectacular disasters have occurred where, say, a bridge has failed or a corporation has collapsed, but also because people – stakeholders – have not been properly considered, which has led to an attrition of confidence and lack of trust so that relationships have been undermined and a 'them and us' situation has evolved. This is between professionals in the built environment and the public, between employers and employees and between the stakeholders of a construction project or property development and the project team or client.

## Education and ethical dialogue

Education is a universal platform by which certain principles might be passed on about ethics using a common parlance and providing an opportunity for vigorous debate. Moral decisions can be made in the safety of the classroom and long-held beliefs and values which have simply been acquired can be challenged by those who have different ones, forcing participants to bring their assumptions into the open.

The disciplines of architecture, civil engineering, quantity and real estate surveying, construction and project management and urban planning are open to conflict with each other, because they are quite different in their traditional approach though there are similarities. Traditionally there have been different values, such as whether:

- Design excellence is associated with art, functionality or sustainability.
- Value for money is associated with capital cost, whole life cost or a user/ business efficiency value.
- Risk assessment is associated with innovation, harm or opportunity.
- Health and safety is associated with the environment, quality, design, construction or use.
- Project management is expressed as people management, stakeholder satisfaction, performance or enhancing client value.
- Quality is associated with specification standard, fit for purpose customer care or efficiency.
- Information technology is associated with control, enabling new technologies or managing knowledge more efficiently.

Because of this uncertainty between process and product there is a difficulty with ethics being understood across disciplines except in very tenuous broad issues such as integrity, trust, confidentiality, transparency and reliability.

There is a need to have interdisciplinary dialogues and to get into each other's shoes so that professionals and clients can respect each other's positions and we can move away from the confrontational ethics of 'I am right and you don't know what you are doing'.

There is plenty of room also to reflect on what the public expects of built environment professionals, the influence of the commercial imperatives on ethical behaviour and the emerging need to consider future sustainable development needs in our design and planning and delivery.

# Index

Note: bold figures = key reference pages; cs = case study; tab = table; fig = figure.

absolute 13, 16, **20**, 123, 152; duty 212; trust 363
ACAS model **133**
accountable(ility) 25, 61, 98; business 74, 79; client's 92, 157; code 105; contructor 112; employer 122; professional 64, **95**, 96; public 101, 194; shared 57
accounting: creative 34; environmental/social **192–3**; false 54, 243, **254**, 264; open book 226; transparent 69
acid test 57
act, intentional 30; moral Kant's formulas 21; utilitarianism 18
administrative corruption 246
agape 23
agnostic 21
Ahrens, C. (2003) 55
altruistic **16–17**, 24, 32, 80, 230
America 10, 34, 35, 37, 88, 110, 143, 171; Airlines 48; Latin/South 196, 275
appeals 48, 125, 171
appraisal, company 82, 83; employee 82, 224
Aquinas, T. **20–1**, 285
archetype 21, 219
architect 9, 10, 16, 43, 63, 87–88, 91–2, 102, 232–3, 238, 308, 347
architecture 9, **10–11**, **45–6**, 121, 367
Aristotle(ian) **13–16**, 285
audit(ing), CABE 26; CCS 353; governance 74, 244; environment 70, 71, 209, 298; financial 63, 74, 239, 361; H&S 325, 328; KPI 298; trail 90, 183
autonomy 65, 91, 124, 139

bad 5, **12–20**, 271, 272
Bagley, C.E. (2003) **54–5**,
Bakunin, M. 23
Badaracco, J.L. Jr. 53
bankruptcy 54, 100
Barings Bank 54
beautiful 15, 16
behaviour(al): business **35–6**, 67, 81, 111, 229, 251; collaborative 113, 217, 290; compensatory 165; competitive 83, 234, 248; conflict 228; contractor 102, 153; contractual 284; corrupt 244; dilemmas **29–31**; discriminatory 117; ethical **12**, 13, 21, 27, 108, 216, 281; exemplary 165; Human 64; moderating 17; personal 254; professional **12**, 37, 55, 82, 87–8 143, 232–3, 360, 362; reasonable 221; risk 236; sacrificing 224; (un)safe 160, 320, 324, 327, 362; standard 16, 57, 94, 239, 366; supportive 217; sustainable 81, 300, 342; trust building **224**, 250; unprofessional 25
beliefs 55, 98, 123, 129, 135, 266; client 232; trust and 92, 217; human rights 118; religious 126, 130
Bentham, J. 17
best practice 103, 116, 128, 151, 191; environmental 206, 210
BIFM 82, **96–9**
BiTC 70, 209
Blockley, D.I. 2005 46, **55–6**
*Bonne Foi* 287
Bowie, N. 24, 25
BP **26**(cs)
BREEAM 70, 75, 183, 200, 204, 205, 210–11, 213, 305, 309

bribe(ery) 22, 73, 102, **243–78**; case study
    263–4; cost of **255**(cs); definition **245**,
    258 (tab); international 250–1;
    legislation against 247–8; OECD code
    250–1, 274; payers index 251, **276**;
    practice 282; professions 96; pact 108;
    planning 144, 172, 172, 257; property
    270, 270–2(cs)
Bristol City Council 69, 357, 358, 359, 386
Broadmead employment (cs) 69
Brumwell, P. 53
Brunel, I.K. 1, 10
Buchanan, D. and Badham, R. **56–7**
builder: bribed 263; chartered 88;
    cowboy 144, 249; master 10, 105;
    house 62, 186, 199–200; small **105**,
    **336–40**
building: building for life 26; managers
    **158**; process 3, 11, **12** 196; permission
    39, 110; users 13
buildings: Citicorp **44–5**(cs); civic 10, 27;
    public 10, 44, 183, 204, 205, 213
built environment 2, 9; business 78;
    corruption in 249, 250, 253–7;
    dilemmas **39–49**; employment 116,
    **140–3**; ethics **24–7**, 58; professional
    conduct in **93–101**, 113; sustainability
    189, **192–3**; trust in **220–8**
business 58, 194, 197; cartels 48, 249,
    266, 268; competition **260–9**;
    compromise 25, 30, 63; ethics 1, **35–6**,
    **61–4**; ethics synergy star **41–2**;
    definition 1; H&S 51; international 58,
    **55–6**; motive 61; ethics model 67;
    situational ethics **14, 15**
business ethics model **41–2, 67–8**

CABE **26–7**(cs) 113, 210
Cadbury 10, 116, **120–2**(cs)
Cadbury Report 74
calculus based trust (CBT) 218
campaign: corruption 250a; H&S 341,
    353; inappropriate developments 63,
    121; labour standards 36, 63, 76, 117;
    planning decisions 172; third world
    medicine 76; site conditions 209
cancer 33
capitalist 41
care: duty of 38, 46, 74, 92, 117, 148, 154,
    155, 156, 290, 341; lack of 161; and
    diligence 97, 98; of 118, 126, 127(cs)
    204, 229
career 17

cartel **48**, 249, 266, 268
categorical imperitive **21**
certainty 18(cs) 189, 273; contractual
    284, 285, 286, 289, 334
chain of custody 36(cs)
challenger space shuttle (cs) 50
change: behaviour 74, 78; climate 55, 62,
    74, 78, 106, 122, 175, 190, 191, **197**(cs)
    203, 206, 212, 306, 334; cultural 151,
    154, 166, 191, 207, 209, 211, 320, 334;
    ethical 165; management 225, 304,
    232, 269, **304–5**, 361; organisation 139,
    328, 329
change driver 56, 57
channel tunnel 38, 50(cs) 227
charter 94; partnering 103–4, **239–40**,
    284, 289, 359; project 218, 225;
    supplier 299
choice 2, 14, 53; business 153, 154; client
    88, 93, 103; employee 83, 118, 119, 127,
    137, 139; ethical 30; professional 91
Christian 20, 130
CIAT 93
CIBSE 93
CIOB 91, 93; professional **94–9**, 105, 108;
    site conditions 161, 209; sustainability
    205; corruption 254, **256–7**, 266
Citicorp building 44–5(cs)
citizenship 55, 75, 144
civic 10
claims 47, 82, 108, 255, 266, 264(cs) 282;
    case study 148–9
client: demand 46, 185, 205; ethics **50–2**,
    55, 57, 58; health & safety 50–1, **154–5**,
    169; leadership 128, 298, 299;
    relationships 5, 55, 113, 185, 294, 360,
    303, **364**, 365; requirements 34, 83, 155;
    sustainability 75–6, 77–8, 79–80(cs)
code: corporate (cs) 62, **65–6**, 66(cs);
    community 105; of conduct **87**;
    definition **61–2, 75–6**; ethical 37, **62**,
    **56–8**, 65–7; framework 65–6; partnering
    **103–4**; of practice 69, 104, 105, 112,
    263; professional 1, 75, **90**, **93–100**, 281;
    small building **105**; structural 44;
    tendering **261**
collaboration 5, 34, 101, **103**, 113, 163,
    218–19, 225–7, 297, 348, 358–64(cs);
    contract 240, 284, 289, 289, **306–7**(cs);
    education 352–4(cs)
combined code 66, 74, 239, 244
commandment 21
common good 10, 37, 189, 228

common sense 40, 262, 325
communications 3, 57, 220, 221;
  employee 125, 128; inter-
  organisational 225; stakeholder 74, 75,
  332–3, 358(cs) **364**
community 2, 13, **26–7**, 58, 67, 195, 219,
  333(cs), 353(cs); CCS 111–12, 299–300,
  353; codes 105; decision model 55;
  Kohlberg 31; marginalised(cs) 138;
  rural 184; social responsibility to
  69–70(cs), 72, 82(tab) 200; worker
  **120–2(cs)** 129
Companies Act 73, 244
competence(cy) **90–3**; employee 138;
  H&S 155, 337; professional **88, 95**;
  small builders 338–442(cs); suppliers
  236, 264
competition(ive) 38, 40, **47–8**, 100;
  advantage 217, 248, 249; and collusion
  **264–6, 311–15(cs)**; and trust 230, 270,
  365; anti 268; distorted 110, 247, 251–2,
  253(cs); value integrity model 267–8;
  law 266; market 187; tendering 103,
  **260**, 317
Competition Act 246, 254, 264, 266, 283,
  311, 312
Competition Tribunal 312
compromise 5, 11, 30, 43, 63, 92, 232, 268;
  development 175, 185; integrity 247;
  sustainability 196 201; trust 224;
  unacceptable 13, 25, 27
concentration: of effort 54; level 197–8
confidence 45, 48
confidential(ity) 22, 37, 43, **89**, 95, 96–9,
  224, 367; client 93, 135, 244;
  commercial 77, 78, 144, 244, 266, 269;
  disclosure 143; professional 244
conflict: of interest 25, 34, 37, **40(cs)**;
  professional 106, 113, 172, 175, 177,
  184, 185, 232–3, 244, 260(cs) 332, 360
conscientious 55, 81, 212
consequential(ist) 13, 15, 16, 17, 19, 30,
  31, 54
Considerate Constructors 69, 71, **79(cs)**,
  **111–12**, 204, 206, 213, 327, **352–3(cs)**
construction: CSR in 78–80; dilemmas
  47–9; image 48; manager 36, 88, 360;
  scope 41; sustainable 77–8
Construction Housing Regeneration Act
  101, 344
consumer product safety commission 51
continuous professional development 92,
  109

continuous improvement: H&S 158, 163;
  virtue ethics 13, 55; partnering 104(fig),
  226, 234–6, 345–6(cs), 360(cs);
  sustainability 79(cs)
contract(ual): administration 93; of
  employment 27, 66, 123, **125–7**; JCT
  design & build 284; JCT framework
  103, 284, 333, 351, 357, 365; partnering
  charter 284; liability 51, 55, 154; NEC
  103, 284; personal 19; prime 103,
  **227–8(cs)**; professional **92**;
  psychological 27, 116, 128, **129**, 168; of
  sale 48; social 13, 19, 41, 113, 193;
  traditional 47, 58, 90
contradiction, ethical 40, 70, 306
corporate: codes 66–7; governance 65,
  71, **73–4**, 244; and trust 63, 239; values
  5, 34, 57
Corporate Social Responsibility (CSR) 11,
  32, **68–71**; drivers 65; good business
  **75–7**; reporting 72–3, 80–2; rationale
  74–5; triple bottom line 68
corruption 219, 243; anti model 267–9;
  cause 248; collusion 264–6; definition
  244–5; examples 249; gifts 110(cs), 256,
  258, 259(cs); international 250–1;
  legislation 247; national 251–2;
  professional/commercial 254–6;
  property 269–72; reasons for 246;
  reform 273–4; survey 257
councillor 26, 40, 171, 173, 175–6, 183,
  213
courage 13–14
covenanted 92
cowboy 48, 105, 144
Cragg, W. 53
Crane, A. and Matten, D. 54
creative 12, 169, 183, 202
credibility 94, 194; eco 210;
  measurement 209; professional 2, 89;
  public 101
Cribbs Causeway 18(cs)
culture(al): context 4, 22, 66; company 54,
  68, 117 127, 133, 138–9, 143, 340;
  corrupt 245, 246, 252, 269;
  development 9–12; differences 65,
  110; employee 123; H&S 151, 162,
  165–7, 324, 327; informal 65; multi
  55; project 168, 209, 219, 225–6,
  303, 307, 319–20, 358–64; social
  norms 55
customer 143, 188, 204, 205; care 24 (cs)
  48, 357, 360, 367; orientated 12;

satisfaction 3, 30; trust 55; business
   62; relationships 232, 234–5, 299, 346

De Archectura 9
Da Vinci, L. 10
decision making 34, 37, 104, 175, 178,
   183–6, 224; models 29, 53–8
defamed 95
Dell Computers 51(cs)
democracy 26, 167, 178, 185; community
   40, 172, 181
deontological 11, 20, 21, 30
dermatological 152
design, functionality 44, 45, 304, 332
designer: dilemma 43; H&S 155–6, 341;
   reputation 46
developing economies 136–7(cs);
   leadership 137; Durban CC 137–8(cs)
development: economics 2; life cycle 2;
   personal 47, 138; property 26
dignity 110
dilemma: classical 38; construction 47–9;
   contractor image 49; designers 43–7;
   doctor's 38–9; planning 39–41;
   prisoners 38; property 41–3; renting 43;
   selling 42
diligence 74, 97, 98
discretion 38, 91, 92, 122–3, 164, 230;
   judgement 289
discrimination 116–43; choice 100;
   direct 230; disability 131–3, 132(cs);
   equal opportunity 98; indirect
   130–1(cs); intellectual 122; legislation
   130–1; preventing 35; worker 142, 146
disciplinary: professional 66, 87, 90, 94,
   95, 100, 108; employment 125, 133;
   H&S 324
diversity: employment 75, 116, 123, 128,
   131(cs) 142, 146; legislation 131;
   procurement 204; of systems 46
DTI 77, 210
duty of care 46, 74, 117; contract 290;
   H&S 149, 154, 155, 156, 341

ecclesiastical 10
eco homes 184, 200, 205, 340
ecological 190, 192; footprint 201–3,
   202(cs) 210
economics 4, 46, 307; advantage 19; of
   climate change 197(cs); and ethics 1, 2;
   of sustainability 196; theory 230
economies: emerging 36, 63–117, 136–7,

191, 250; developed 46; transitional
   252; western 3
Egan 12, 58, 73, 101, 206
egoism 10, 11, 16–17, 73, 228
employability 123, 135
employment 124; absenteeism 140;
   contract 125–7; full 124; informal 124;
   law 123; presenteeism 140; retention
   134–5; self 141
energy: efficiency 35, 192, 204, 206;
   embodied 137, 203, 238; energy saving
   77, 186; renewable 77, 206
engineer(ing) 2, 9, 11, 44–5(cs), 55 58, 62,
   88, 106, 122; civil 10, 55, 367; military
   10; contracts 287; reputation 46–7;
   structural 346
Enron 52, 62, 63, 73, 239, 244
environment(al) 60, 61, 86, 98;
   management 138, 206(cs) 219; working
   128, 129(cs) 317
equal: distribution 10; opportunities
   82–3, 84, 119, 209; rights 19; opps.
   recruitment 134–5
Equality Act 130
equality 116–17; commission 130; living
   conditions 303; pay 124; planning 175,
   178, 185; worker 307(cs)
ethical: acceptance 57; autonomy 124;
   awareness 53, 5; behaviour 13, 21, 25,
   27, 37, 90–1, 100, 108, 229, 223, 254,
   290, 341, 366; business 14, 72;
   challenges 33; code 25, 37, 62,, 65–7,
   87–101, 93(cs) 220, 281; conduct 61;
   criteria 56; decision making 5–6, 37,
   53–7; dilemma 38–53, 229; driver
   11(fig); frameworks 12–24, 64, 110,
   135; gatekeepers 113; guidelines 54, 94,
   174, 366; investment 75, 111; leadership
   108–10, 117, 123–4, 137; nihilism 23;
   partnering code 103–4; policy 52, 81;
   relativism 23; subjectivism 23; trading
   70, 73, 83
ethics: at work 33; definition 1, 12, 20, 37,
   64; macro 30; micro 30; personal 2,
   29, 30, 233
eudaimonia 14
eurhythmy 9
Europe 10, 37, 55, 110, 124, 126, 130, 211;
   Eastern 252; Western 275
European Council 244
European Council for Civil Engineering
   55

executive pay 35, 62
expert(ise) 88, **91–2**, 113, 252; advice 176, 187; knowledge 139

facilities manager 57, 99
fair(ness) 95, 96, 122, 140, 231, 244, 285 ; advertisement 10; bounded 361; business 67, 269; competition 97, 264; pay; timescales 332; trading 66; trial 118; valuation 362; working conditions 118
Fairclough, J. 12
Fair Labour Association 70
faith 224, 233
false 24; accounting 63, 243, 254; advertising 51(cs); claims 254, 257; employment **316–21**; promises 109; value 245
Fan, V. 36, 81, 254
fat cat 64
Federation of Master Builders 105
feeling 14, 17
financial: advantage 48; decisions 107; gain 138; health 72; incentives 12, 95, 193, 205
fire 50(cs), 51(cs) 150, 159
Fletcher, J. 23
flexible working 119, 127(cs) 132, 140
Ford 10, 11
Forestry Stewardship 36
forms (Socrates) 13
Forum for the Future 67, 70(cs) 76
freedom 16, 24, 32, **118**; of thought 122; of speech 137; from harm 153
Freedom of Information Act 177
Friedman, M. 1, 67
Friends of the Earth (FOE) 200
FTSE 62, 70, 74
FTSE4Good 6

generation gap **34–5**; Baby boomer **34**; X,Y **34–5**; young 27
generosity 14
gifts 30, 65, 81, 95, 97(tab), **110**(cs) 256, **258–60**
Global Reporting Initiative 72, 299
Golden Rule 17
good 9, **12–13**; absolute 14; client's 92; life 1; neighbour 79, 111; relative 14; of society 10, 13; supreme 14
good faith **281–91**; concept 113, 174, 218; definition 286; duty of 285, 289;

European Contract law 288; guiding principles 289; in France 287; in Germany 287; Latham 290; legal meaning of 288; partnering and 283–4
good practice **293–365**
Gothic 10
Greek 10, 14, 15
Greenbury Report 74
Groak, S. (1992) 12
guilty 20, 270; companies 266, 311; conscience 179

happiness 13, 14, 15, 16, **17–19**
Harrison, M.R. (2005) 5, 16, 17, 54
health and safety **149–69**, 322(cs), 337(cs); building managers 158–9; contractor 158; co-ordinator 156–7; decision approach 57; developing economies 136; ethical approach **152–4**; migrant workers 142; policy 65, 322–4; principle contractor 157–8; professional 94; refurbished buildings 159; reporting 151–2; respect for people 128; responsibilities 154–8; safe work see *code*; self employment 141; small builders 105, 337; survey results 81–2
Health and Safety Executive (HSE) 57, 151
Heathrow T5 141, 294, 319–21(cs)
Higgs Report 74
Hippocratic Oath 39
Hitler 23
Hollis, M. 228
honesty 9, 30, 36, 65, 82, 91, 96, 111, 175, 360; in planning **180**(cs) 185; tendering codes 261
Housing Grants and Construction Act 102
housing: affordable 121, 175, 204, 330–4(cs); CSR comparison 297–9(cs); Focus 270–1(cs); supply **186**; sustainable code 186, 200; value 249(cs); worker 121
human rights **117–20**; declaration of 118; discretion 122–3; professional discretion 122–3
Human Rights Act 1998 120

ICE 95
identity based trust (IBT) 218–19
IFAC 110(cs)

ignorance (veil of)  19
ILO  36, 37, 63; labour costs 136–7(cs)
image: business  53, 138; contractor 48;
  male  116; project 359(cs); worker 209
impartial: decisions  55, 74; planning  171,
  234; professional  97, 99, 103; theory
  11, 13
impartiality  48, 85
independence(ent): ethic  21, 31;
  professional 90, 99, 247 ; role 36(cs), 74,
  145, 268, 273–4, 282, 320(cs)
Industrial Revolution  10, 122, 197
inequality(ies)  12, 65, 201, 273
Institute of Business Ethics (IBE); Survey
  35, 62, 111
integrated  2, 12, 143, 149, 167; teams 192,
  223, 319(cs)
integrity: case study 303(cs), 345(cs),
  360–2(cs) 367; pacts 267 268;
  professional  95, 96
INTERCAPE  33
intercultural  20
interdisciplinary  58, 367
interest: community  90; self  21, 31, 228,
  230, 246, 286
international business ethics  5, 63, 64
investment: ethical  75; for the future  192;
  long term  73
irrational  14, 22
Islam  10
Italian  10, 48, 270

JCT constructing excellence  103, 294
Jones, Eleanor  39
judgement: consequential  31; ethical  24;
  moral  64, 90; rule based  31
justice: ethic  189; good faith  285; social
  195, 201; universal  31; theory  14,
  19–20, 24, 180

Kant(ian)  11, 12, 21–2, 24, 29, 32, 43, 55,
  113, 117, 152, 165, 202, 266, 272(cs) 310
knowledge: body of  87–8, 91; based trust
  (KBT  218
Kohlberg, L.  30–3, 90, 113
KPIs: business  63, 73; CSR 73, 72, 75,
  81–2, 194, 210, 298–300(cs);
  performance  123; quality 164,
  303–4(cs)

labour: cheap  45; dilemmas 45–7; force
  140–1; migrant 46–7; standards 55, 62

land use  39, 41, 171–2, 175, 203;
  decisions  177, 187; policy  173
language: common  55, 163, 178, 217, 219,
  366; Foreign  20, 53, 53, 142
Latham, M.  12, 58, 73, 101(cs)  234, 284,
  290
leadership  14; client  128; ethical 108–10,
  117, 123–4, 137; market  32, 210;
  professional  109
lean production  46
learning 80(cs), 142, 222, 306, 320(cs),
  351–2(cs); curve  204, 213, 259, 303;
  organisation  138–40; life long  95, 128
legal  10, 19; compliance  1, 12, 32, 56,
  117, 322; duties  282, 341; requirements
  46, 64, 84, 93, 127; rights  281; system
  284, 285, 288
legal cases: Birse v St David  289; Picture
  Library v Stilleto Visual Programme
  Ltd  288; Sutcliffe v Thackrah  282;
  Walford v Miles  289
Legionnaires  46, 159
Leesen, N  54, 55
LeMessurier  44
level playing field: equality  19, 164, 264,
  267; trade  55, 249, 250, 261, 264, 274;
  sustainability  190, 196
Lewicki, R.J. and Bunker, B.B.  218
liberalism  16
lie  5, 23, 34
local authority  69, 171, 231, 330
logic(al)  14, 102; competing  194–5, 201
logistics  49, 141
love  21, 23
loyalty: client  43, 103; worker  83, 142,
  146, 229; project team  221, 319(cs),
  364–5(cs); and trust  216, 217

manmade  2
manslaughter 51(cs)
manufacturing  10, 12, 53, 132, 162,
  344–6(cs); offsite  104
materialism  24
Maxwell, R.  63, 73, 145, 239
McIntyre, A.  13, 16
Medici  10
meta ethics  5, 27
methodology, book  4
Michaelangelo  10
middle class  89
Millennium Bridge 46, 183(cs)
Mill, J.S.  17, 18

minimalism(t) 30, 123
misconduct 100, 289
moderation 10
money laundering 252, 272–3
Mont Blanc tunnel 51(cs)
moral: absolutes 20; development 78;
    dilemma 31, 47; distance 54, 55; hot air
    160–1; imagination 162–4 123;
    intensity 54; law 21–2; reasoning 30–3;
    solution 43; standard 20; scepticism
    23; judgements 90; responsibility 55,
    64–5, 88, 154, 162, 323
morality: conventional 31–2; and justice
    19; language of 163; post conventional
    31–2; pre conventional 31–2; virtual
    160–1; religious 21
morals 16, 22, 30, 228, 259, 266
Myers, D. 62, 105

National Audit Office 205, 210
Natural Step 70(cs) 299
neighbourhood 208
Neo Aristotilism 13, 16
Nestlé 63
New Engineering Contract 103
Nicomachean Ethics 13, 14
Nietzche, F. 24
ninety nine percent 49
no blame 109, 145, 225, 240, 284,
    323–8(cs), 364(cs)
non consequentialist 12, 20–2, 32
Northern Rock 63
Nozick, R. 19

offshoring 52–3
openness 66(cs), 82, 109, 185, 226,
    298–9(cs)
ownership 26, 201, 307(cs), 324(cs),
    356(cs)
objectivity 91, 95, 96
Office of Fair Trading (OFT) 48, 111, 256,
    263–4(cs), 266, 311–15(cs)

Patankar, M.S. (1994) 57
partnership 27, 66(cs), 69(cs); code 103–5;
    educational 333(cs) 348–55(cs); PPP
    227; specialist 235(cs), 351–5(cs),
    348(cs); supply chain 224–5
payment, late 101–2(cs)
peer review 65
perfect 13, 20
personal ethics 2, 29, 30, 34, 233

philosopher 12, 32
Piaget, J. 30
place 12, 201; sense of 26(cs) 46
planner 26, 58, 88, 168, 171–84; roles
    181–2, 182, 185, 201, 213
planning: appeal 171, 332; application 11,
    172, 177–8, 203, 206, 330(cs); decisions
    179–81 (cs's, 333–5(cs); morals 184–5,
    213; policy statements 173, 181, 332(cs)
    343; urban space 22–3
Plato 13–14
policy: development control 330;
    diversity 131; equal opportunity 83;
    ethical 52, 81; safety 322,(cs);
    sustainable 5, 353
political: affiliations 40; party 172, 176
pollution: airborne 202, 208; control 201,
    202; creating 192, 200, 250; effects 150;
    prevention 76(cs) 112; noise 208 ;
    waterborne 202
positivism 25
postmodernism 122
potential entrants 81–3, 84, 171, 354
pragmatism 24
pre payment 22
pressure groups 63, 64, 70, 75, 76(cs), 106,
    176, 181(cs), 194, 237(cs)
prime contract 103, 227(cs)
Privy Council 94
probability 49, 54, 153, 160, 191
procurement 204, 246; ethics 262;
    directive 5, 261–5
production 5, 10, 12; lean 40
professional: advertising 84 ; audit 26(cs);
    behaviour 12, 37. 82, 233, 360; code
    95–100; competence 90–3; conduct 55,
    81, 96–9(tab); corruption 107–8; ethics
    25–6; exclusiveness 88–9; fees 97, 98,
    100; indemnity insurance 25, 94;
    institutions 37, 46, 94–100;
    membership 232; practice 35–7, 107,
    122; rights 91–3; rules 93–100;
    standards 95, 96; sustainability 106–7;
    trust 232–3; & sustainability 93–4, 95
profit: balance 1, 42; motive 61, 124
promises, false 106
property: developer dilemma 41–3,
    development 9, 41–2, 78, 139; renting
    43; selling 42; stakeholder 41; value 41
proportionalism 22
psychological 17, 161, 105; contract 27,
    116, 128, 129, 168

public 29, 41; interest 85, 87; opinion 65; disclosure 144
Public Disclosure Act 144

Quaker 10, 122
Quakers Friars 356
quality, building 41; control 43; defects 3, 82, 83, 162, 163, 290, 303–4(cs); definition 149; and distortion 164–5; of life 12, 19, 22, 121(cs), 127, 206(cs), 194–5, 199(cs), 237, 306(cs); manufacturing 294, 344–6(cs); & moral imagination 162–4; poor 255(cs); product 30, 52; and trust 217; workers 142, 317(cs) 323

rational(ism) 14, 24, 83
Rawls, J. 19, 32, 180
Ray, N. (2002) 45
recruitment 134–5
recycling 46, 76(cs) 192; Harbourside 78(cs); plasterboard 79–80(cs); CSR reporting 82; eco homes 205; tax 206; measurement 211; case study 353
register: CCS site 79; CHAS 339; client 51; companies 105; land 269; practitioner 37; risk 157, 305, 353, 301–2
relationships: contractor–client 347(cs); employees 65; interpersonal 95; professional 78, 83; project 364(cs); trading 3, 101, 344(cs); and trust 232–3
relativism 14, 23, 25
reliance 43, 362(cs)
religion(ious) 8, 10, 17–18, 20–1, 22, 98, 123, 126, 130(cs) 275
rent 43; fair 43; tenant 43, 58, 144(cs); uplift 43
reputation: business 49–52, 56, 76, 100, 145, 244; construction industry 62; contractor 102(cs) 111; decision making model 56–7; definition 56; LeMessurier 45; professional 46–7, 86, 100; and trust 220
respect 55; codes 65, 96, 206; human right 117; for people 128, 223, 327(cs), 334(cs), 347(cs), 364(cs)
responsibility 2; client 154; contractor 68, 79; corporate 68, 326(cs); designer 155; environmental 35, 55, 61, 70, 95; moral 55, 64–5, 88, 154, 323(cs);

professional 47, 244; social 69, 71, 80(cs), 298(cs), 351(cs)
retrofitting cost 207–8
RIBA 91, 93, 96–9, 107, 233
RICS 91, 93, 95, 96–9, 265
RIDDOR 1995 151, 322
right 20
risk 49–52, 57, 72, 236–9; assessment 78, 160–2; ethics 153; contractor 347–9(cs); management 301–2, 305(cs)
Robin Hood 22
role conflict 34
Roman 10
Royal Academy of Engineering 106
RTPI 93, 95, 96–9
rule utilitarianism 19

SARS 39
satisfaction: customer 3, 365; employee 117, 129, 299, 342; job 83, 138; stakeholder 56, 367
Section 106: 69, 108, 176–7, 213, 330–5(cs)
Securities and Exchange Commission 61
Sedgewick, P. 19
self interest 21, 31, 228, 230, 246, 286
self knowledge 53
selfish 27, 91, 231, 237
selling 42, 200, 248, 270
sensitive(ity) 23, 27, 34, 49, 340
shareholder 17 20, 30, 54, 57, 67, 69, 72, 73–4, 76(cs) 77; pressure groups 63, 64, 65, 75
Singer, I. 10
situational ethics 22–3; applications 24–5
skills: ethical codes 53; business ethics 83; CSCS 324; employment 80 (cs), 135; professional 92 108
social: accountability code 70, 105; expenditure 72
social contract 13, 19, 41, 113, 193, 195
Society for Construction Law 281
Soil Association 36
soul 14, 15
stakeholder 2–3, 15, 17, 20, 25, 27, 30, 41; CSR 68, 72, 75, 81, 82, 192, 297–9(cs); engagement 76(cs), 79(cs), 358–9(cs&tab) 364; governance 74; indecisions 53, 54, 56, 332(cs); management 171, 356(cs); model 71(fig)
Stansbury, N. 108, 254, 255, 256, 261, 282
standards, absolute 21, 152

state capture 246
Stern Report 197–9
Stock Exchange 66, 74
stress: bolts 44; conflict 34, 150, 152, 158, 168, 334
supply(iers) 12; agreement 319(cs); chain 36(cs), 52, 102(cs), 141, 157–8, 207, 224–5 360–4(cs); code 261; partnering 225, 227(cs), 234–6, 345(cs)
surveys: apprenticeship 317; business behaviour 35, 62, 97; codes of ethics 62; construction managers 36, 47; corruption 73, 108, 252, 257, 276; CSR reporting 62, 80–2; ethical perception 34, 35, 111; flexitime 127; morality 34; migrant labour 53; personal ethics 29; QS morals 36; sustainability 205; trustworthiness 76–7; US contractors 41; Y generation 35
sustainable(ity) 70, 189–213; 3 areas 192; client 154–5; competing logics 195–6; construction 77, 79(cs) 205–7; development 189–92, 307–9(cs); economics 196–200, 207–8; good business 75–6; housing 78, 186–7; implementation 195–6; measuring 209; & planning 173, 183–4; & professionals 106–7, 113; procurement 203–4; social 194, 200, 206(cs); reporting 72, 80–2; & trust 89; UN principles 106; urban (eco) footprint 201–3
Sydney Opera House 44

tax: aggregate levy 206; avoidance 243, 316; business rate 176; climate change levy 206; development 176; green 194; incentives 199, 205; landfill 32, 75, 78, 206
Taylor, F. W. 10
Teleological 14, 124,
terms of engagement 98, 100
TGWU 141
Thatcher, M. 34
theory: Aristotelian 190, 25; egoism 10, 11, 13, 16–7, 73, 228; justice 13, 19–21, 117, 189, 202; Kantian 11, 12, 20, 21–2, 24, 29, 30, 32, 41, 43, 55, 101, 113, 117, 152, 162, 202, 213, 266, 272, 310; natural law 13, 20, 20–1, 22, 41, 43; neo Aristotelian 13, 16; post utilitarian 19; proportionalism 22; situational 7, 22–3,

24; utilitarian 13, 16, 17–19, 29, 41, 43, 86, 101, 165, 180, 183, 190, 195, 266, 272, 309, 310; virtue 13–14, 19, 24, 55–6, 88
tolerance 16, 33, 100, 123, 208, 221, 362; zero 166, 320; public 252
town planning 9, 11, 93
training 62, 73, 138–9, 338–40(cs)
transparency 101–7; development control 175; governance; professional 308(cs); relationships 12, 25, 97; reporting 80–1; and risk 301–2; trust 89, 192, 231, 234, 239, 244
Transparency International 3, 4, 5, 73, 77, 78, 108, 245, 251, 252, 253, 254, 261, 262, 267, 269, 274, 275, 276, 277, 282, 344
triple bottom line 68, 192
trust 216–40; accordingly 217; contracts 216–17; contracting 346–8(cs); in construction 220–8; definition 217–18; delegation 221, 222(cs); & efficiency 230–1; & ethics 229(cs); & governance 239; levels of 218–20; inter organisational 224; joint venture 227–8; partnering 225–7, 301, 235(cs); & planning 233–4, 237(cs); in practice 233–236; professional 8, 232–3; project 224, 358–63(cs), 361(fig); supplier relations 234–6; and risk 220, 236, 238; team building 221
Trustmark 91–92
trustworthy(ness) 42, 88, 92, 135
truth, Aristotle 14–15; economical with 44, 62; Kant 21, 23; survey 89

UCATT 141
Uff 281
UNCAC 219, 250
UN Global Compact (environment) 106, 108
unfair: advantage 244, 258; dismissal 125; distribution 251; influence 251, 254; profit gain 258, 263, 266
universal law (Kant) 21
urban 367; design 205, 210, 121(cs); ecological footprint 201–3; landscape 27; planning 2, 27, 201–2 367; pollution 46; renewal 182; scam 271; space 26–7
utilitarianism 17–19; act 18; post 19; rule 19

value 2, 13, 23, 29, 48; archetypal 219;
  client 44, 163; community 37, 135;
  house 237, 271(cs); of life **152**;
  organisational 5, 34, 57, 230, 259, 298;
  objective v.23; personal 17, 34, 176,
  183, 224
value for money 13, 72, 187, 231–2, 262,
  **302–3**(cs) 305, 367
Vardy, P. and Grosch, P. 15
Vetruvius 9, 10
virtue ethics **13–14**, 19, 24, 55, 88, 196,
  213
virtues 13–14, 24 123, 196, 290

wants 14
war 11, 22, 34, 250 252, 287
warrant 56–7
waste management 209, 213
Watt. J. 10
whistle-blowing 81, 126, **143–4**(cs) 266,
  267
white collar 146

Whitney, E. 10
wisdom 13, 14, 16, 189
women 32, 35, 116, 124, 127, 130, 131
work: ethic **11**; life balance **11**; UN safe
  code 31
workers: agency 142; contract 142;
  migrant 142; temporary 141
Working Well Together (WWT) 340
World Wildlife Fund (WWF) 70, 80,
  299
Worldcom 63, 73, 239
wrong 9, **12**, **20–2**, 156, 163

X generation 34
X contractor 313

Y Generation 34, 35
Yorkshire Water Co 24 (cs)

zero: accidents 151, 153, 163, 164, 290;
  carbon 78, 184, 186, 191, 199, 204, 207;
  defects 149; tolerance 166, 320(cs)